Asian Agriculture in a Climate Change-driven and Risky World

CABI Climate Change Series

Climate change is a major environmental challenge to the world today, with significant threats to ecosystems, food security, water resources and economic stability overall. In order to understand and research ways to alleviate the effects of climate change, scientists need access to information that not only provides an overview of and background to the field, but also keeps them up to date with the latest research findings.

This series addresses many topics relating to climate change, including strategies to develop sustainable systems that minimize impact on climate and/or mitigate the effects of human activity on climate change. Coverage will encompass all areas of environmental and agricultural sciences. Aimed at researchers, upper level students and policymakers, titles in the series provide international coverage of topics related to climate change, including both a synthesis of facts and discussions of future research perspectives and possible solutions.

Titles Available

1. Climate Change and Crop Production
Edited by Matthew P. Reynolds

2. Crop Stress Management and Climate Change
Edited by José L. Araus and Gustavo A. Slafer

3. Temperature Adaptation in a Changing Climate: Nature at Risk
Edited by Kenneth Storey and Karen Tanino

4. Plant Genetic Resources and Climate Change
Edited by Michael Jackson, Brian Ford-Lloyd and Martin Parry

5. Climate Change Impact and Adaptation in Agricultural Systems
Edited by Jürg Fuhrer and Peter J. Gregory

6. Livestock Production and Climate Change
Edited by Pradeep K. Malik, Raghavendra Bhatta, Junichi Takahashi, Richard A. Kohn and Cadaba S. Prasad

7. Climate Change and Insect Pests
Edited by Christer Björkman and Pekka Niemelä

8. Climate Change and Agricultural Water Management in Developing Countries
Edited by Chu Thai Hoanh, Vladimir Smakhtin and Robyn Johnston

9. Climate Change Challenges and Adaptations at Farm-level: Case Studies from Asia and Africa
Edited by Naveen Prakash Singh, Cynthia. Bantilan, Kattarkandi Byjesh and Swamikannu Nedumaran

10. Climate Change Impacts on Urban Pests
Edited by Partho Dhang

11. Climate Change Impacts and Sustainability: Ecosystems of Tanzania
Edited by Pius.Z. Yanda, Claude G. Mung'ong'o and Edmund B. Mabhuye

12. Climate, Ticks and Disease
Edited by Pat Nuttall

13. Climate Change, Adaptation and Gender: Policy, Practice and Methodological Underpinnings
Mamta Mehar and Narayan Prasad

14. Climate Change and Sustainable Agro-ecology in Global Drylands
Edited by Adel El-Beltagy, Rattan Lal and Kauser Malik

15. Climate Change and Food Security
Edited by Sukanta Mondal, Pankaj Singh and Ram Lakhan Singh

16. Epigenetics for Climate-Smart and Sustainable Agriculture
Edited by Jen-Tsung Chen

17. To be confirmed.

18. Asian Agriculture in a Climate Change-driven and Risky World
Edited by Hermann Waibel, Sabine Liebenehm and Ernah

Asian Agriculture in a Climate Change-driven and Risky World

Edited by

Hermann Waibel

Sabine Liebenehm

and

Ernah

CABI

CABI is a trading name of CAB International

CABI
Nosworthy Way
Wallingford
Oxfordshire OX10 8DE
UK

Tel: +44 (0)1491 832111
E-mail: info@cabi.org
Website: www.cabi.org

CABI
200 Portland Street
Boston
MA 02114
USA

Tel: +1 (617)682-9015
E-mail: cabi-nao@cabi.org

A catalogue record for this book is available from the British Library, London, UK.

ISBN-13: 9781800627314 (hardback)
9781800627321 (ePDF)
9781800627338 (ePub)

DOI: 10.1079/9781800627338.0000

Commissioning Editor: Ward Cooper
Editorial Assistant: Emma McCann
Production Editor: James Bishop

Typeset by Straive, Pondicherry, India
Printed and bound by CPI Group (UK) Ltd, Croydon, CR0 4YY

Contents

Acknowledgements

As is the case with most books like this one, there are many more people who have contributed to its completion, other than the editors and the chapter authors. All of them deserve recognition, acknowledgement and thanks.

First of all, we acknowledge the cooperation, administrative support and flexibility of the Faculty of Economics at Leibniz University (LUH), Hannover. They allowed me (Waibel) to use the remaining funds of the Institute of Development and Agricultural Economics after retirement, to proceed with this project. The project started with a mini-conference in Bandung, Indonesia, in June 2023, financed by the LUH funds. The conference was organized by one of us (Ernah) for which she and her student assistants, Putri Maudi Kusnaedi, Athhar Izzan Kautsar and Ravena Risna Annisa deserve great appreciation. Most of the chapter authors were present or had joined online and with their help and active participation, the first outline of the book was drafted.

A big appreciation has to go to CABI, our UK publishing house, for their confidence and trust. Hereby, we would like to thank Dr. David Hemming for his vital role in the inception phase and for his continued support and endless patience in the long gestation period of this book. At the same time, we express our thanks to Ali Thomson, who was our editorial partner at the start. Then she went on maternity leave. When she came back, the book manuscript was just about to finish. In between, Kin Yunn Ho took care of us in the best possible way.

Along the same line, a special appreciation goes to Guillermo Morán, now a doctoral student at Hochschule der Bundeswehr in Munich, for being our main editor, helping to weed out many formal and formatting errors of the chapters. Without his help the manuscript perhaps would never have been completed.

There are a number of renowned experts on Asian agricultural and economic development who offered their comments and made excellent suggestions. As external reviewers, they pointed out weaknesses in individual chapters of the book. We list them randomly by country.

In Thailand, it is Prof. Dr. Nipon Poapongsakorn, former President and now Distinguished Fellow at the Thailand Development Research Institute (TDRI). Furthermore, Associate Professor Dr. Wannapong Durongkaveroj from Ramkhamhaeng University, Bangkok. We thank them both for their thorough reading of the Thailand chapter and their great help to make it better. Now living in Down Under, as a wineyard owner and sheep farmer, we have to thank Dr. Rainer Adam, formerly Director of the Regional Office of the Friedrich Naumann Foundation, in Bangkok. We must thank him for sharing his in-depth knowledge of the Agricultural Policies in Asian countries. We also want to say thank you very much to Dr. Agnes Quisumbing, Senior Research Fellow with the International Food Policy Research Institute (IFPRI). Her thorough reading of the Philippines chapter is most appreciated.

In Germany, we highly appreciate the valuable comments given by Prof. Dr. Michael Grimm, Development Economist, Passau University for the Indonesia chapter. Likewise, we thank Prof. Dr. Erich Schmidt, Leibniz University (LUH) Hannover and Dr. Matthias Waibel, Research Institute of Organic Agriculture (FiBL) in Frick, Switzerland, who gave helpful comments on various chapters and structural questions of the book.

Last but not least, we thank all the contributors, authors and co-authors of the book for writing the chapters and contributing their ideas, as well as for playing a major role in the shaping of this book. In addition to submitting their own chapters, many authors and co-authors also thoroughly reviewed other chapters and we gratefully acknowledge this extra input. We highly appreciate their time and efforts invested in the preparation of this book.

Ernah
Sabine Liebenehm
Hermann Waibel

Contributor list

Badrun Nessa Ahmed, Bangladesh Institute of Development Studies (BIDS), Ministry of Planning, Government of Bangladesh; https://orcid.org/0000-0002-4726-8755. Email: tamanna@bids.org.bd

Suputtra Boriwut, Faculty of Economics, Ramkhamhaeng University, Bangkok, Thailand. Email: suputta.aou@gmail.com

Eko Ruddy Cahyadi, Department of Management, Faculty of Economics and Management, IPB University. Email: ekocahyadi@apps.ipb.ac.id

Bezawit Beyene Chichaibelu, Center for Development Research (ZEF), University of Bonn, ORCID: 0000-0001-6196-6261. Email: bchichai@uni-bonn.de;

Xinyi Du, School of Economics, Singapore Management University, Singapore; ORCID iD: https://orcid.org/0009-0004-0950-4763. Email: xydu.2025@phdau.smu.edu.sg

Ernah, Department of Agricultural and Social Economics, Faculty of Agriculture, Padjadjaran University. Email: ernah@unpad.ac.id

Tomoki Fujii, School of Economics, Singapore Management University, Singapore; ORCID iD: https://orcid.org/0000-0002-9879-4835. Email: tfujii@smu.edu.sg

Bernd Hardeweg, Faculty of Agriculture/Environment/Chemistry, University of Applied Sciences, Dresden, Germany, https://orcid.org/0009-0008-5551-4531. Email:bernd.hardeweg@htw-dresden.de

Gesa Horstkotte-Wesseler, Independent Researcher, Hoeilaart, Belgien. E-mail: gwesseler@yahoo.com Email: Gesa.WESSELER@ec.europa.eu

Dan Huang, College of Economics and Management, Huazhong Agricultural University, Wuhan, China. Email: 3453804700@qq.com

Shaoze Jin, Research Center for Rural Economy,Ministry of Agriculture and Rural Affairs, China; Email: szjin_rcre@126.com

Thanaporn Krasuaythong, Faculty of Economics, Kasetsart University, Bangkok, Thailand. Email: thanaporn_k@hotmail.com

Sabine Liebenehm, Agroscope, Confederation Swiss, Ettenhausen, Switzerland. Email: sabine.liebenehm@agroscope.admin.ch

Myriam Al Lily, College of Business and Management, American University of Bahrein, https://orcid.org/0000-0003-4890-7026. Email: miriam.allily@gmail.com

Shi Min, College of Economics and Management, Huazhong Agricultural University, Wuhan, China. Email: min@mail.hzau.edu.cn

Chomponuch Nantajit, Faculty of Economics, Ramkhamhaeng University, Bangkok, Thailand. Email: chompunuch.nantajit@gmail.com

Somkid Naprom, Faculty of Management Science, Udon Thani Rajabhat University, Udon Thani, Thailand. Email: napromm@yahoo.com

Priyanka Parvathi, Leibniz University Hannover, Germany. Email: priyanka.parvathi@gmail.com

Phouvong Phami, Wuhan University, China. Email: phouvongphami@whu.edu.cn

Diemuth Pemsl, Independent Researcher, Seattle, USA, ORCID: 0000-0002-0290-8632. Email: dpemsl@gmx.net

Jonathan Pincus, Fulbright School of Public Policy and Management, Fulbright University Vietnam. Email: jonathan.pincus@fulbright.edu.vn

Sirman Purba, Independent Researcher; Jakarta, Indonesia. Email: sirmanpurba34@gmail.com

Zhongming Shi, Chair of Architecture and Building Systems, ETH Zurich, Switzerland; ORCID iD: https://orcid.org/0000-0002-3744-1649. Email: shi@arch.ethz.ch

Marc Völker, Institute for Population and Social Research, Mahidol University, Salaya, Nakhon Pathom, Thailand, https://orcid.org/0000-0001-6185-5043. Email: marc.voe@mahidol.edu

Christoph Waibel, Construction Robotics and Innovative Building envelopes (CRIB), Flemish Institute for Technological Research (VITO), 2400 Mol, Belgium; ORCID iD: https://orcid.org/0000-0001-6077-1411. Email: christoph.waibel@vito.be

Hermann Waibel, Faculty of Economics and Management, Leibniz University Hannover, Hannover, Germany, https://orcid.org/0000-0002-3811-1975. Email: waibel@ifgb.uni-hannover.de

Justus Wesseler, Agricultural Economics and Rural Policy Group, Section Economics, Social Science Department, Wageningen University and Research, the Netherlands, https://orcid.org/0000-0002-6009-8543. Email: justus.wesseler@wur.nl

Haowen Zhuang, Zhejiang Economics and Information Center, Hangzhou, China. Email: haowen1001@hotmail.com

Wu Lifeng, CAC Nantong Chemical Co., Ltd., Shanghai, China. Email: wwllff@sohu.com

List of Authors and Affiliations

1 **Introduction and Overview**
Ernah[1], *Sabine Liebenehm*[2] *and Hermann Waibel*[3]
1) Department of Agricultural and Social Economics, Faculty of Agriculture, Padjadjaran University, ernah@unpad.ac.id
2) Agroscope, Confederation Swiss, Ettenhausen, Switzerland email: sabine.liebenehm@agroscope.admin.ch
3) Faculty of Economics and Management, Leibniz University Hannover, Hannover, Germany; Email: waibel@ifgb.uni-hannover.de; https://orcid.org/0000-0002-3811-1975

2 **Climate Change and Weather Extremes in Asia**
Sabine Liebenehm[1], *Shi Min*[2] *and Hermann Waibel*[3]
1) Agroscope, Confederation Swiss, Ettenhausen, Switzerland; email: sabine.liebenehm@agroscope.admin.ch
2) College of Economics and Management, Huazhong Agricultural University, Wuhan, China; email: min@mail.hzau.edu.cn
3) Faculty of Economics and Management, Leibniz University Hannover, Hannover, Germany; Email: waibel@ifgb.uni-hannover.de; https://orcid.org/0000-0002-3811-1975

3 **A Review of Risk and Risk Preference in Asian Agriculture: Methodology and Empirical Evidence**
Sabine Liebenehm[1] *and Miriam Al Lily*[2]
1) Agroscope, Confederation Swiss, Ettenhausen, Switzerland; email: sabine.liebenehm@agroscope.admin.ch
2) College of Business and Management, American University of Bahrein; email: miriam.allily@gmail.com; https://orcid.org/0000-0003-4890-7026

4 **Agricultural Development and Policy in Thailand – With a Case Study of the Vegetable Subsector**
Marc Völker[1], *Bernd Hardeweg*[2] *and Hermann Waibel*[3]
1) Institute for Population and Social Research, Mahidol University, Salaya, Nakhon Pathom, Thailand, https://orcid.org/0000-0001-6185-5043, email: marc.voe@mahidol.edu;
2) Faculty of Agriculture/Environment/Chemistry, University of Applied Sciences, Dresden, Germany, Email: Hardeweg, Bernd bernd.hardeweg@htw-dresden.de; https://orcid.org/0009-0008-5551-4531;

3) Faculty of Economics and Management, Leibniz University Hannover, Hannover, Germany; Email: waibel@ifgb.uni-hannover.de; https://orcid.org/0000-0002-3811-1975

5 Agricultural Development and Policy in Vietnam: Also a Success Story?

Jonathan Pincus[1] and Hermann Waibel[2]
1) Fulbright School of Public Policy and Management, Fulbright University Vietnam; Email: jonathan.pincus@fulbright.edu.vn; ORCID: (lead author)
2) Faculty of Economics and Management, Leibniz University Hannover, Hannover, Germany; Email: waibel@ifgb.uni-hannover.de, ORCID: https://orcid.org/0000-0002-3811-1975.

6 The Philippines – Increasing Productivity in a Risky Environment

Hermann Waibel[1], Gesa Horstkotte-Wesseler[2] and Justus Wessler[3]
1) Faculty of Economics and Management, Leibniz University Hannover, Hannover, Germany; Email: waibel@ifgb.uni-hannover.de, ORCID: https://orcid.org/0000-0002-3811-1975.
2) European Commission, Brussels; email: Gesa.WESSELER@ec.europa.eu
3) Agricultural Economics and Rural Policy Group, Section Economics, Social Science Department, Wageningen University and Research, The Netherlands, email: justus.wesseler@wur.nl;https://orcid.org/0000-0002-6009-8543

7 Agricultural Development and Policy in China – with Two Case Studies from South-west China

Dan Huang[1], Shi Min[2], Shaoze Jin[3] and Haowen Zhuang[4]
1) College of Economics and Management, Huazhong Agricultural University,Wuhan, China
2) College of Economics and Management, Huazhong Agricultural University, Wuhan, China; email: 闵师 min@mail.hzau.edu.cn (Lead Author)
3) Research Center for Rural Economy, Ministry of Agriculture and Rural Affairs, PRC, Beijing, China; email: szjin_rcre@126.com
4) Zhejiang Economics and Information Center, Hangzhou, China, email: haowen1001@hotmail.com

8 Indian Agricultural Policy Developments: A Review

Priyanka Parvathi
Leibniz University Hannover, Germany; email: priyanka.parvathi@gmail.com

9 Review of Agricultural Development and Policy with a Case Study on the Role of Homestead Fish Ponds for Food Security in Bangladesh

Badrun Nessa Ahmed
Bangladesh Institute of Development Studies (BIDS), Ministry of Planning, Government of Bangladesh; https://orcid.org/0000-0002-4726-8755; mailto:tamanna@bids.org.bd

10 Agricultural Development and Policy in Indonesia since Independence with a Case Study of the Oil Palm Sector

Eko Ruddy Cahyadi[1] and Ernah[2]
1) Department of Management, Faculty of Economics and Management, IPB University, ekocahyadi@apps.ipb.ac.id;
2) Department of Agricultural and Social Economics, Faculty of Agriculture, Padjadjaran University, ernah@unpad.ac.id

11 Agricultural Development and Policy in Laos

Phouvong Phami[1], Priyanka Parvathi[2] and Hermann Waibel[3]
1) Wuhan University, email: phouvongphami@whu.edu.cn
2) Leibniz University Hannover, Germany; email: priyanka.parvathi@gmail.com
3) Faculty of Economics and Management, Leibniz University Hannover, Hannover, Germany; Email: waibel@ifgb.uni-hannover.de, ORCID: https://orcid.org/0000-0002-3811-1975

12 The Role of Biotechnology in Chinese Agriculture with a Case Study of Bt Cotton in Shandong Province
Shi Min[1], Wu Lifeng[2] and Diemuth Pemsl[3]
1) College of Economics and Management, Huazhong Agricultural University,Wuhan, China; Email: min@mail.hzau.edu.cn (lead author)
2) CAC Nantong Chemical Co., Ltd., Shanghai, China. Email: wwllff@sohu.com
3) Seattle, USA, Email: dpemsl@gmx.net; ORCID: 0000-0002-0290-8632.

13 The Mekong River and Village Development in Thailand
Suputtra Boriwut[1], Thanaporn Krasuaythong[2], Chomponuch Nantajit[3] and Somkid Naprom[4]
1) Faculty of Economics, Ramkhamhaeng University, Bangkok, Thailand; email: suputta.aou@gmail.com; (lead author)
2) Faculty of Economics, Kasetsart University, Bangkok, Thailand; email: thanaporn_k@hotmail.com
3) Faculty of Economics, Ramkhamhaeng University, Bangkok, Thailand; email: chompunuch.nantajit@gmail.com
4) Faculty of Management Science, Udon Thani Rajabhat University, Udon Thani, Thailand; email: napromm@yahoo.com

14 Exploring Differences in Rural Household Debt between Thailand and Vietnam: Economic Environment versus Household Characteristics
Bezawit Beyene Chichaibelu
Center for Development Research (ZEF), University of Bonn; Email: bchichai@uni-bonn.de; ORCID: 0000-0001-6196-6261

15 Food Self-sufficiency and Building-integrated Urban Agriculture: Lessons from Singapore
Tomoki Fujii[1], Christoph Waibel[2], Xinyi Du[3] and Zhongming Shi[4]
1) School of Economics, Singapore Management University, Singapore; Email: tfujii@smu.edu.sg, ORCID iD: https://orcid.org/0000-0002-9879-4835; (lead author)
2) Construction Robotics and Innovative Building envelopes (CRIB), Flemish Institute for Technological Research (VITO), 2400 Mol, Belgium; ORCID iD: https://orcid.org/0000-0001-6077-1411
3) School of Economics, Singapore Management University, Singapore; ORCID iD: https://orcid.org/0009-0004-0950-4763
4) Chair of Architecture and Building Systems, ETH Zurich, Switzerland; ORCID iD: https://orcid.org/0000-0002-3744-1649

16 Synthesis
Ernah[1], Sabine Liebenehm[2], Sirman Purba[3] and Hermann Waibel[4]
1) Department of Agricultural and Social Economics, Faculty of Agriculture, Padjadjaran University; Email: ernah@unpad.ac.id
2) Agroscope, Confederation Swiss, Ettenhausen, Switzerland; Email: sabine.liebenehm@agroscope.admin.ch
3) Jakarta, Indonesia; Email: sirmanpurba34@gmail.com
4) Faculty of Economics and Management, Leibniz University Hannover, Hannover, Germany; Email: waibel@ifgb.uni-hannover.de, ORCID: https://orcid.org/0000-0002-3811-1975

17 Outlook
Hermann Waibel
Faculty of Economics and Management, Leibniz University Hannover, Hannover, Germany; Email: waibel@ifgb.uni-hannover.de, ORCID: https://orcid.org/0000-0002-3811-1975

1 Introduction and Overview

Ernah, Sabine Liebenehm and Hermann Waibel

Abstract

This chapter introduces the purpose of the book and gives a quick overview of the chapters. The relevance of climate change and the generally riskier world are exemplified with some key factors. Thereafter, the structure of the book, with its four main parts, is explained. Within each part, the respective chapters are briefly outlined. There are eight Asian countries covered by the book.

Motivation and Purpose

This book is about agricultural development in major Asian countries. It deals with the past, the present and the expected future situation. The countries included in the book are India, China, Indonesia, Bangladesh, Philippines, Vietnam, Thailand, Laos and Singapore, when ranked by size of population. Most of the chapters were written by authors who are nationals of these countries and who were PhD students of the first editor of the book (Waibel), at Leibniz University Hannover, Germany. Other authors were German PhD students who carried out their PhD or postdoctoral research, using or collecting data in one of the countries. The third group of authors are long-term research partners adding interesting aspects to the book (i.e. Laos and Singapore). The book also builds upon the work of the first editor who has over 30 years' research experience in these countries, including a 3-year stay in the Philippines and a 7-year stay in Thailand, where he carried out major projects on vulnerability to poverty (e.g. Klasen and Waibel, 2013, 2015).

The starting point of the book was a small research conference in June 2023 in Bandung, Indonesia, which was organized by one of the book editors (Ernah) and attended by most of the chapter authors.

Rationale of the Book

There are several good reasons to write a book on agriculture in Asian countries. First, economically, Asia, to date, is the heart of the world. Its share of global GDP is higher than that of the USA and the EU together. The two most populous countries in the world, India and China produce almost 30% of the world's economic output (IMF, 2025). Given the political transition ongoing in the USA and the security threats in Europe, it is

 Asian Agriculture in a Climate Change-driven and Risky World (H. Waibel, S. Liebenehm and Ernah eds)
DOI: 10.1079/9781800627338.0001

likely that the economic dominance of Asia will grow further. Second, while over half the world population lives in Asia, its share of land area is much less. This poses a challenge as regards agriculture and food security. Third, Asia is responsible for two-thirds of global agricultural GDP and agricultural commodities, significantly contributing to their export earnings. For example, China is a major exporter of pork products and is the world's largest producer and consumer of fertilizers. India, aside from being the world's top rice exporter, also exports wheat and spices, such as black pepper and cardamom. Vietnam and Thailand rank second and third as the world's leading exporters of rice, and Indonesia, being the country with the largest area under oil palm plantations, exports palm oil and related products (FAO, 2023). Agriculture in Asia is also a major employer. Labor shares are at least double the GDP shares, which are now mostly below 10%.

Changes in the natural and political environment have exacerbated the challenges for agriculture in Asia. Ongoing climate change, protectionist tendencies and trade barriers caused by the policy shifts in the USA, require new conceptual thinking and policy actions, away from business as usual. Hence, there are good reasons to produce a book on Asian agriculture. While many research volumes on Asian agricultural development have been written, including especially on the green revolution (e.g. Hayami and Ruttan, 1985; Evenson and Gollin, 2003) and recently by Otsuka and Fan (2021), what is missing is a comprehensive and comparative country analysis that takes up some of the most pressing issues on agricultural development and rural transformation in Asia. This book aims to fill this gap.

Key Issues

There are several key topics that deserve to be studied. Foremost, this includes the structural transformation, including land and labor markets in rural areas (e.g. Meng, 2012), rural education (Li *et al.*, 2005), new technologies such as digital technologies (Deichmann *et al.*, 2016), water-saving techniques (Won *et al.*, 2005) and modern biotechnologies (Pinnstrup-Andersen and Cohen, 2000), demography and population aging and, above all, a comprehensive rural development strategy that defines the future role of Asia's farmers in the future.

A major issue in the structural transformation debate is the stagnation or even decline in farm sizes across Asian countries. This has raised concerns among agricultural economists that Asian agriculture will not be able to compete with the large-scale farms of western countries, such as the USA, Canada, Europe and Australia. For example, already 20 years ago, Fan and Chan-Kang (2005) noted that most farms in Asia are small, asking the question whether "is *small really beautiful*"? Felipe (2012) has attributed the decline in farm sizes in Bangladesh and China to land fragmentation and population growth. Otsuka *et al.* (2016) submit that small farms can never reach the production efficiency of the large farms. They question the classic inverse plot size-yield relationship and demand a policy shift towards large farms. However, as highlighted by Briones and Felipe (2013) the more important issue is to promote long-term productivity growth in agriculture and facilitate the modernization and upgrading of small farms side-by-side with agro enterprises who are the core of global value chains. The question of how agricultural policies have evolved, how the agricultural sector has performed and how farming will be organized in the future are topics that rank prominently throughout this book. This also includes the question regarding the future of the rural villages where most farming is taking place. How can rural villages become attractive places to live and work? (Rigg *et al.*, 2020).

Other topics highly relevant for Asian agriculture include the location- and crop-specific impacts of climate change (e.g. Mendelsohn, 2014) and the increasing importance of adopting climate-smart agricultural practices (Yuan *et al.*, 2025). Similarly, a chapter is devoted to the debate about genetically modified crop varieties (e.g. Whitty *et al.*, 2013) taking China as a case study.

In sum, the profound transformations in the economic and environmental conditions shaping agriculture in Asia call for comprehensive and forward-looking policy adjustments. Hence, there are lessons that can be learned from investigating and comparing the development path taken by different countries in Asia.

Structure of the Book

The book has 17 chapters, organized over four parts. Part I establishes the methodological framework that underpins the book's title and objectives. Part II presents historical reviews of agricultural development and policies in seven Asian countries. Using a framework that distinguishes key time periods, the chapters analyze general economic policies, sector-specific agricultural policies and the performance of the agricultural sector over time. Each country chapter also contains an in-depth case study based on previous, mostly PhD research. In Part III, special topics are being discussed, including biotechnology in China, village development in the Mekong Basin, high-tech farming in Singapore and rural household debt. Part IV includes a synthesis chapter that derives some lessons learned from the papers presented in the book. Finally, an outlook chapter with some subjective viewpoints about the future of agriculture and rural areas in Asia completes the book.

In the following, a brief summary of the chapters is provided, in order to give the reader some orientation and quick access to the contents of the book.

Chapter 2 presents an introductory overview of climate change and extreme weather events and their implications for agriculture in Asia. It sets the stage for the country-specific policy chapters in Part II as well as for some of the special topics in Part III of the book. The chapter begins with a review of recent scientific findings on climate change in Asia, including regionally differentiated trends in temperature, precipitation, and the frequency and intensity of extreme weather events, drawing on key sources such as the Sixth Assessment Report of the Intergovernmental Panel on Climate Change (IPCC). Also, the agricultural sector's vulnerability towards these climatic changes and their impact on agricultural productivity and food security are discussed. Finally, the chapter outlines examples of climate-smart agricultural practices and adaptation strategies, with a particular focus on policies and interventions that have shown promise in enhancing the resilience of farming systems in the region. Recommendations from the IPCC are integrated throughout, insofar as they pertain to agricultural development and risk management in Asia.

Chapter 3 introduces the topic of risk in agriculture, in accordance with the underlying hypothesis of the book that the world is becoming a riskier place. The chapter highlights the risk categories which are typical for agriculture, as well as methods to measure farmers' risk preferences, and the empirical evidence of risk management decisions in Asian agriculture. A range of production, market, institutional and personal risks are analyzed that influence agricultural decision-making. Furthermore, the chapter presents approaches to measuring risk preferences, including survey-based methods, lottery choice experiments and simulation games, highlighting their strengths and limitations in a readily accessible and generally understandable way. The chapter also highlights some empirical study results with contradictory findings. For example, in China it was found that a higher degree of risk aversion can reduce technology adoption, while studies in India and in Cambodia found the opposite, for risk-reducing technologies. Given that climate and market risks are increasing in the foreseeable future, understanding farmers' risk preferences is essential for designing effective policies that enhance resilience and promote sustainable agricultural practices. Insofar, the chapter can make a contribution for a better understanding of farmers' behavior in an increasingly risky world.

Chapters 4 to 11 focus on the country policy reviews including case studies, mostly based on the authors' PhD or postdoctoral research. Starting with Thailand in Chapter 4, a historical review of development and agricultural policies is presented, organized by the UN development decades. It is shown how general development theories and policies have shaped agricultural policies and how this has formed the structure of agriculture. Bringing significant events into the picture, the performance of the agricultural sector in each decade is reviewed. Thus, an easily accessible overview of the evolution of agriculture in the country is provided. Current constraints and opportunities are outlined and entry points for further policy debate and subsequent target-orientated research is offered. To further illustrate the process of agricultural development in Thailand, the review is complemented with a case study of the vegetable sector, which is based on a thorough and in-depth

PhD study. The paper ends with an outlook of how to facilitate the transition to farming systems that can make the rural areas more prosperous and allow sustainable agricultural development.

Chapter 5 is on Vietnam. The review of economic and agricultural policies starts with the Vietnam War period, covers the post-war phase, the economic reform period and subsequent development decades. The connection between general development policies, agricultural policies, agricultural transformation and the performance of the agriculture is discussed in detail. A case study of a rural household drawing from a unique panel data base, complements the analysis and a summary of the major issues in relation to agricultural and rural transformation is added. Essential statistics are included to support the conclusion and outlook for Vietnam's agriculture in the near future.

The Philippines is discussed in Chapter 6. This country case study is particularly interesting because the Philippines is special in several regards. First, it is among the most densely populated countries in Asia whose population has been growing rapidly during the past six decades. Second, economic growth in the Philippines since the 1960s had its ups and downs which seems to be related to the respective government administrations. Third, since around 2010, growth was more stable as the political environment became more steady. Fourth, there are two case studies included in the chapter. One of them is about the ongoing debate on the use of modern biotechnology, namely the genetically modified golden rice variety. The second case study deals with the banana industry, a major export earner of the Philippines. The chapter draws conclusions as regards the current performance of the Philippine agricultural sector and ends with an outlook.

Chapter 7 is dedicated to China with a comprehensive analysis of China's agricultural and land reform. Since 1949, China's agricultural sector has undergone significant transformations through continuous adjustments in land policies and rural development strategies. This chapter reviews the evolution of China's agricultural institutions and policies across four distinct periods, namely privatization-to-agricultural-cooperatives between 1949 and 1956, the cooperatives-to-people's-communes approach from 1957 to 1977, the household-contract-responsibility system, starting in 1978 until 2012, and the new era of agricultural-modernization-and-rural revitalization since 2013 until to date. The authors illustrate how each period is marked by key policy changes and corresponding impacts on agricultural productivity, rural economies, and farmers' livelihoods. Furthermore, two case studies from Southwest China illustrate these dynamics. First, the carbon balances of rubber plantations in Xishuangbanna, highlighting ecological trade-offs between economic benefits and sustainability; and second, the development of off-farm labor markets, and their contribution to the income of rubber farmers.

In Chapter 8, Indian agricultural policy developments during post-independence are traced. The effect of national development theories and policies on Indian agricultural policies is analyzed. The author discusses five distinct phases of Indian agriculture, from the 1950s–2020. This ranges from the green revolution, encompassing both its pre- and post- phases and extends to the 1991 economic reforms and India`s growth story thereafter. The description of the five phases identifies patterns of structural changes in agriculture. A case study on organic and fair trade pepper is included. The chapter provides insights into existing constraints and opportunities for a productive policy debate and target-orientated subsequent research.

Chapter 9 outlines structural transformations and policy reforms in the agriculture sector over the years in Bangladesh, starting with independence from Pakistan and continuing to the present. Over the past 35 years, agriculture in Bangladesh has advanced considerably and achieved significant progress. Using historical data, this chapter analyses patterns of structural changes, highlights challenges and opportunities and offers a foundation for impactful policy analysis. A case study on homestead pond aquaculture in Bangladesh is included as an example of pro-poor development efforts.

Chapter 10 takes the reader to the world's fourth most populated country, Indonesia. It reviews the evolution of agricultural development, corresponding policy and policy changes in Indonesia. The time frame ranges from independence in 1945 to the present. It traces key policy phases, including post-independence food security efforts, the Green Revolution under the

New Order, and the decentralization and liberalization of the reform period. Recent policies have emphasized sustainable practices, digital transformation, and global market competitiveness, reflecting a shift from food security to food sovereignty. The chapter also addresses emerging challenges and their implication on future directions, including climate change, the digitalization of agriculture, and possible disruptions in global supply chains due to geopolitical tensions. A case study on the oil palm industry, derived from the PhD theses of the two chapter authors, highlights the sector's significance to Indonesia's economy, while examining its environmental impacts and social complexities.

Part II of the book ends with Chapter 11, with the history and analysis of economic and agricultural development in Laos. The starting point is the time of independence from France. The transition thereafter from a command and control economic system towards a communist governed market economy, is analyzed, using a historical framework. While mining products and hydropower electricity have become important parts of the Laos economy, it is still an agricultural country, where the vast majority of households are engaged in farming. A case study of natural resource extraction underlines the importance of forests for Lao farmers.

In sum, the policy chapters are the core of the book. They give a unique overview of what has happened to agriculture in eight Asian countries which make up almost half of the world's population. The extensive literature reviews and exclusive case studies are a rich source of information for those dealing with Asian general economic and agricultural sector development.

Part III of the book presents special issues of agriculture in Asia. Chapter 12 analyses biotechnology in China. The country has invested early and substantially in the development of biotechnology and today has the largest R&D capacity in this field outside the USA. Bt cotton was approved in China in 1997 as one of the first genetically modified (GM) crops and celebrated as a major success. Nevertheless, the country picked a notably cautious approach in regulating agricultural biotechnology, only recently approving additional GM crops such as maize and soybean for commercial production. The importance of this chapter stems from the large number of impact studies of Bt cotton. Also, the editor of the book has carried out a 12-year in-depth case study research in one county in Shandong province in China.

Chapter 13 is taking a look at village development in Thailand of villages along the Mekong River using household and village level data. It is based on a 2023 survey of 110 villages, along the Mekong River, out of which 26 were in Thailand. Furthermore, data come from the long-term household survey, namely the Thailand Vietnam Socioeconomic Panel (TVSEP), covering a 10-year period. In addition, in-depth interviews with 10 Mekong villages during the Covid phase are included. The analysis is largely based on descriptive statistics. The chapter provides a point of departure for further research of village development in the Mekong River Basin.

Chapter 14 takes up the issue of rural household debt which is an increasingly pressing and rapidly growing problem in many Asian countries. In some areas, household debt has recently reached unprecedented levels. In this chapter cross-country differences in credit market participation, the level of household debt and the degree of over-indebtedness between rural households in Thailand and Vietnam are analyzed. Using the panel data set of the Thailand Vietnam Socioeconomic Panel, and by applying a decomposition model, the socio-economic determinants of indebtedness, including respective economic environments, household characteristics and behavioral parameters are identified. Significant differences are found between Vietnam and Thailand, with the latter country facing higher risks of indebtedness. Conversely, for households in Vietnam at the upper end of the debt distribution, it is household and behavioral characteristics, rather than credit market conditions that explain indebtedness.

In Chapter 15, a special case study on urban, high-tech farming is presented. The case of Singapore, with its policy aim to achieve 30% food self-sufficiency by 2030, is analyzed from an economic and technical perspective. Analyzing Singapore's changing landscape for urban agriculture and its popular 30-by-30 goal, key challenges such as land constraints, high production costs and resource limitations are highlighted. Building-integrated agriculture (BIA), which utilizes building surfaces such as rooftops, façades and balconies for food production within urban environments, is examined as a potential way to increase self-sufficiency through a simulation. Despite the BIA's potential, practical issues – including regulatory concerns and infrastructure limitations – have been worked out clearly in this unique write-up.

Part IV of the book contains two chapters, i.e. synthesis and outlook. Chapter 16 undertakes a synthesis of the major findings which serves as a basis for policy recommendations. Hereby, the focus is on the policy chapters. These allow to draw uniformities and differences in terms of development of the agricultural sector in the eight Asian countries covered in this book, including the urban farming model in Singapore. The issues highlighted are the degree of structural transformation, including especially the evolution of farm size distributions, the diversification of the sector, including emerging dualistic tendencies with a subsistence agriculture on the one hand and high tech, modern farming on the other hand. The role of technology and the questions of farmer entrepreneurs is highlighted. Finally, the role of rural villages as they have been treated over the development decades in the seven Asian countries is emphasized. The synthesis chapter paves the way for the last chapter that presents some outlook for the future.

The last chapter of the book, builds upon the synthesis chapter and looks into the future, i.e. until say 2050, the middle of the century. The basis for this outlook is an expert opinion survey among the authors of the policy chapters. They were asked for statements on how they see agriculture in their respective country by 2050. The results of this survey, together with some relevant literature is synthesized in this last chapter.

References

Briones, R. and Felipe, J. (2013) *Agriculture and Structural Transformation in Developing Asia: Review and Outlook*. Asian Development Bank Economics Working Paper Series, Manila, Philippines. p. 363.

Deichmann, U., Goyal, A. and Mishra, D. (2016) Will digital technologies transform agriculture in developing countries? *Agricultural Economics* 47(S1), 21–33.

Evenson, R.E. and Gollin, D. (2003) Assessing the impact of the Green Revolution, 1960 to 2000. *Science* 300(5620), 758–762.

Fan, S. and Chan-Kang, C. (2005) Is small beautiful? Farm size, productivity, and poverty in Asian agriculture. *Agricultural Economics* 32(1), 135–146. DOI: 10.1111/J.0169-5150.2004.00019.X

FAO (2023) World Food and Agriculture – Statistical Yearbook 2023. Rome. Available at: https://doi.org/10.4060/cc8166en (accessed 05 March 2026).

Felipe, J. (2012) Inclusive growth: Why is it important for developing Asia. *The CADMUS Journal* 1(4), 36–58.

Hayami, Y. and Ruttan, V.W. (1985) *Agricultural Development*. Johns Hopkins University Press, Baltimore.

IMF (2025) Available at: https://www.imf.org/external/datamapper/PPPSH@WEO/EU/CHN/USA (accessed 13 September 2025).

Li, Q., De Brauw, A., Rozelle, S. and Zhang, L. (2005) Labor market emergence and returns to education in rural China. *Review of Agricultural Economics* 27(3), 418–424.

Mendelsohn, R. (2014) The impact of climate change on agriculture in Asia. *Journal of Integrative Agriculture* 13(4), 660–665.

Klasen, S. and Waibel, H. (2013) *Vulnerability to Poverty: Theory, Measurement and Determinants, with Case Studies from Thailand and Vietnam*. Palgrave Macmillan, London, UK.

Klasen, S. and H. Waibel (2015) Vulnerability to poverty in South-East Asia: Drivers, measurement, responses, and policy issues. *World Development* 71(special issue), 1–3.

Meng, X. (2012) Labor market outcomes and reforms in China. *Journal of Economic Perspectives* 26(4), 75–102.

Otsuka, K. and Fan, S. (2021) *Agricultural Development: New Perspectives in a Changing World*. International Food Policy Research Institute, Washington DC, USA.

Otsuka, K., Liu, Y. and Yamauchi, F. (2016) The future of small farms in Asia. *Development Policy Review* 34(3), 441–461.

Pinstrup-Andersen, P. and Cohen, M. J. (2000) *Modern Biotechnology for Food and Agriculture: Risks and Opportunities for the Poor. Agricultural Biotechnology and the Poor*. Consultative Group on International Agricultural Research, Washington DC, USA, pp. 159–172.

Rigg, J., Phongsiri, M., Promphakping, B., Salamanca, A. and Sripun, M. (2020) Who will tend the farm? Interrogating the ageing Asian farmer. *Journal of Peasant Studies* 47(2), 306–325.

Whitty, C.J., Jones, M., Tollervey, A. and Wheeler, T. (2013) Africa and Asia need a rational debate on GM crops. *Nature* 497(7447), 31–33.

Won, J.G., Choi, J.S., Lee, S.P., Son, S.H. and Chung, S.O. (2005) Water saving by shallow intermittent irrigation and growth of rice. *Plant Production Science* 8(4), 487–492.

Yuan, F., Ospina, R., Perumal, A.B., Noguchi, N., He, Y. *et al.* (2025) Smart agriculture in Asia. *Plant Communications* 6, 101377.

2 Climate Change and Weather Extremes in Asia

Sabine Liebenehm, Shi Min and Hermann Waibel

Abstract

This chapter presents a foundational overview of climate change and extreme weather events and their implications for agriculture across Asia. It sets the stage for the country-specific policy chapters in Part II and some of the special topics in Part III. The chapter begins with a review of recent scientific findings on climate change in Asia, including regionally differentiated trends in temperature, precipitation, and the frequency and intensity of extreme weather events, drawing on key sources such as the Intergovernmental Panel on Climate Change (IPCC) Sixth Assessment Report. Thereafter, it discusses the agricultural sector's vulnerability towards these climatic changes, highlighting the impacts on agricultural productivity and food security. Finally, the chapter outlines examples of climate-smart agricultural practices and adaptation strategies, with a particular focus on policies and interventions that have shown promise in enhancing the resilience of farming systems in the region. Recommendations from the IPCC are integrated throughout, insofar as they pertain to agricultural development and risk management in Asia.

The Current Situation: Climate Change and Weather Extremes in Asia

The continuous rise in global temperatures is expected to lead to more frequent and more intense extreme weather events. These include heatwaves, intensive rainfall, prolonged droughts, storms and typhoons, as well as wildfires. Such events pose severe negative impacts for agriculture and rural livelihoods. Their impacts go beyond primary production and include disruptions in non-farm employment, transportation and storage of food products, market access, food prices and food safety (Agnolucci *et al.*, 2020; Randell *et al.*, 2022; Behera *et al.*, 2024).

At the same time, agriculture itself is a major driver of climate change (World Bank, 2024). About one-third of all emissions is attributable to the global agrifood system. Key contributors include methane from livestock and nitrous oxide from synthetic fertilizer use, both of which are sources of non-CO_2 emissions in agriculture (FAO, 2014; Behera *et al.*, 2024). These are compounded by biodiversity loss, and water overuse, often driven by the expansion of food production to meet the demands of a growing population and rising per capita food consumption, especially of animal-based products (Fuglie, 2021; World Bank, 2024). Historical reliance on land expansion has also led to widespread deforestation. Additionally, food waste – especially in urban areas – adds to the

(H. Waibel, S. Liebenehm and Ernah eds)
DOI: 10.1079/9781800627338.0002

sector's indirect climate footprint (FAO, 2013; Fuglie, 2021; Behera *et al.*, 2024).

With about 60% of the world's population but only about one-third of the world's arable land and a similar share of its freshwater resources, Asia is particularly vulnerable to climate change impacts. The IPCC Sixth Assessment Report (2022) emphasizes the following challenges for Asian agriculture:

- South Asia is projected to experience more frequent and intense heatwaves and greater variability in monsoon rainfall.
- Southeast Asia is likely to face fewer but more extreme tropical cyclones, rising flood levels and prolonged inundation in the Mekong Delta.
- East Asia is expected to see more intense droughts and a rising number of strong typhoons.
- In mountainous regions such as the Hindu Kush Himalayas, glacier retreat and reduced snow cover threaten both irrigation and river-fed agriculture.

These climatic shifts have tangible agricultural impacts. For example, under a warming scenario of 1–4°C, rice and maize yields in India are projected to fall by 10–30% and 25–70%, respectively. Additionally, pests such as the "golden apple snail" are thriving under the new climate conditions, threating rice-producing countries, including those covered in this book, namely China, India, Indonesia, Bangladesh, Vietnam, Thailand and the Philippines.

Extreme events are already taking a toll. Panda (2021), in an ADB background paper on agricultural insurance in the Asia Pacific Region, points at the huge costs of flooding events in the recent past due to climate change. Globally, flooding was said to account for some 40% of all loss-related natural catastrophes since 1980, with losses worldwide totaling more than US$1 trillion (ADB or Panda, 2021). The 2011 flood in Thailand, mentioned in Chapter 4, resulted in exorbitant losses, totaling US$43 billion. Similar vulnerabilities are compounded by climate variability linked to El Niño and La Niña phenomena, which affect rainfall patterns critical for rainfed agriculture (Montesclaros and Teng, 2021).

In summary, climate risks to agriculture in Asia are mounting. The intersection of environmental fragility, demographic pressure and socio-economic dependencies on agriculture creates a high-stakes situation. In the following section, we turn to empirical studies that assess the agricultural impacts of climate change, with a focus on productivity and food security in the Asian context.

Vulnerability and Impact on Agricultural Production and Food Security

The Sixth Assessment Report (AR6) of the IPCC (IPCC, 2023) demonstrates that East Asia (e.g. China), South Asia (e.g. Bangladesh, India) and Southeast Asia (e.g. Thailand, Vietnam), are all highly vulnerable to extreme heatwaves, including more frequent and intense heatwaves. Risks from heavy precipitation are also increasing, though these are more geographically heterogeneous and generally of lower magnitude. The reports states that "*climate change has reduced food security and affected water security due to warming, changing precipitation patterns, reduction and loss of cryospheric elements, and greater frequency and intensity of climatic extremes, thereby hindering efforts to meet sustainable Development Goals*" (IPCC, 2023, p. 50).

Several studies have shown that the increasing frequency of extreme weather events has contributed to rising food prices, disproportionately affecting low-income households and exacerbating nutritional insecurity (Iizumi and Ramankutty, 2016; Fanzo *et al.*, 2018). These events also disrupt food supply chains by damaging infrastructure such as roads, storage facilities and transportation networks thereby reducing food availability, accessibility and quality (Barrios *et al.*, 2008).

Building on IPCC–AR6, Clarke *et al.* (2022a, b) compiled a database of attribution statements for individual events and linked specific extreme weather events to local impacts. Their analysis offers more localized insights at smaller spatial scales than the regional aggregations typically used by the IPCC. One example is the 2015 drought in Indonesia triggered by the occurrence of El Niño during the dry season. The combination of high temperatures and extremely low precipitation affected over 100 000 hectares of crops, resulting in significant income loss, elevated food prices and increased poverty.

The event also triggered forest and peat land fires leading to air pollution with detrimental health effects for millions of people across Indonesia, Malaysia and Singapore. In some areas, food scarcity even led to wildlife intrusion, including reports of long-tailed monkeys raiding villages in search of food (Rohmah, 2015).

There is little doubt that crop productivity in many parts of Asia will be negatively affected by climate change. Montesclaros and Teng (2021) emphasize the sensitivity of crop growth to both daytime and nighttime temperatures. Higher night temperatures increase plant respiration rates, potentially reducing yields and grain quality, while sustained higher daytime temperatures cause increased spikelet sterility and consequently reduce yields in wheat and rice.

Using a global dataset covering 18 crops and 70% of crop production by area, Agnolucci *et al.* (2020) show that countries most affected by climate change are also those that are the most food insecure – with the lowest caloric food supply and average crop yield. They find that irrigation is the most effective adaptation strategy to counteract the negative implications of temperature increases on crop yields.

While attribution of climate change impacts to agricultural productivity remains scientifically challenging, new methods are improving our understanding. For example, Ortiz-Bobea *et al.* (2021) developed an econometric model, combining historical weather data and counterfactual climate scenarios, to show that climate change has reduced global agricultural total factor productivity by 21% between 1961 and 2020. Their work underscores that long-term agricultural productivity growth is increasingly vulnerable to ongoing climate shifts.

Ozdemir *et al.* (2022) use an asymmetric dynamic panel model to investigate the short- and long-term effects of climate change on agricultural productivity in Asia over the period of 1980–2016. They found that CO_2 emissions have a short-term positive effect on productivity –likely due to the carbon fertilization effect – but this turns negative in the long run. The authors emphasize that rainfed agriculture, which covers 80% of the world's cultivated land and produces about 60% of global food, is especially at risk under climate change.

Beyond direct impacts on crop and livestock productivity, climate change also threatens human health, particularly in countries with limited medical infrastructure. Hashim and Hasim (2016) highlight that typhoons have more severe health impacts in countries such as Bangladesh and Laos, compounding agricultural and economic vulnerability.

In sum, this selection of empirical observations of climate change effects for agriculture in Asia, demonstrates that climate change is already undermining food systems across Asia. These impacts are expected to intensify, posing a serious threat to agricultural development, poverty reduction and food security in Asia, and, in particular, in the countries covered by this book.

This situation requires both urgent policy action and a rethinking of conventional theoretical models of agricultural transformation. As nicely pointed out by Barret *et al.* (2023), conventional models of agricultural transformation do not capture climate change well because they ignore the connection between economic processes and changes in the natural environment and therefore often predict that, in equilibrium, factor productivity in absolute terms does not decrease over time. Additionally, the models assume that natural capital stocks are constant over time. Yet climate change directly affects the quality-adjusted stocks of land, labor and capital, and reshapes technological pathways. Innovations such as renewable energy (solar, wind, geothermal), carbon capture storage (CCS), and drought-resistant crops offer promise – but they also carry potential risks and trade-offs. The key challenge is to leverage technological progress to enhance resilience and sustainability, while minimizing negative externalities and ensuring that agricultural transformation is both inclusive and climate smart.

The following section will examine adaptation strategies – drawing on recent empirical work and conceptual thoughts, including contributions from this book's authors.

Adaptation and Mitigation

A wide range of strategies exists to reduce both the vulnerability of agricultural systems to climate change and their contribution to it. These strategies include mitigation measures, which aim to reduce greenhouse gas emissions from agriculture, and adaptation measures, which help farmers adjust to the impacts of climate variability and extremes.

Agriculture can contribute to climate mitigation by reducing greenhouse gas emissions and enhancing carbon sinks. Key examples include soil carbon sequestration through zero-tillage practices; agroforestry and the expansion of perennial crops (as long as grown following sustainability principles); improved nutrient and fertilizer management; and more efficient use of energy and water inputs.

Adaptation measures are diverse, highly context-specific and span multiple scales. Examples include crop cultivar improvements; on-farm water storage and soil moisture conservation; artificial irrigation, agroforestry and farm-level diversification; efficient irrigation systems, which, according to the IPCC (2022), represent the most widely documented category of water-related adaptation in agriculture. However, adaptation success depends not only on technology but also on local context, farmers' perceptions and access to knowledge and support systems. Several studies emphasize the importance of integrating local knowledge with scientific tools.

In addition, indirect measures such as reducing food loss and waste in production, processing and consumption as well as the change to more sustainable diets, especially with lower meat consumption, are essential. Fanzo *et al.* (2018), uses a food systems approach to explore the two-way relationship between climate change and nutrition, identifying actionable entry points along the food value chain. Their proposed strategies include promoting the production of nutrient-dense crops (e.g. fruits, vegetables, legumes), improving processing and labeling of nutritious foods, and enhancing consumer education. The authors emphasize the role of public institutions in setting enabling conditions, including communication, incentives and regulation.

Waibel *et al.* (2013) highlight the importance of farmers' perceptions towards climate change. Using rural household panel data from 2007 and 2008 in Thailand and Vietnam, they show that rural households are aware of changing climatic conditions and that their subjective experiences can inform the design of locally tailored agricultural adaptation strategies and tactics. Their follow-up paper (Waibel *et al.*, 2018), using panel data from 2007 and 2013, confirms that while most farmers do perceive climate change, their interpretations vary by location and are shaped by socio-economic characteristics and recent experiences with shocks. The study highlights that adaptation decisions are often based on perception rather than scientific guidance and calls for improved information, extension and education services for small-scale farmers in Thailand and Vietnam to cope with climate change.

The role of education and information with regards to farmers' perceptions of climate change and their adoption of climate-smart and environmentally friendly practices was shown in a study of small-scale rubber producers in Southwest China by Min *et al.* (2020). Despite observable increases in average annual temperature, almost 40% of farmers did not perceive any change. The authors link this discrepancy to differences in education and shock exposure – emphasizing the importance of climate literacy in promoting climate-smart agricultural practices.

In the three Asian mega-deltas (Mekong in Vietnam; Ganges–Brahmaputra in Bangladesh; and Ayeyarwady in Myanmar), which are the major rice-producing areas of their respective countries, Schneider and Asch (2020) identify serious challenges such as salinity intrusion and water scarcity. The study shows that the most frequently used agricultural adaptation strategies are planting salt-tolerant varieties, the adjustment of cropping calendars, and the use of water-saving technologies which are applied on a case-by-case basis. Yet there is still no integrated adaptation framework across these vulnerable rice-growing regions.

In Pakistan, Habib-ur-Rahman *et al.* (2022) integrate crop growth models with climatic and economic models to formulate adaptation strategies for rice–wheat cropping systems. These systems are not only prevalent in Pakistan but also widespread across India. Their recommendations include modifications in crop management such as optimizing sowing time, adjusting plant density, enhancing early warning systems and fostering more transdisciplinary research, including ICT-based technologies, applicable to agriculture.

Liebenehm *et al.* (2023) use the same household panel data set as those mentioned in the two previous sections, i.e. Waibel *et al.* (2013; 2018), albeit for a much longer time horizon, i.e. 2007 to 2017. In addition, they combine survey data with satellite-based historical precipitation and temperature data. This enabled the authors to assess the impact of drought events on climate-relevant agricultural management decisions. The model captured the interrelatedness and simultaneity of drought-coping decisions with household-specific time-invariant heterogeneity, possibly correlated with drought exposure and coping decisions. They find that drought shocks significantly influence fertilizer use and labor decisions, with clear cross-country differences. For instance, Thai farmers tend to reduce fertilizer and switch to machinery services, while Vietnamese farmers increase spending on hired labor and inputs. While the efficiency of these responses remains uncertain, the study highlights the location-specific and dynamic nature of adaptation behavior. While it is not clear to what extent farmers' responses to extreme weather events are efficient (or climate-smart), the study underlines the context and location-specificity of climate adaptation.

In Mindanao, the Philippines, Chandra *et al.* (2017) evaluate the effectiveness of the popular Farmer Field Schools (FFS) approach as a tool to promote climate-smart agriculture. Using a combination of field observations, household interviews and focus group discussions, they find that collective actions at community level such as communal seed banks, are key to successful adaptation. They also observe that climate-smart technologies are highly location-specific, technically rigorous and involve knowledge-intensive processes. Hence, experiential learning, guided by capable trainers, is a good way to foster knowledge and capacities of local farming communities and other service partners for safeguarding local food production with income and other livelihood goals. This case underscores that bottom-up approaches, when embedded in supportive institutional environments with peer-to-peer and cross-scale farmer planning and engagement, can foster sustainable and climate-resilient livelihoods.

Conclusions

To conclude this introductory chapter on climate change in Asia and its implications for agriculture, three overarching lessons are highlighted.

First, assessing agriculture's dual role as a driver of climate change and as a source of mitigation requires clarity on the basis of comparison. For example, Min *et al.* (2019), in an interdisciplinary study on the carbon balance effects of rubber expansion in Southwest China, demonstrate that when rubber plantations replace virgin forests, the net carbon impact is negative. In contrast, when established on former cropland such as rice or maize, the carbon balance can be positive. This underscores the critical question: could some climate damage have been avoided or minimized through more strategic and sustainable development planning in the first place and what institutional conditions need to be put in place to avoid such misguided decisions in the future?

Second, the successful adoption of climate-smart agricultural practices depends on an enabling institutional setting and policy environment. Incentive structures must support transitions to more sustainable systems. Bottom-up approaches – anchored in local communities – play a key role, particularly in fostering climate-related education for the next generation of farmers. Public investments in land use changes, soil health improvements and water retention infrastructure are most effective when coordinated at the village level through collective actions. Furthermore, adaptation and mitigation strategies must look beyond agriculture to encompass rural employment opportunities in environmentally sustainable local sectors such as renewable energy and digital services.

Third, the management of global crises, whether within or outside the agricultural sector, requires international coordination aligned with the Sustainable Development Goals. The 2008 global food price crisis illustrates the dangers of uncoordinated national responses and the emergence of a global prisoner's dilemma. As Timmer (2009) explains, countries such as India implemented rice export bans to protect domestic consumers, while major exporters such as Thailand and Vietnam discouraged exports through fiscal measures to benefit from rising global prices. As a result, import-dependent countries such as the Philippines faced severe supply constraints, even at tripled prices.

Similar dynamics were observable during the Covid-19 pandemic, where unilateral national policies at times undermined global solidarity and response effectiveness. As Asia's international influence continues to grow, the region has an opportunity – and also a responsibility – to draw lessons from these crises. Enhancing regional cooperation, strengthening multilateral institutions and prioritizing sustainable, climate-resilient development agricultural technologies and pathways will be essential to navigating future shocks under increasingly uncertain climate scenarios.

References

Agnolucci, P., Rapti, C., Alexander, P., De Lipsis, V., Holland, R.A. *et al.* (2020) Impacts of rising temperatures and farm management practices on global yields of 18 crops. *Nature Food* 1(9), 562–571.

Barrett, C.B., Ortiz-Bobea, A. and Pham, T. (2023) Structural transformation, agriculture, climate, and the environment. *Review of Environmental Economics and Policy* 17(2), 195–216.

Barrios, S., Ouattara, B. and Strobl, E. (2008) The impact of climatic change on agricultural production: is it different for Africa? *Food Policy* 33(4), 287–298. DOI: 10.1016/j.foodpol.2008.01.003

Behera, B., Haldar, A. and Sethi, N. (2024) Agriculture, food security, and climate change in South Asia: a new perspective on sustainable development. *Environment, Development and Sustainability* 26, 22319–22344. DOI: 10.1007/s10668-023-03552-y

Chandra, A., Dargusch, P., McNamara, K.E., Caspe, A.M. and Dalabajan, D. (2017) A study of climate-smart farming practices and climate-resiliency field schools in Mindanao, the Philippines. *World Development* 98, 214–230.

Clarke, B., Otto, F., Stuart-Smith, R. and Harrington, L. (2022a) Extreme weather impacts of climate change: an attribution perspective. *Environmental Research: Climate* 1(1), 012001.

Clarke, H., Nolan, R.H., De Dios, V.R., Bradstock, R., Griebel, A. *et al.* (2022b) Forest fire threatens global carbon sinks and population centres under rising atmospheric water demand. *Nature Communications* 13(1), 7161.

Fanzo, J., Davis, C., McLaren, R. and Choufani, J. (2018) The effect of climate change across food systems: implications for nutrition outcomes. *Global Food Security* 18, 12–19. DOI: 10.1016/j.gfs.2018.06.001.

FAO (2013) *Food Wastage Footprint: Impacts on Natural Resources*. Food and Agriculture Organization of the United Nations, Rome.

FAO (2014) *Agriculture, Forestry and Other Land Use Emissions by Sources and Removals by Sinks: 1990–2011 Analysis*. Food and Agriculture Organization of the United Nations, Rome.

Fuglie, K.O. (2021) Climate change upsets agriculture. *Nature Climate Change* 11, 293–299. DOI: 10.1038/s41558-021-01017-6.

Habib-ur-Rahman, M., Ahmad, A., Raza, A., Hasnain, M.U., Alharby, H.F. *et al.* (2022) Impact of climate change on agricultural production: issues, challenges, and opportunities in Asia. *Frontiers in Plant Science* 13, 925548.

Hashim, J.H. and Hashim, Z. (2016) Climate change, extreme weather events, and human health implications in the Asia Pacific region. *Asia Pacific Journal of Public Health* 28(2_suppl), 8S–14S.

Iizumi, T. and Ramankutty, N. (2016) Changes in yield variability of major crops for 1981–2010 explained by climate change. *Environmental Research Letters* 11(3), 34003. DOI: 10.1088/1748-9326/11/3/034003.

IPCC (2022) Climate Change 2022: Impacts, Adaptation and Vulnerability. Contribution of Working Group II to the Sixth Assessment Report of the Intergovernmental Panel on Climate Change [H.-O. Pörtner, D.C. Roberts, M. Tignor, E.S. Poloczanska, K. Mintenbeck, A. Alegría, M. Craig, S. Langsdorf, S. Löschke, V. Möller, A. Okem, B. Rama (eds.)]. Cambridge University Press. Cambridge University Press, Cambridge, UK and New York, NY, USA, 3056 pp. DOI:10.1017/9781009325844

IPCC (2023) Sections. In: Climate Change 2023: Synthesis Report. Contribution of Working Groups I, II and III to the Sixth Assessment Report of the Intergovernmental Panel on Climate Change [Core Writing Team, H. Lee and J. Romero (eds.)]. IPCC, Geneva, Switzerland, pp. 35–115. DOI: 10.59327/IPCC/AR6-9789291691647.

Liebenehm, S, Huong, J. and Hermann, W. (2023) Extreme weather and agricultural management decisions among smallholder farmers in rural Thailand and Vietnam. *Agricultural Economics* 54(5),593–622.

Min, S., Jikun Huang, J., Hermann Waibel, H., Yang, X. *et al.* (2019) Rubber boom, land use change and the implications for carbon balances in Xishuangbanna, Southwest China. *Ecological Economics* 156, 57–67.

Min, S., Xiaobing, W., Shaoze, J., Hermann, W. and Jikun, H. (2020) Climate change and farmers' perceptions: impact on rubber farming in the Upper Mekong region. *Climate Change* 163, 451–480.

Montesclaros, J.M.L. and Teng, P.S. (2021) Agriculture and food security in Asia. In *Climate Change, Disaster Risks, and Human Security: Asian Experience and Perspectives*. Springer Singapore, Singapore, pp. 137–168.

Ortiz-Bobea, A., Ault, T.R., Carrillo, C.M., Chambers, R.G., and Lobell, D.B. (2021) Anthropogenic climate change has slowed global agricultural productivity growth. *Nature Climate Change* 11(4), 306–312.

Ozdemir, D. (2022) The impact of climate change on agricultural productivity in Asian countries: a heterogeneous panel data approach. *Environmental Science and Pollution Research* 29, 1–13.

Panda, A. (2021) *Climate Change and Agricultural Insurance in the Asia and the Pacific Region. Asian Development Outlook 2021 Update: Transforming Agriculture in Asia*. Background Paper.

Randell, H., Gray, C. and Shayo, E.H. (2022) Climatic conditions and household food security: Evidence from Tanzania. *Food Policy* 112, 102362.

Rohmah, A. (2015) Drought hits most Indonesian provinces due to El Nino. Available at: www.aa.com.tr/en/world/drought-hits-most-indonesian-provinces-due-to-el-nino/22867 (accessed 13 July 2025).

Schneider, P. and Asch, F. (2020) Rice production and food security in Asian Mega deltas – a review on characteristics, vulnerabilities and agricultural adaptation options to cope with climate change. *Journal of Agronomy and Crop Science* 206(4), 491–503.

Timmer, P. (2009) *Rice Price Formation in the Short Run and the Long Run: The Role of Market Structure in Explaining Volatility*. Working Papers (Center for Global Development) (172). Available at: https://www.cgdev.org/sites/default/files/1422136_file_Timmer_Rice_Price_FINAL.pdf (accessed 17 July 2025).

Waibel, H., Tongruksawattana, S. and Voelker, M. (2013) *Voices of the poor on climate change in Thailand and Vietnam. The Environments of the Poor in Southeast Asia, East Asia and the Pacific*. Asian Development Bank, Singapore, 170–186.

Waibel, H, Pahlisch. T.H. and Voelker, M. (2018) Farmers' perceptions of and adaptations to climate change in Southeast Asia: The case study from Thailand and Vietnam. In: Zilberman, D., Lipper, L., McCarthy, N., Asfaw S. and Branca, G. (eds) *Climate Smart Agriculture – Building Resilience to Climate Change, Vol. 52*. Springer Natural Resource Management and Policy Book Series, Rome, pp. 137–160.

World Bank (2024) Climate-smart agriculture. Available at: https://www.worldbank.org/en/topic/climate-smart-agriculture (accessed 17 July 2025).

3 A Review of Risk and Risk Preference in Asian Agriculture: Methodology and Empirical Evidence

Sabine Liebenehm and Miriam Al Lily

Abstract

This chapter examines the risk categories faced by farmers, methods to measure their risk preferences and empirical evidence linking these preferences to risk management decisions in the Asian agricultural context. First, we review the diverse range of production, market, institutional and personal risks that influence agricultural decision-making. Next, we explore approaches to measuring risk preferences, including survey-based methods, lottery choice experiments and simulation games, highlighting their strengths and limitations.

Empirical evidence from studies across Asia suggests that higher risk aversion generally reduces technology adoption in China, while in India and Cambodia it encourages the adoption of risk-reducing technologies. Risk-averse farmers are also more likely to purchase insurance, adjust input use, diversify crops and engage in contract farming.

With increasing climate and market risks, understanding farmers' risk preferences is essential for designing effective policies that enhance resilience and promote sustainable agricultural practices. This chapter synthesizes empirical patterns and methodological insights to support future research and deepen understanding of farmer behavior under uncertainty in the Asian agricultural context.

Introduction: Risk in Agriculture

The objective of this chapter is to review current methodologies and empirical evidence in the study of risk preferences in agriculture.

Risk in agriculture is a complex challenge that affects both the stability of food production and the livelihoods of farmers. Unlike many other industries, agriculture is uniquely susceptible to a wide range of unpredictable and uncontrollable factors. In general, these risks can be categorized into five main types, namely production risk, market risk, institutional risk, personal risk and financial risk (Hardaker *et al.*, 2015; Komarek *et al.*, 2020).

Production risk in agriculture refers to the uncertainty in crop and livestock production outcomes, driven by a variety of environmental factors. Agricultural production is influenced by numerous uncontrollable natural elements, including weather conditions, pests, diseases and soil quality (Rosenzweig *et al.*, 2014). These factors contribute to the variability of production outcomes, creating significant challenges for farmers who rely on consistent yields to sustain their livelihoods.

Even slight deviations from expected weather can lead to significant variations in yields (Porter and Semenov, 2005). The frequency of extreme weather events such as droughts, floods and

DOI: 10.1079/9781800627338.0003

hurricanes has increased in recent years due to climate change (Habib-ur-Rahman *et al.*, 2022).

Crop pests thrive under changing environmental conditions and can lead to large-scale crop losses if not managed effectively (Deutsch *et al.*, 2018; Savary *et al.*, 2019). Likewise, livestock diseases can lead to animal mortality and reduced productivity (Tiwari *et al.*, 2013).

Another source of production risk is dysfunctional soil management leading to poor soil health (Dias *et al.*, 2015; Yang *et al.*, 2020). Also, growing crops on soils with excessive heavy metals, poor drainage or high salinity can augment yield uncertainty (van Es and Karlen, 2019).

Market risks in agriculture arise from fluctuating prices, changing costs and uncertain access to markets, all of which significantly impact farm profitability. A key contributor to market risk is price volatility in agricultural commodities, influenced by multiple interconnected factors (Tomek and Peterson, 2001). For example, weather-related impacts on crop yields can reduce supply, driving prices up, while bumper harvests in favorable years can lead to oversupply and price drops.

On the input side, energy price shocks can lead to a sudden increase in production costs. Increasingly agriculture is energy-intensive, thus, any volatility in energy markets has a cascading effect on the overall cost structure for farmers, resulting in fluctuating profitability (Sands and Westcott, 2011).

Information asymmetry also plays a critical role in market risk (Perloff and Rausser, 1983; Liao and Chen, 2017). Farmers with limited access to real-time market data may be disadvantaged when making decisions about when and where to sell their products. Moreover, market access depends on various factors, including infrastructure, logistics networks, trade policies and even the bargaining power farmers hold within the supply chain (Satyasai and Pereira, 2019).

Institutional risks are due to unpredictable changes in policies and regulations from government and private institutions (e.g. trade partners). Environmental regulations, which may impose limits on water usage, chemical application or land use, require farmers to adopt new practices to remain compliant (Du *et al.*, 2021). Institutional risks also encompass uncertainties around land use and property rights, which are essential for stable agricultural operations.

Personal risks: Specific to individuals, these risks involve human health and relationships, such as injuries, family health issues, and disease transmission between livestock and humans, reducing the stability of farm operations and income.

Financial risks: Associated with farm financing, these risks arise from variability in cash flow due to fixed financial obligations, such as loans. Changes in interest rates, credit availability or credit conditions further compound financial risk.

As outlined above, each type of risk carries unique implications and challenges. How farmers respond to these risks is influenced by their individual risk preferences and subjective perceptions of the likelihood and impact of each risk. Recognizing these preferences can lead to the development of more tailored and supportive risk management tools, policies and resources.

Measurement of Risk Preferences

Risk cannot be eliminated but it can be managed by a diverse set of tools. The adoption of these risk management strategies is strongly influenced by the decision-makers attitudes toward taking risks (Ullah *et al.*, 2016). As a result, a substantial body of research has focused on measuring farmers' risk preferences, which is essential for designing effective interventions and policies.

Risk preferences can be measured using three primary approaches: (i) self-reporting through survey (e.g. Dohmen *et al.*, 2011), (ii) incentivized lottery choice experiments (e.g. Binswanger, 1980, 1981), and (iii) econometrics and mathematical models or simulations (e.g. Merrill *et al.*, 2019; Clark *et al.*, 2020). Each method has its advantages and problems but generally provides unique insights into farmers' risk behaviors, in different contexts.

Survey-based Measurement

Survey-based measures are a common and practical way to assess individuals' risk preferences.

One widely adopted approach is the question developed for the German Socio-Economic Panel by Dohmen *et al.* (2011). Participants rate their willingness to take risks on an 11-point Likert scale using the following prompt:

> *"Are you generally a person who is fully prepared to take risks, or do you try to avoid taking risks? Please choose a number on a scale from 0 (unwilling to take risk) to 10 (fully prepared to take risks).".*

A variation of this method is the Domain-Specific Risk-Taking Scale (DOSPERT) developed by Weber *et al.* (2002). This psychometric tool assesses risk-taking across five specific domains: financial decisions, health and safety, recreational activities, ethical considerations and social decisions. With both a 40-item full scale and a reduced 30-item version (Blais and Weber, 2006), DOSPERT provides a nuanced view of individuals' risk preferences across different contexts.

While survey-based measures are sometimes criticized for not directly reflecting risk aversion in the concavity of the utility function (Pratt *et al.*, 1964; Arrow, 1971), they have gained increasing acceptance in large-scale surveys. Recent evidence supports their validity. For example, Falk *et al.* (2023) confirmed the internal validity of survey-based measures and demonstrated that they align well with experimental measures in describing individual risk preferences. Similarly, Vieider *et al.* (2015) found significant correlations between survey responses and incentivized lottery choices across 30 countries. Furthermore, in a study with rural households in Thailand, Hardeweg *et al.* (2013) found that the survey measure is a cost-effective and reasonably accurate alternative to experiments.

Above all, studies using the German Socio-Economic Panel (SOEP) data (Dohmen *et al.*, 2011; Lönnqvist *et al.*, 2015) show that survey-based measures often outperform experimental ones in predicting actual risky behavior. On a broader scale, Falk *et al.* (2018) demonstrated the behavioral validity of survey measures across 76 countries, reinforcing their usefulness for cross-country comparisons.

However, survey-based measures have at least two major limitations. First, they do not capture risk aversion through the curvature of the utility function. Second, survey measures are not incentive-compatible, meaning there is no cost to misrepresenting one's true preferences, which may be particularly problematic if responses are influenced by social desirability bias. On the other hand, survey-based measures are cost-effective and supply information from a large and diverse set of respondents. In contrast, incentivized elicitation methods with actual payoffs are designed to encourage truthful revelation of risk preferences. These methods are typically designed as multiple lottery choice sets and are time-consuming and costly. We will introduce these in the next section.

Lottery Choice Experiments

One of the earliest empirical applications of lottery choice experiments was developed by Binswanger (1980, 1981) in India. He offered respondents a set of six lottery choices with increasing payoffs and variance. For example, selecting Choice 1, which guarantees a fixed payoff, indicates strong risk aversion. Conversely, choosing Choice 6 – with a high payoff but extreme variability – suggests a greater willingness to take risks. Inefficient choices were included in the design to assess the sincerity and understanding of respondents.

A refinement of this design has been developed by Eckel and Grossmann (2008). It offers participants a single choice among six gambles, each with a 50% probability of winning a higher prize. The gambles vary in riskiness, allowing respondents to be classified into one of six risk categories. In Eckel and Grossmann's (2008) original version, the experiment is conducted across three frames: abstract with losses; abstract without losses; and investment with losses. In the loss frames, participants face the possibility of losing up to $6, potentially forfeiting their entire endowment if they choose the riskiest gamble. In the no-loss frame, all payoffs are shifted upward to ensure the minimum outcome is zero, removing the risk of losses. The investment with losses frame maintains the payoff structure of the abstract/loss design, except that all decisions are described as investment. By comparing these treatments, researchers can distinguish between risk aversion and loss aversion and assess whether respondents are sensitive to a gamble vs. investment framing.

Another key influential study, and probably one of the most commonly used measures of risk preferences was developed by Holt and Laury (2002). They introduced the accept/reject lottery experiments based on multiple price lists (MPL). Instead of eliciting a certainty equivalent, the goal here is to elicit a probability equivalent, i.e. the point at which the respondent is indifferent between a safer option and a riskier one. The Holt and Laury (2002) design involves 10 decisions. Each requires a choice between two lotteries: Option A, a safer binary gamble with lower variability, and Option B, a riskier binary gamble with more variable outcomes. As the choices progress, the expected payoff for the riskier option increases, making it more attractive. At some point, most subjects switch from Option A to Option B, which serves as a measure of their risk aversion.

Individuals who switch to the riskier option early are classified as risk-loving, while those who switch later demonstrate risk-averse behavior. This simple yet effective method provides a clear assessment of participants' risk preferences.

A simplified version of the Holt and Laury (2002) MPL design was developed by Dohmen *et al.* (2011). Respondents face 20 decisions, choosing between a constant lottery (50% chance of winning 300 € or 0 €) and a safe amount that increases by 10 € per row, from 0 € to 190 €. With monotonic preferences, participants typically switch from preferring the lottery to the safe option at a certain point. Participants' switching point – the row at which they opt for the safe option over the lottery – serves as a measure of their risk aversion. Since the lottery's expected value is 150 €, risk-averse individuals tend to choose the safe option when it is smaller or equal to 150 €, while risk-loving individuals may continue choosing the lottery even when the safer option is greater than 150 €.

While these lottery choice experiments assess risk aversion in the gain domain and assume linear probability weighting as per Expected Utility Theory, Tanaka *et al.* (2010) proposes a relatively simple MPL design to account for both loss aversion and non-linear probability weights, as suggested by Kahnemann and Tversky (1979) and Tversky and Kahnemann (1992). The experiment consists of three series of paired lotteries: the first two series involve different probabilities and positive payoffs, while the third series introduces losses.

A drawback of the above-mentioned MPLs is that they may be too complex for use in countries/contexts where respondents have lower levels of education, leading to a higher proportion of erroneous responses and potential bias in risk preference measurement (Dave *et al.*, 2010; Cook, 2015; Crosetto and Filippin, 2016; Filiz *et al.*, 2020). One effective solution is the visualization of lottery pairs e.g. with colored chips, allowing respondents to easily understand probabilities and rewards. Illustrating choices on single picture cards that simulate an actual lottery draw enhances comprehension and ultimately improves the accuracy of risk preference assessments (Liebenehm and Waibel, 2014; Al Lily and Liebenehm, 2024).

While the experiments suggested by Binswanger (1980, 1981) or Eckel and Grossmann (2008) as described above appear easier than the several MPLs (Holt and Laury, 2002; Tanaka *et al.*, 2010; Dohmen *et al.*, 2011), yet another intuitive approach is the Bomb Risk Elicitation Task (BRET) developed by Crosetto and Filippin (2013). In the static version of BRET, subjects make a single decision: how many of 100 boxes to collect, with each collected box earning 10 euro cents. However, one of the boxes contains a bomb. After participants make their choice, a number between 1 and 100 is drawn from an urn. If the drawn number is less than or equal to the number of collected boxes, the bomb explodes, and the participant loses all earnings. If the number is greater than the number of collected boxes, the participant receives the corresponding payment. The participants thus have to weigh how much risk they are willing to tolerate relative to the likelihood of the bomb exploding and zero payment.

Finally, Filiz *et al.* (2020) propose another simple measure of risk preferences using two decks of cards: a low-risk deck (Deck A) with lower variability and a high-risk deck (Deck B) with greater variability. Both decks have the same expected value, and participants are informed of the differences. Participants can either draw a card from Deck A – indicating risk aversion; draw a card from Deck B – indicating risk-seeking; or express indifference – indicating risk neutrality.

Simulation Games

In addition to survey questions and lottery choice experiments, a couple of studies have measured risk preferences through simulation games. One of the most popular simulations is the Balloon Analogue Risk Task (BART) developed by Lejuez *et al.* (2003). In this simulation participants are provided with a computer-based simulation in which they inflate a series of virtual balloons. Each balloon has a fixed but unknown probability of bursting, which increases with the number of pumps. If the balloon bursts, participants lose the accumulated value for that balloon. If participants stop inflating before the balloon bursts, they can "bank" their earnings and move on to the next balloon.

Another simulation game, closely aligned with farmers' real-life decision-making, was employed by Merrill *et al.* (2019) and Clark *et al.* (2020). In this digital simulation, participants managed a virtual livestock facility and faced decisions about investing in biosecurity measures when confronted with disease outbreak scenarios. The researchers systematically varied the scenarios by altering the likelihood and severity of disease outbreaks. Participants' investment choices were recorded across these varying scenarios, providing insights into their willingness to invest in biosecurity under different conditions of uncertainty. This approach aimed to capture the dynamics of decision-making in a realistic agricultural context.

With risk elicitation measures described above varying in complexity, the choice of method depends on the research question, sample characteristics, and the desired level of precision. Using a measure that participants struggle to understand can introduce noise and lead to biased estimations. It's also important to recognize that seemingly inconsistent behavior – such as expressions of indifference or multiple switches – may not indicate poor understanding. In some cases, these outcomes are a deliberate design choice to capture nuanced decision-making. In the next section, we'll explore the empirical evidence on the use of different approaches in the context of various agricultural risks in Asia.

Empirical Evidence on Risk Preferences in Asian Agriculture

This section synthesizes empirical research linking farmers' risk preferences to their decision-making in the context of Asian agriculture. Table 3.1 provides a structured overview of the key studies that examine the relationship between risk preferences and various risk-related activities in Asian agriculture. To facilitate analysis, these studies are categorized according to the primary type of risk: (A) production risks, (B) market risks, (C) institutional risk and (D) personal risks.

In Panel A, the majority of studies in the context of production risk focus on how risk aversion is associated with technology adoption. In China, there is a consistent pattern in which higher risk aversion correlates with lower technology adoption across different agricultural activities, including cotton, rice, wheat and hog farming. Examples include studies by Liu (2013), Hui *et al.* (2023) and Wu and Li (2023). In India, the pattern is more nuanced. Patil and Veettil (2024) show that risk-averse farmers are less likely to adopt mechanized rice transplanting, favoring more traditional and predictable methods despite lower efficiency. However, risk-averse farmers are more willing to adopt low-capital, risk-reducing technologies, such as stress-tolerant rice varieties. Similarly, Ward and Singh (2015) find that Indian farmers with higher risk aversion are more likely to adopt drought-tolerant rice seeds, which are perceived as reducing exposure to climate-related production risks. In Cambodia, Shimamoto *et al.* (2018) report that risk-averse rice farmers are more likely to adopt post-harvest technologies, such as moisture meters, which help minimize losses and uncertainty in grain quality. The country-specific differences may be less likely to stem from the measurement method, as most of the Chinese studies as well as the Indian and Cambodian studies use the Tanaka *et al.* (2010) approach. Instead, differences in market conditions, government support, and the perceived reliability of new technologies may explain the varying adoption patterns. Risk aversion is also linked to increased insurance uptake. For example, Jianjun *et al.* (2015) and Jin *et al.* (2016) find that Chinese farmers with higher risk aversion

Table 3.1. Risk preference estimates in Asian countries. (Authors' own table).

Study	Country	Measure	Risk-related activity	Sample size	Direction of effect
Panel A: Production risks					
Liu (2013)	China	Tanaka *et al*. (2010)	Technology adoption	320 cotton farmers	Higher risk and loss aversion → Lower likelihood to adopt Bt cotton seeds
Wu and Li (2023)	China	Holt and Laury (2002)	Technology adoption	688 rice and wheat farmers	Higher risk aversion → Lower likelihood to adopt water-saving seeds
Patil and Veettil (2024)	India	Eckel and Grossmann (2008)	Technology adoption	3,517 rice farmers	Higher risk aversion → less likely to adopt mechanization Higher risk aversion → more likely to adopt stress tolerant rice variants
Gao *et al*. (2022)	China	Dohmen *et al*. (2011)	Technology adoption	48 hog farmers	Higher risk aversion → Lower likelihood to adopt African Swine Fever resistant semen
Hui *et al*. (2023)	China	Tanaka *et al*. (2010)	Technology adoption	1,038 rice farmers	Higher risk and loss aversion → Lower likelihood to adopt low carbon technologies
Ward and Singh (2015)	India	Tanaka *et al*. (2010)	Technology adoption	491 rice farmers	Higher risk and loss aversion → Higher likelihood to adopt drought-tolerant rice seeds
Shimamoto *et al*. (2018)	Cambodia	Tanaka *et al*. (2010)	Technology adoption	142 rice farmers	Higher risk aversion → Higher likelihood to adopt post-harvest technology
Jianjun *et al*. (2015)	China	Dohmen *et al*. (2011)	Technology adoption + Insurance	200 farmers	Higher risk aversion → Lower likelihood to adapt in terms of planting new crop varieties and adopting new technology Higher risk aversion → Higher likelihood to purchase weather index crop insurance

Jin *et al.* (2016)	China	Dohmen *et al.* (2011)	Insurance	200 wheat and soybean farmers	Higher risk aversion → Higher likelihood to purchase weather index crop insurance
Senapati (2024)	India	Holt and Laury (2002)	Insurance	400 farmers	
Liu and Huang (2013)	China	Tanaka *et al.* (2010)	Pesticide use	320 cotton farmers	Higher risk aversion → More pesticide use Higher loss aversion → Less pesticide use
Gong *et al.* (2016)	China	Holt and Laury (2002)	Pesticide use	300 subsistence and semi-subsistence farmers	Higher risk aversion → More pesticide use
Hou *et al.* (2020)	China	Tanaka *et al.* (2010)	Pesticide use	298 rice farmers	Higher loss aversion → More pesticide use
Salazar and Rand (2020)	Vietnam	Survey Questions	Pesticide use	2,205 farmers	Higher risk aversion → Less pesticide use
Qiao and Huang (2021)	China	Tanaka *et al.* (2010)	Fertilizer use	306 cotton farmers	Higher risk aversion → More fertilizer use only if elasticity of fertilizer use is high
Khor *et al.* (2018)	Vietnam	Holt and Laury (2002) and Dohmen *et al.* (2011)	Fertilizer use	294 maize farmers	Higher risk aversion → Less fertilizer use among poorer farmers
Sarwowosri and Mußhoff (2020)	Indonesia	Holt and Laury (2002)	Crop diversification	437 rubber and 199 oil palm farmers	Higher risk aversion → Higher likelihood to diversify
Khanal *et al.* (2019)	India	Eckel and Grossmann (2008)	Crop diversification	643 onion farmers	Higher risk aversion → Higher likelihood to diversify aside improved seed adoption, preventive measures against diseases and pests

Continued

Table 3.1. Continued.

Study	Country	Measure	Risk-related activity	Sample size	Direction of effect
Yu *et al.* (2021)	China	Tanaka *et al.* (2010)	Cooperative membership	385 vegetable farmers	Higher risk aversion → Higher uptake of cooperative membership
Panel B: Market risks					
Peng and Xu (2024)	China	Dohmen *et al.* (2011)	Price risk	409 large-scale grain farmers	Higher risk aversion → Avoidance of price risk through contract farming and diversification
Mohan (2020)	Nepal	Eckel and Grossmann (2008)	Organic certification	270 tea farmers	Higher risk aversion → Higher likelihood to be certified organic
Panel C: Institutional risk					
Wang *et al.* (2020)	China	Dohmen *et al.* (2011) and Weber *et al.* (2002)	Subsidy for pro-environmental production	712 hog farmers	No association
Panel D: Personal risks					
Nainggolan and Rommel (2023)	Indonesia	Tanaka *et al.* (2010)	Willingness to become a farmer	577 students at agricultural university	No association
Si *et al.* (2021)	China	Survey Question	Exposure to environmental pollution	714 rural female Chinese pig breeders	Higher risk aversion → Lower level of environmental exposure

Note: [a] → indicates associations rather than causal relationships.

are more likely to purchase weather index crop insurance. Likewise, Senapati (2024) found that that risk aversion does affect insurance decisions, but not on its own. Instead, it matters when it interacts with how much loss the farmer expects. This implies that risk-averse farmers are more likely to buy insurance if they believe they could lose a lot.

A similar association is observed in the context of pesticide use. Studies in China (Liu and Huang, 2013; Gong *et al.*, 2016) find that higher risk aversion is linked to greater pesticide use. However, the effect of loss aversion on pesticide use remains inconclusive, with Liu and Huang (2013) finding that higher loss aversion leads to less pesticide use, while Hou *et al.* (2020) report the opposite. One reason for this ambiguity could be explained by the fact that excessive pesticide use makes production riskier (Salazar and Rand, 2020).

With respect to fertilizer use, Khor *et al.* (2018) suggests that poorer Vietnamese upland maize farmers tend to reduce their fertilizer intensity as their risk aversion increases. However, no such relationship is found among wealthier farmers. In China, Qiao and Huang (2021) find that the association between risk aversion and fertilizer use depends on the output elasticity, i.e. when the elasticity is high, farmers apply more fertilizer with increased risk aversion. These findings that risk aversion is associated with increased pesticide use and lower fertilizer use, especially among poorer farmers, aligns with the classification of pesticides as damage control inputs and fertilizer as yield-enhancing inputs, which often depend on liquidity (Jagnani *et al.*, 2021; Liebenehm *et al.*, 2023).

Lastly, crop diversification decisions as a coping strategy to manage production risks are associated with higher risk aversion in Indonesia (Sarwosri and Mußhoff, 2020) and India (Khanal *et al.*, 2019), as farmers spread their risk across multiple crops.

In Panel B, under market risks, there is growing evidence that farmers with higher risk aversion tend to adopt a range of strategies aimed at mitigating price volatility – a major source of income instability in agricultural markets. These strategies typically include contract farming, crop diversification, cooperative membership and organic certification, each of which serves to reduce uncertainty in output pricing or improve bargaining power.

In China, for instance, Peng and Xu (2024) report that risk-averse grain farmers are more likely to enter into contractual agreements with buyers or processors to avoid price risks. By locking in prices ahead of the production cycle, these arrangements help insulate farmers from market fluctuations at harvest time. In Nepal, Mohan (2020) finds that more risk-averse farmers are more likely to seek organic certification for their products. This decision is influenced not only by the opportunity to access a price premium but also by the perception that organic markets – often smaller and more specialized – may offer greater price stability and long-term buyer relationships. Yu *et al.* (2021) found that risk-averse farmers are more likely to join a cooperative.

In Panel C, under institutional risks, given the unpredictability and sometimes limited accessibility of formal institutional support, farmers have developed various strategies to mitigate institutional risk. One such strategy is political engagement and advocacy. Empirical findings suggest that the relationship between risk aversion and engagement with institutional programs is not always straightforward. For instance, Wang *et al.* (2020) observe no significant association between risk aversion and participation in pro-environmental subsidy programs among Chinese hog farmers. Finally, in Panel D, under personal risk, Nainggolan and Rommel (2023) find that risk aversion does not affect students' willingness to become farmers. This suggests that personal career decisions may be influenced by factors beyond risk preference alone in Indonesia. In contrast, Si *et al.* (2021) shows that Chinese farmers with higher risk aversion were more inclined to adopt measures to protect their health. This behavior reflects a proactive approach to mitigating personal health risks associated with livestock farming.

In summary, higher risk aversion generally leads to lower technology adoption in China, while in India and Cambodia, it is linked to greater adoption of risk-reducing technologies. Risk-averse farmers in China are also more likely to purchase insurance and increase pesticide use, with fertilizer use rising when output elasticity is high. In contrast, poorer farmers in Vietnam reduce fertilizer use as their risk aversion increases. Additionally, higher risk aversion is associated with greater crop diversification in India and Indonesia, as well as price risk avoidance through contract farming in China and securing price premiums via organic certification in Nepal.

Conclusion and Outlook

As this chapter has shown, much has been learned about farmer's risk behavior in different geographic, ecological and socioeconomic contexts. Since farming, per se, is a risky business, agricultural decision makers have learned to cope with shocks and risks. While we can conclude that generally, farmers are moderately risk averse, there is a large variation in how they react to uncertain and probabilistic events across cultural and socioeconomic conditions. It is reasonable to assume that the likelihood of co-variate and idiosyncratic shocks in the near future stemming from climate, biology as well as economic and political factors will increase (Aryal *et al.*, 2020). Thus, the question is, how decision-makers will react. Will farmers react more carefully with new technologies, designed to increase productivity or will they be more prepared to take into account the possibility of failure? In other words, are farmers in Asia willing to take up higher risks or will they become more risk averse? Some possible answers can be found in some literatures. For example, Gloede *et al.* (2015) in a sample of some 4000 rural households in Thailand and Vietnam, found that experiences with shocks leads to higher risk aversion. Liebenehm (2018), using the same panel data set, found significant changes in risk attitudes over time among rural households in Thailand and Vietnam with differences between the two countries as regards the type of shocks. In a recent paper Liebenehm *et al.* (2023) analysed how individual risk aversion changes in response to shocks and its effect on yields and prices. The authors found that rainfall shocks increase individuals' risk aversion, especially among households that are net buyers of food. As pointed out by Hallegattet *et al.* (2018) climate change can lead to a significant increase in poverty. If economic political crises come in addition, poverty traps may re-emerge on a large scale in Asia.

Given an increasingly risky environment, it seems mandatory that additional insurance mechanisms must be put in place by governments. Aside from new, climate-smart technologies these should also include community-based emergency funds as well as effective warning systems with real-time information.

References

Al Lily, M. and Liebenehm, S. (2024) Risk and time preferences among the urban poor in Saudi Arabia. *Review of Development Economics* 29, 1613–1633. DOI: 10.1111/rode.13178.

Arrow, K.J. (1971) The firm in general equilibrium theory. In: *The Corporate Economy: Growth, Competition, and Innovative Potential*. London: Palgrave Macmillan, pp. 68–110.

Binswanger, H.P. (1980) Attitudes toward risk: experimental measurement in rural India. *American Journal of Agricultural Economics* 62(3), 395–407.

Aryal, J.P., Sapkota, T.B., Khurana, R., Khatri-Chhetri, A., Rahut, D.B., *et al.* (2020) Climate change and agriculture in South Asia: adaptation options in smallholder production systems. *Environment, Development and Sustainability* 22(6), 5045–5075.

Binswanger, H.P. (1981) Attitudes toward risk: theoretical implications of an experiment in rural India. *The Economic Journal* 91(364), 867–890.

Blais, A.-R. and Weber, E.U. (2006) A Domain-Specific Risk-Taking (DOSPERT) scale for adult populations. *Judgment and Decision Making* 1(1), 33–47. DOI: 10.1017/S1930297500000334.

Clark, E.M., Merrill, S.C., Trinity, L., Bucini, G., Cheney, N., *et al.* (2020) Using experimental gaming simulations to elicit risk mitigation behavioral strategies for agricultural disease management. *PLoS ONE* 15(3), e0228983. DOI: 10.1371/journal.pone.0228983.

Cook, J.H. (2015) Confusion in risk aversion experiments in low-income countries (SSRN Working Paper No. 2617701). Social Science Research Network. https://doi.org/10.2139/ssrn.2617701.

Crosetto, P. and Filippin, A. (2013) The "bomb" risk elicitation task. *Journal of Risk and Uncertainty* 47, 31–65.

Crosetto, P. and Filippin, A. (2016) A theoretical and experimental appraisal of four risk elicitation methods. *Experimental Economics* 19(3), 613–641. DOI: 10.1007/s10683-015-9457-9.

Dave, C., Eckel, C.C., Johnson, C.A. and Rojas, C. (2010) Eliciting risk preferences: when is simple better? *Journal of Risk and Uncertainty* 41(3), 219–243. DOI: 10.1007/s11166-010-9103-z.

Deutsch, C.A., Tewksbury, J.J., Tigchelaar, M., Battisti, D.S., Merrill, S.C., *et al.* (2018) Increase in crop losses to insect pests in a warming climate. *Science* 361(6405), 916–919.

Dias, T., Dukes, A., and Antunes, P.M. (2015) Accounting for soil biotic effects on soil health and crop productivity in the design of crop rotations. *Journal of the Science of Food and Agriculture* 95(3), 447–454. DOI: 10.1002/jsfa.6565.

Dohmen, T., Falk, A., Huffman, D., Sunde, U., Schupp, J., *et al.* (2011) Individual risk attitudes: measurement, determinants and behavioral consequences. *Journal of the European Economic Association* 9(3), 522–550. DOI: 10.1111/j.1542-4774.2011.01015.x

Du, S., Liu, J., and Fu, Z. (2021) The impact of village rules and formal environmental regulations on farmers' cleaner production behavior: new evidence from China. *International Journal of Environmental Research and Public Health* 18(14), 7311. DOI: 10.3390/ijerph18147311.

Eckel, C.C. and Grossman, P.J. (2008) Forecasting risk attitudes: an experimental study using actual and forecast gamble choices. *Journal of Economic Behavior & Organization* 68(1), 1–17. DOI: 10.1016/j.jebo.2008.04.006.

Falk, A., Becker, A., Dohmen, T., Enke, B., Huffman, D., *et al.* (2018) Global evidence on economic preferences. *Quarterly Journal of Economics* 133(4), 1645–1692. DOI: 10.1093/qje/qjy013.

Falk, A., Becker, A., Dohmen, T., Huffman, D. and Sunde, U. (2023) The preference survey module: A validated instrument for measuring risk, time, and social preferences. *Management Science* 69(4), 1935–1950. DOI: 10.1287/mnsc.2022.4455.

Filiz, I., Nahmer, T., Spiwoks, M. and Gubaydullina, Z. (2020) Measurement of risk preference. *Journal of Behavioral and Experimental Finance* 27, 100355. DOI: 10.1016/j.jbef.2020.100355.

Gao, S., Grebitus, C. and Schmitz, T. (2022) Effects of risk preferences and social networks on adoption of genomics by Chinese hog farmers. *Journal of Rural Studies* 94, 111–127. DOI: 10.1016/j.jrurstud.2022.05.018.

Gloede, O., Lukas, M. and Hermann, W. (2015) Shocks, individual risk attitude, and vulnerability to poverty among rural households in Thailand and Vietnam. *World Development* 71(July), 54–78.

Gong, Y., Baylis, K., Kozak, R., and Bull, G. (2016) Farmers' risk preferences and pesticide use decisions: evidence from field experiments in China. *Agricultural Economics* 47(4), 411–421. DOI: 10.1111/agec.12240.

Habib-Ur-Rahman, M., Ahmad, A., Raza, A., Hasnain, M.U., Alharby, H.F., *et al.* (2022) Impact of climate change on agricultural production: issues, challenges, and opportunities in Asia. *Frontiers in Plant Science* 13, 925548. DOI: 10.3389/fpls.2022.925548.

Hallegatte, S., Fay, M. and Barbier, E.B. (2018) Poverty and climate change: Introduction. *Environment and Development Economics* 23(3), 217–233.

Holt, C.A. and Laury, S.K. (2002) Risk aversion and incentive effects. *American Economic Review* 92(5), 1644–1655.

Hardaker, J.B., Lien, G., Anderson, J.R. and Huirne, R.B.M. (2015) *Coping with Risk in Agriculture: Applied Decision Analysis* (3rd ed.). CABI, Wallingford, UK.

Hardeweg, B., Menkhoff, L. and Waibel, H. (2013) Experimentally validated survey evidence on individual risk attitudes in rural Thailand. *Economic Development and Cultural Change* 61(4), 859–888.

Hou, L., Liu, P., Huang, J. and Deng, X. (2020) The influence of risk preferences, knowledge, land consolidation, and landscape diversification on pesticide use. *Agricultural Economics* 51(5), 759–776. DOI: 10.1111/agec.12590.

Hui, M.A.O., Quan, Y.R. and Yong, F.U. (2023) Risk preferences and the low-carbon agricultural technology adoption: evidence from rice production in China. *Journal of Integrative Agriculture* 22(8), 2577–2590.

Jagnani, M., Barrett, C.B., Liu, Y. and You, L. (2021) Within-season producer response to warmer temperatures: defensive investments by Kenyan farmers. *The Economic Journal* 131(633), 392–419. DOI: 10.1093/ej/ueaa063.

Jianjun, J., Yiwei, G., Xiaomin, W. and Nam, P.K. (2015) Farmers' risk preferences and their climate change adaptation strategies in the Yongqiao District, China. *Land Use Policy* 47, 365–372. DOI: 10.1016/j.landusepol.2015.04.028.

Jin, J., Wang, W. and Wang, X. (2016) Farmers' risk preferences and agricultural weather index insurance uptake in rural China. *International Journal of Disaster Risk Science* 7(4), 366–373. DOI: 10.1007/s13753-016-0108-3.

Kahneman, D. and Tversky, A. (1979) Prospect theory: an analysis of decision under risk. *Econometrica* 47(2), 263–291. DOI: 10.2307/1914185.

Khanal, A.R., Mishra, A.K. and Kumar, A. (2019) *Risk Preference and Adoption of Risk Management Strategies: Evidence from High-Value Crop Production in Emerging Economy*. Agricultural and Applied Economics Association, Wisconsin, USA.

Khor, L.Y., Ufer, S., Nielsen, T. and Zeller, M. (2018) Impact of risk aversion on fertiliser use: evidence from Vietnam. *Oxford Development Studies* 46(4), 483–496. DOI: 10.1080/13600818.2018.1445212.

Komarek, A.M., De Pinto, A. and Smith, V.H. (2020) A review of types of risks in agriculture: What we know and what we need to know. *Agricultural Systems* 178, 102738. DOI: 10.1016/j.agsy.2019.102738.

Lejuez, C.W., Aklin, W.M., Zvolensky, M.J. and Pedulla, C.M. (2003) Evaluation of the balloon analogue risk task (BART) as a predictor of adolescent real-world risk-taking behaviours. *Journal of Adolescence* 26(4), 475–479. DOI: 10.1016/S0140-1971(03)00036-8.

Liao, C. and Chen, Y. (2017) Farmers' information management in developing countries – a highly asymmetric information structure. *Production and Operations Management* 26(6), 1207–1220. DOI: 10.1111/poms.12678.

Liebenehm, S. and Waibel, H. (2014) Simultaneous estimation of risk and time preferences among small-scale cattle farmers in West Africa. *American Journal of Agricultural Economics* 96(5), 1420–1438. DOI: 10.1093/ajae/aau056.

Liebenehm, S. (2018) Temporal stability of risk attitudes and the impact of adverse shocks – a panel data analysis from Thailand and Vietnam. *World Development* 102, 262–274.

Liebenehm, S., Schumacher, I. and Strobl, E. (2023) Rainfall shocks and risk aversion: evidence from Southeast Asia. *American Journal of Agricultural Economics* 106(1), 145–176.

Liu, E.M. (2013) Time to change what to sow: risk preferences and technology adoption decisions of cotton farmers in China. *The Review of Economics and Statistics* 95(4), 1386–1403. DOI: 10.1162/REST_a_00295.

Liu, E.M. and Huang, J. (2013) Risk preferences and pesticide use by cotton farmers in China. *Journal of Development Economics* 103, 202–215. DOI: 10.1016/j.jdeveco.2012.12.005.

Lönnqvist, J.-E., Verkasalo, M., Walkowitz, G. and Wichardt, P.C. (2015) Measuring individual risk attitudes in the lab: task or ask? An empirical comparison. *Journal of Economic Behavior & Organization* 119, 254–266. DOI: 10.1016/j.jebo.2015.08.003.

Merrill, S.C., Koliba, C.J., Moegenburg, S.M., Zia, A., Parker, J., *et al.* (2019) Decision-making in livestock biosecurity practices amidst environmental and social uncertainty: evidence from an experimental game. *PLoS ONE* 14(4), e0214500. DOI: 10.1371/journal.pone.0214500.

Mohan, S. (2020) Risk aversion and certification: evidence from the Nepali tea fields. *World Development* 129, 104903. DOI: 10.1016/j.worlddev.2020.104903.

Nainggolan, L.B. and Rommel, J. (2023) An experiment on the link between risk preferences and the willingness to become a farmer. *Journal of the Agricultural and Applied Economics Association* 2(4), 686–702. DOI: 10.1002/jaa2.85.

Patil, V. and Veettil, P.C. (2024) Farmers' risk attitude, agricultural technology adoption and impacts in Eastern India. *Agriculture & Food Security* 13(1), 50.

Peng, P. and Xu, Z. (2024) Subjective preferences, liquidity constraints and price risk management under large-scale farm management. *China Agricultural Economic Review* 16(1), 76–96. https://doi.org/10.1108/CAER-12-2022-0287

Perloff, J.M. and Rausser, G.C. (1983) The effect of asymmetrically held information and market power in agricultural markets. *American Journal of Agricultural Economics* 65(2), 366–372. DOI: 10.2307/1240894.

Porter, J.R. and Semenov, M.A. (2005) Crop responses to climatic variation. *Philosophical Transactions of the Royal Society B: Biological Sciences* 360(1463), 2021–2035. DOI: 10.1098/rstb.2005.1752.

Pratt, J.W. (1964) Risk aversion in the small and in the large. *Econometrica* 32(1/2), 122–136. DOI: 10.2307/1913738.

Qiao, F. and Huang, J. (2021) Farmers' risk preference and fertilizer use. *Journal of Integrative Agriculture* 20(7), 1987–1995. DOI: 10.1016/S2095-3119(20)63450-5.

Rosenzweig, C., Elliott, J., Deryng, D., Ruane, A.C., Müller, C., *et al.* (2014) Assessing agricultural risks of climate change in the 21st century in a global gridded crop model intercomparison. *Proceedings of the National Academy of Sciences USA* 111(9), 3268–3273. DOI: 10.1073/pnas.1222463110.

Salazar, C. and Rand, J. (2020) Pesticide use, production risk and shocks: the case of rice producers in Vietnam. *Journal of Environmental Management* 253, 109705.
Sands, R. and Westcott, P. (2011) *Impacts of Higher Energy Prices on Agriculture and Rural Economies*. United States Department of Agriculture, Economic Research Report Number 123, Washington DC, USA.
Sarwosri, A.W. and Mußhoff, O. (2020) Are risk attitudes and time preferences crucial factors for crop diversification by smallholder farmers? *Journal of International Development* 32(6), 922–942. DOI: 10.1002/jid.3483.
Satyasai, K. and Pereira, C. (2019) Can better market access lead to higher incomes for farmers? *Agricultural Economics Research Review* 32(conf), 113–122. DOI: 10.5958/0974-0279.2019.00021.1.
Savary, S., Willocquet, L., Pethybridge, S.J., Esker, P., McRoberts, N., *et al.* (2019) The global burden of pathogens and pests on major food crops. *Nature Ecology & Evolution* 3(3), 430–439.
Senapati, A.K. (2024) Do farmers value insurance against extreme droughts and floods? Evidence from Odisha, India. *Global Business Review* 25(3), 561–584.
Shimamoto, D., Yamada, H. and Wakano, A. (2018) The effects of risk preferences on the adoption of post-harvest technology: Evidence from rural Cambodia. *The Journal of Development Studies* 54(10), 1819–1837. DOI: 10.1080/00220388.2017.1329527.
Si, R., Zhang, X., Yao, Y. and Lu, Q. (2021) Risk preference, health risk perception, and environmental exposure nexus: evidence from rural women as pig breeders, China. *Social Indicators Research* 162, 1–28.
Tanaka, T., Camerer, C.F. and Nguyen, Q. (2010) Risk and time preferences: Linking experimental and household survey data from Vietnam. *American Economic Review* 100(1), 557–571. DOI: 10.1257/aer.100.1.557.
Tiwari, R., Sharma, M.C., Mishra, K.K. and Singh, B.P. (2013) Economic impacts of infectious diseases of livestock. *The Indian Journal of Animal Sciences* 83(3), 316–320.
Tomek, W.G. and Peterson, H.H. (2001) Risk management in agricultural markets: a review. *Journal of Futures Markets* 21(10), 953–985. DOI: 10.1002/fut.2004.
Tversky, A. and Kahneman, D. (1992) Advances in prospect theory: cumulative representation of uncertainty. *Journal of Risk and Uncertainty* 5(4), 297–323. DOI: 10.1007/BF00122574.
Ullah, R., Shivakoti, G.P., Zulfiqar, F. and Kamran, M.A. (2016) Farm risks and uncertainties: Sources, impacts and management. *Outlook on Agriculture* 45(3), 199–205. DOI: 10.1177/0030727016665440.
Van Es, H.M. and Karlen, D.L. (2019) Reanalysis validates soil health indicator sensitivity and correlation with long-term crop yields. *Soil Science Society of America Journal* 83(3), 721–732. DOI: 10.2136/sssaj2018.09.0338.
Vieider, F.M., Lefebvre, M., Bouchouicha, R., Chmura, T., Hakimov, R., *et al.* (2015) Common components of risk and uncertainty attitudes across contexts and domains: evidence from 30 countries. *Journal of the European Economic Association* 13(3), 421–452. DOI: 10.1111/jeea.12102.
Wang, J., Yang, C., Ma, W. and Tang, J. (2020) Risk preference, trust, and willingness-to-accept subsidies for pro-environmental production: An investigation of hog farmers in China. *Environmental Economics and Policy Studies* 22(3), 405–431. DOI: 10.1007/s10018-020-00262-x.
Ward, P.S. and Singh, V. (2015) Using field experiments to elicit risk and ambiguity preferences: behavioural factors and the adoption of new agricultural technologies in rural India. *Journal of Development Studies* 51(6), 707–724. DOI: 10.1080/00220388.2014.989996.
Weber, E.U., Blais, A. and Betz, N.E. (2002) A domain-specific risk-attitude scale: Measuring risk perceptions and risk behaviors. *Journal of Behavioral Decision Making* 15(4), 263–290. DOI: 10.1002/bdm.414.
Wu, H. and Li, J. (2023) Risk preference, interlinked credit and insurance contract and agricultural innovative technology adoption. *Journal of Innovation & Knowledge* 8(1), 100282. DOI: 10.1016/j.jik.2022.100282.
Yang, T., Siddique, K.H.M. and Liu, K. (2020) Cropping systems in agriculture and their impact on soil health – a review. *Global Ecology and Conservation* 23, e01118. DOI: 10.1016/j.gecco.2020.e01118.
Yu, L., Chen, C., Niu, Z., Gao, Y., Yang, H., *et al.* (2021) Risk aversion, cooperative membership and the adoption of green control techniques: evidence from China. *Journal of Cleaner Production* 279, 123288.Table 3.1. Continued.

4 Agricultural Development and Policy in Thailand – With a Case Study of the Vegetable Subsector

Marc Völker, Bernd Hardeweg and Hermann Waibel

Abstract

In this chapter a historical review of agricultural development in Thailand is provided. It is shown how general development theories and policies have shaped agricultural policies and how this has formed the structure of agriculture. Bringing significant events into the picture, the performance of the agricultural sector in over several development decades, following the UN classification, is reviewed. Thus, an easily accessible overview of the evolution of agriculture in the country is delivered. This can help to better understand current constraints and opportunities of the agricultural sector and offer some entry points for a productive policy debate and target-orientated subsequent research. To further illustrate the process of agricultural development in Thailand, a case study of the vegetable sector, which is based on an in-depth PhD research study, is presented. The paper concludes with an outlook for agriculture and rural development in Thailand.

Introduction

Similar to other countries in Asia, Thailand has been an agricultural country in the past. It is now an emerging service economy and has reached the status of an upper middle-income country. Economic growth has been facilitated by labour and capital transfer from the agricultural to the industrial and service sectors of the economy. Development and agricultural policies have been a major driver of these changes. Hence, this chapter takes a historical approach to analyse the structural transformation, or lack thereof, by development decades, as these were defined by the United Nations. Using five categories, namely (i) general development policy, (ii) agricultural policies, (iii) structure of agriculture, (iv) significant events and (v) the performance of the agricultural sector, the history of sectoral development, is presented in Table 4.1, using keywords. A detailed analysis by development decade follows. Formally, the first development decade starts in the 1960s, however, the analysis includes the decade before (1950–1960) and ends with the current decade.

To briefly summarize the results of the development and agricultural review upfront, a couple of key points can be highlighted. First, Thailand is an economic development success story. For several decades, it had impressive rates of economic growth and has been outstandingly successful to curb excessive population growth. Thus, it has reached the status of an emerging market economy and upper middle-income country with the prospect of eventually becoming an industrialized country. Thailand has shown to be fairly resilient toward numerous ecological, economic and political crises over the last seven

DOI: 10.1079/9781800627338.0004

Table 4.1. Development overview by decades. (Author's own table).

Decades	General Development Policy	Agricultural Policies	Structure of Agriculture	Significant Events	Performance of Agriculture
Prior Decade (1950–1960)	State enterprises and investments in infrastructure; economic nationalism	Irrigation projects, Rice breeding programs; forest concession policy, opening up of new agricultural land and rural settlements	Dominated by rice production and peasant-based small-scale farming	Growing influence of USA in development assistance, partly through multi-lateral aid agencies	Output increased; stagnant or declining productivity
First decade (1961–1970)	Less support of state enterprises; liberal economic policy; free market approach, 1st and 2nd national economic development plan; compulsory primary education	Promote crop diversification away from rice; establishment of the BAAC	Expansion of field crops (e.g. maize, cassava) but rice remains the major crop; agricultural GDP declines, labor shares remain high.	US military bases and main roads because of Vietnam War; counter-insurgency but less direct rural development assistance	Output increase due to land expansion, yields declined
Second decade 1971–1980	Public investments in infrastructure and defense financed by foreign capital	Promotion of agribusiness through public incentives; irrigation dams, rural roads and electrification	Expansion of agribusiness, high-value/profit agricultural commodities (poultry, vegetables)	Oil price crisis; increased demand from Japan for fruits and vegetables	Crop productivity increased
Third decade 1981–1990	Structural adjustment, deregulation and liberalization, open economy, infrastructure investments; import substitution while export promotion; high rates of economic growth	Taxation of agriculture (rice premium) until 1985; insufficient land tenure security	GDP share and export share of agriculture declined; adoption of new technologies low; dualistic structures in agriculture emerged	Low agricultural prices, structural adjustment; man-made pest outbreaks; environmental concerns in agriculture emerged	Area expansion reached its peak; agricultural output still increasing, agriculture as the engine of growth
Fourth decade 1991–2000	Deregulation of financial markets, inflow of large amounts of foreign capital; FDI in retail markets	Trade liberalization (Uruguay Round) domestic input subsidies continued	GDP share and labor share decline further; out-migration and mechanization	High economic growth at the beginning, negative growth after Asian Financial Crisis (1997)	Rice yields and fertilizer use increased moderately; still below other Asian countries

Continued

Table 4.1. Continued.

Decades	General Development Policy	Agricultural Policies	Structure of Agriculture	Significant Events	Performance of Agriculture
Fifth decade (2001–2010)	Thaksin government with strong support for rural electorate, ended by military coup in 2006; followed by political and economic turbulences;	Thailand Village Fund, paddy pledging price significantly increased; debt moratorium for farmers; further subsidy increase in post-Thaksin governments	Adoption and diffusion of mechanical technology for small-scale farmers; average farm size stagnant; land transfers made more difficult due to changes in inheritance practices; slowdown in rural-urban migration; shock and age-related return migration	Financial crisis continued until early decade, credit constraints, drop in off-farm income; food price crisis	Agricultural Value Added increased; minor increase in crop yields
Sixth decade (2011–2020)	Yingluck Government terminated by military coup in 2014; military-backed Prayut government beyond 2020	Rice Pledging Scheme until 2014; area-based subsidy (limited to 20 rai); income guarantee program	The GDP share of agriculture fell below 10%, labor share still 30%. Arable land, picked up again; deforestation continued; average farm size remains constant	Country-wide flood; military coup put an end to pro- poor rural development policies	Input intensity declined; crop production, productivity stagnant; livestock production increased
Beyond 2020	End of Prayut government in 08/2023; political instability with six different prime ministers after 2023 elections	Climate Change Action Plan for the Agricultural Sector (CCAPA) 2023–2027	Further crop diversification; more mechanization; intro-duction of digital tools	Severe drought in 2020 with impact on off-season rice and corn pro-duction; climate vulnerability increasing	GDP growth around 2 %; negative growth in agriculture, 2020 and 2024; GDP share agriculture 8.7%, labor shares still the same

decades. Yet, the national policy environment, with a continuously strong influence of the established institutions, is adding domestic uncertainty to the risks that are already imminent from overall world development. Second, Thailand has been lucky when it comes to food production and agriculture. Relative to its population, land resources are bountiful, and in many commodities, total output exceeds domestic demand. Hence, Thailand has become a lead exporter in several commodities, especially for rice. However, agricultural productivity is low. For example, rice yields are only a fraction of those of its major competitors, such as India and Vietnam. They have remained almost stagnant to date. Third, in spite of various programs to subsidize small-scale farmers, to date, there has not been a comprehensive policy of rural development, enabling agricultural structural transformation towards higher value-added production. The agricultural sector had been taxed and its resources transferred for developing the industrial and service sector. Prior to the election of Thaksin, in 2001, the political elite did not pay serious attention to the rural electorate. When the military finally took over again in 2014, populist policies were abandoned and agricultural policies shifted back to the subsidy style, incapable to trigger significant change. Most importantly, average farm size stayed stagnant over all development decades, with less than 4 ha per household. Less than 1% of farms are of the size that could generate sufficient income for a family. Hence, practically all farm households must earn their income from multiple sources with a decreasing share of income from their own agriculture. Thus, agriculture in Thailand is caught in a "small-scale farmer trap." Fourth, dualistic structures are emerging in agriculture. On the one hand, there is the overwhelming share of millions of small-scale farms, who in effect, are all part-time farmers. While they have modernized to some extent, through scale-specific mechanical technology, their productivity remains low. On the other hand, there is the world of the agribusiness conglomerates such as Charoen Pokaphand, Betagro and Centaco who are often engaged in contract farming with small-scale farmers. Some commodities, like pork, chicken and milk are produced in large-scale farms while the majority of crops remain with small-scale farmers. Although there exists a 20-year (2017–2036) plan by the Ministry of Agriculture and Cooperatives to develop the agricultural sector, little is said there about structural transformation and farm size development targets.

In this chapter, we will describe how development and agricultural policies have evolved over the decades, what has changed and what has not. We end the chapter with some outlook as regards the prospects for Thai agriculture.

The 1950s

Thailand's general development policies in the 1950s were characterized by an attempt to establish state enterprises in various sectors such as glass, paper, sugar and textiles (Robinson *et al.*, 1991). This was largely due to the government's and the Thai–Chinese business community's fear of foreign influence when the USA had become the dominant political power in Southeast Asia. At the same time, there was a growing influence of the local communist party (Akrasanee and Ajanant, 1986, p. 80, in Booth, 2016). However, the economic nationalist approach of the early 1950s proved to be ill-fated due to the inefficiency of state enterprises. Towards the end of the decade, a new government came to power and changed the nation's development strategy to the promotion of the private sector and large-scale World Bank-financed investments in infrastructure (Robinson *et al.*, 1991).

In terms of agricultural policies, agricultural development in the 1950s focused mainly on irrigation projects. The first large-scale irrigation schemes in the Central Plain region were completed to support rice production and exportation (Kermel-Torrès, 2004) and rice breeding programs were implemented (Doner, 1974). The Thai government also promoted rice breeding and the use of mineral fertilizer (ECAFE/FAO Agriculture Division, 1955). A special project was the support for domestic production of kenaf, as major input for state-owned rice sack factories. To protect local production an import ban on kenaf was established (Goss and Burch, 2001). Towards the end of the decade, Thailand introduced taxes on agricultural commodities and exports to fund subsequent investments in agriculture (Kermel-Torrès, 2004).

The structure of agriculture in the 1950s was heavily dominated by rice production, which accounted for 92% of the area under crop cultivation. Rice production alone contributed around 25% to the country's GDP while approximately 86% of the total working-age population was engaged in agriculture (ECAFE/FAO Agriculture Division, 1955; Robinson *et al.*, 1991). Exports of agricultural products were a major source of income for the government through export taxes and state trading. Peasant-based small-scale farming was a widespread practice and the vast majority of farmers owned their land. Apart from rice, other common crops included maize, sugarcane, peanuts, cotton, tobacco and vegetables. The area of agricultural land under irrigation increased considerably. The area available for further agricultural development during that time was substantial, as only around one-third of the total land was cultivated (ECAFE/FAO Agriculture Division, 1955).

With regards to significant events, while there are no reports about major weather shocks, the broader geo-political situation had influenced agricultural development in Thailand. A large part of public development investments was financially supported by the USA which saw agricultural aid to Thailand as a means to counter the spread of communism in Southeast Asia. Agricultural experts were sent to the country via multi-lateral aid agencies such as the World Bank and bilateral agencies such as USAID and the Rockefeller Foundation. Foreign experts promoted the advancement of breeding programs to increase rice productivity (Doner, 1974) and emphasized the need for diversification of the Thai agricultural sector.

Agricultural performance in the 1950s was characterized by a lack of productivity increases and in some areas even declining yields of rice. Increase in rice production mainly took place through area expansion in forest land. Labor productivity was low as mechanization was still in its infancy (Goss and Burch, 2001).

The First Decade

Development policies in the 1960s were characterized by a shift away from the support of state enterprises and promoting a more private sector approach instead. The Board of Investment (BoI) was established which fostered private investment while public involvement in commerce and industry became more restricted. Other noticeable developments during that decade included the repeal of laws restricting private land ownership, the fixation of the exchange rate to the US dollar. Notably, the nation's first economic development plan was launched (Viksnins, 1973; Robinson *et al.*, 1991; Goss and Burch, 2001).

Among the main goals of agricultural policies in the 1960s was diversification, first through expansion of crops such as cassava, kenaf, maize and sugar, which were needed as inputs for the industrial sector. In the second half of the decade, diversification efforts were accelerated through financial support by the then newly founded BAAC (Bank of Agriculture and Agricultural Cooperatives) (Goss and Burch, 2001). Moreover, the nation's first two Economic Development Plans, which were launched in the 1960s, strongly promoted the development of agricultural irrigation through the construction of large multi-purpose dams, which particularly benefitted rice production in the Central Plains (Kermel-Torrès, 2004). Furthermore, farmers started to adopt labor-saving technology such as power tillers (Shintani, 2003; Rieber, 2013).

As a result of the country's agricultural diversification efforts, the structure of agriculture had changed. The share of land allocated to the cultivation of crops other than rice, such as cassava, maize, kenaf and sugarcane, increased substantially, though rice still remained the dominant crop. At the same time, the proportion of agriculture in GDP declined markedly due to the relative strength of the manufacturing and construction sectors (Goss and Burch, 2001).

The establishment of Thailand as the primary operating base and staging area for the US military in its involvement in the Vietnam war marks a significant event in the 1960s. The importance of foreign support from the USA increased dramatically due to the war. Not only did Thailand receive substantial direct foreign aid during that time, but further benefitted from investments in military and strategic infrastructure, the spending of US military personnel in the country and an expansion of exports to the US-supported Republic of Vietnam. Japanese direct investment also increased strongly (Goss and Burch, 2001). However, during the course of that decade, the focus of US aid to Thailand shifted from rural development towards counterinsurgency and security activities (Thaxton, 1973).

With regards to agricultural performance, agricultural output increased but productivity continued to decline due to the continuous substantial expansion into less fertile land (Goss and Burch, 2001; Nidhiprabha, 2019).

The Second Decade

Thailand's development policies in the 1970s were dominated by Keynesian-based theories of state interventions with capital intensive, all-inclusive development approaches. It increasingly relied on foreign debt to finance its growing public investments in infrastructure and defense (Robinson *et al.*, 1991). Similar to many other developing countries, the nation pursued an import-substitution, domestic industrialization strategy to reduce its reliance on imported goods. Priority was given to the local production of capital-intensive goods such as automobiles, pharmaceuticals and textiles (OECD, 2021). A range of protectionist policies such as the use of tariffs and other trade barriers as well as export bans of selected agricultural products and price controls of important inputs were implemented (Robinson *et al.*, 1991). Market access of foreign firms was restricted across all sectors, but exceptions were granted for foreign investment in sectors that required foreign capital and more generally to American companies due to a bilateral treaty between the two countries (OECD, 2021). A significant and far-reaching policy, during the 1970s, was the launching of the National Family Planning Program with the aim to bring down high population growth. The program was a big success especially with the establishment of the Population and Community Development Association (PDA) by Meechai Viravaidya (popularly called "*Mr. Condom*") in 1974. Fertility went down from over 6 children per woman during the 1960s to about 1.32 in the 2020s. Population growth dropped from 3% to below 0.2% during the same period.

The 1970s saw a pivotal shift in Thailand's agricultural policies with the promotion of agribusiness through a range of incentives provided by the BoI to companies located outside of the Greater Bangkok area (Goss and Burch, 2001). In terms of the continuing development of irrigation systems, more emphasis was given towards smaller structures in the underdeveloped north-east of the country (Kermel-Torrès, 2004). Moreover, agricultural technology transfers to Thailand, in particular in the form of modern rice genetic materials, were facilitated through the International Rice Research Institute (Suphannachart and Warr, 2011). In 1977, the World Bank-financed National Agricultural Extension Project (NAEP) was introduced. The project applied the popular Training &Visit approach with the establishment of a system of village extension workers (Kaset Tambon) and so-called contact farmers (Cernea, 1983; Benor *et al.*, 1984).

In terms of the structure of agriculture in the 1970s, an important shift could be observed towards the expansion of agribusiness and the growing importance of high-profit agricultural commodities, such as poultry, fruits, vegetables and fresh flowers, and export (Goss and Burch, 2001; Singhapreecha, 2014). Price and wage guarantees were introduced for the first time during that decade (Kermel-Torrès, 2004). The dominance of rice-milling and sugar refining groups declined whereas agribusiness companies which vertically integrated production, marketing, trade and retail, such as the Charoen Pokphand (CP) Group became more important (Goss *et al.*, 2000; Goss and Burch, 2001). In the south of Thailand, oil palm plantations started to replace the less profitable rubber trees or land expansion took place through deforestation (Kermel-Torrès, 2004). Technological progress in the agricultural sector continued through further mechanization (e.g. the introduction of mechanical threshers) and the increased use of fertilizer (Shintani, 2003; Rieber, 2013). Moreover, modern varieties of rice were increasingly adopted by farmers as a result of the Green Revolution reaching the developing world (Rieber, 2013).

The first OPEC oil shock was a significant event in the 1970s and led to a primary commodities price boom. Moreover, during that decade, increasing Japanese food import demand and investment positively affected the agri-food industries in Thailand (Goss and Burch, 2001).

With regards to agricultural performance, the 1970s marked a turning point. Crop yields started to increase in the second half of the decade (Nidhiprabha, 2019).

The Third Decade

In development policies during the 1980s, a turn to neoclassical and neoliberal development strategies took place. These included structural adjustment, with the liberalization and deregulation of markets and the promotion of an open economy. The principle was that free markets will generate the best welfare outcomes, provided that strong institutions were in place, effectively monitored by the government. During this period, Thailand followed open economy philosophies and provided conducive conditions for an expansion of the private sector, including low taxes, low entry barriers and at the same time extensive investment in agricultural-based industries. Foreign Direct Investment (FDI) increased. Also, Thailand became a major recipient of Official Development Assistance (ODA). As a result, infrastructure investments such as roads, ports and export processing zones were undertaken. During this decade, Thailand was on its way to join the "Asian Tigers," like South Korea and Taiwan.

Agricultural policies were dominated by government interventions in the agricultural sector, including in commodity markets. In spite of structural adjustment policies, deregulation and liberalization, significant export premiums for rice and rubber were in place, thus lowering farm gate prices. Hence, productivity-enhancing input use and introduction of innovations were compromised. Domestic protection of industries like the fertilizer industry prevailed which led to poor quality and inappropriate fertilizer products. In pest and disease control, the government had in place a \$10 million annual dole-out system for chemical pesticides to be distributed among farmers or applied via aerial spraying, in case of pest outbreaks. Unsurprisingly, the system was susceptible to misuse and corruption (Waibel, 1991). Furthermore, land tenure regulations suffered from a lack of imprecision in titling, resulting in limited acceptance of land as loan collateral. Agricultural technologies were mainly delivered by a multitude of government agencies, such as the Department of Agriculture (DOA) for plant research and the Department of Agricultural Extension (DOAE) for technology transfer. The structure of government departments largely prevented the implementation of a need-based, problem-solving approach and resulted in inefficiencies. Besides, most of the staff were deployed in Bangkok head offices, causing a lack of personnel in the rural areas. Government cooperation with the private sector was poor, for example, in areas like germplasm improvement and seed multiplication. The rice mortgage program was first initiated in 1981 with rice millers as the core agents. The program offered farmers a government-sponsored loan, valued at a predicted market price for their paddy (Mittal, 2021). In theory, once the market rebounded, farmers would sell their paddy and repay the loan. In the opposite case, farmers could default on the loan and leave their rice with the government. The financial crisis in the last phase of the decade had practically put an end to that scheme.

The structure of agriculture during the 1980s was largely the result of government policies. Following general development patterns with an inter-sectoral reallocation of resources, agriculture's GDP share continued to decline rapidly and had reached a level of 12.5% by 1990. Labor shares, however, declined at much lower rates. By the end of the decade, still some 60% of labor was in the agriculture sector resulting in low labor productivity of agriculture. However, in reality, labor shares in agriculture could be lower and labor productivity higher than reported in the statistics for a number of reasons. Firstly, many of the rural laborers are engaged in home production by unpaid work on family farms of which output is not accounted for. Secondly, in labor surveys, respondents will report agriculture as their first occupation but actually they spend more time in non-agriculture. Thirdly, agricultural wage labor is often in seasonal employment patterns, (e.g. in plantation crops) which are prone to measurement errors. Finally, employment outside agriculture is often temporary without formal contract, thus underreporting is likely.

As regards the structure of production, the share of crops in agriculture's GDP remained at close to 80%, with little change during the decade. At the same time, rice remained the dominant crop in spite of its continuous decline where the country was still the world's leading exporter.

In terms of significant events, the first part of the decade was marked by a worldwide decline in agricultural commodity prices. A major cause was the debt crisis which caused a drop in economic growth, decline in international trade, and an increase in interest rates. The debt crisis paved the way for the austerity and structural adjustment policies promoted by the International Monetary Fund (IMF). However, as the crisis was overcome, the economy grew at unprecedented high rates and the agricultural sector benefited from this growth. To a large extent, Thai agriculture during the 1980s, was spared from major natural disasters like floods and drought which occasionally occurred. However, pest outbreaks during the second half of the 1980s, caused by over- and misuse of chemical pesticides in rice and other crops with hazards to consumers and the environment, were events that raised the attention of researchers and policy analysts (Poapongsakon *et al.*, 1998). The decline of rain forests, especially in southern Thailand was a major factor that had led to severe flood events in October 1988. This prompted the government to impose a nationwide ban on logging and to set a minimum target for forest coverage of 40%.

Performance of agriculture during the third decade is shown by the growth in agricultural output, still mainly driven by the expansion of agricultural land, which, by the end of the decade, had reached its plateau (Siamwalla, 1995). On the other hand, value added in agriculture, more than doubled between 1981 and 1990, albeit at lower rates during the first half of the decade, due to the low prices. As a result of its ample land resources and relatively favorable climatic conditions, Thailand was a surplus country in many agricultural crops. Aside from rice, cassava, sugar, natural rubber and fruits became the main export commodities (Poapongsakorn *et al.*, 2006). Rural poverty which had increased during the first half of the decade, quickly went back to pre-crisis levels, attributable to a generally well-performing agricultural sector. For example, a steep increase in the livestock production took place between 1986 and 1990. By and large, however, the agricultural sector was moving into a kind of dualistic structure. One the one hand, vertically integrated, agro-industry corporations such as the Charoen Pokphand Foods and Betragro, companies engaged in poultry, dairy products, seafood and high quality vegetables had taken root. On the other hand, a largely traditional small-scale farming sector, with rice as the dominant crop remained. Most of agriculture in Thailand during the 1980s still belonged to the latter group, i.e. farming with low yields, hardly increasing over time and low levels of external inputs, especially chemical fertilizers.

The Fourth Decade

In development policies during the early 1990s, Thailand extended its open market policies to the financial sector. This had significant consequences. While prior to 1990, the Bank of Thailand (BoT) kept a tight grip on financial capital movements into and out of Thailand, these controls were largely lifted during the early 1990s. There was the hope that Bangkok could replace Hong Kong as a regional financial center (Warr, 1999). This further accelerated the growth boom of the 1980s well into the next decade. In fact, at the turn of the 1990s, Thailand registered impressive two-digit economic growth rates. At the same time, however, the "seeds" for the crisis to come were sown. Growth during this period was fueled by foreign capital inflow, by far exceeding domestic investments. To curb domestic lending, the BoT kept the exchange rate constant and raised interest rates. This led to high rates of return for foreign investors in Thailand and hereby further accelerated foreign capital inflow. Most of the capital investment went into real estate and commercial office space in Bangkok, eventually creating a bubble economy leading to declining investment quality. In 1997, the BoT devalued the Thai Baht and adopted floating exchange rates. As a result, investor confidence collapsed, capital flows reversed, exports declined, imports became expensive and consequently the economy went into a deep recession. The tourist sector which had been continuously expanding since the 1980s, benefited from a low Baht exchange rate (Zhang *et al.*, 2009) but it was unable to compensate for the slump in the economy. As submitted by many analysts, it was not only financial policies that contributed to the crisis. In other important policy areas, the Government was too cautious and hesitant. For example, there was a lack of investment in education, especially in rural areas. There were insufficient policy incentives for the SME sector and there was almost no broad-based rural development strategy. Notwithstanding the financial crisis, towards the end of the decade, Thailand had transited from an agriculture-based economy to a newly industrialized country (NIC). A negative side effect of Thailand's rapid growth during the 1990s has been an increase in environmental problems. It was only in the Seventh National Economic and Social Development Plan (1992–96) that the need to promote effective use of natural resources protection and environmental conservation was mentioned.

For agricultural policies, during the fourth decade, there weren't many specific policies for agriculture, just as in the case before. Government policy was focused on industrial growth and the reallocation of resources, especially labor from agriculture to industry and the service sector. Food security was never really an issue in Thailand; agricultural output was sufficient due to ample land resources, relative to the population. Aside from some minor price support schemes and somewhat "cosmetic" input subsidies, the only area where the government exercised significant investment in agriculture was irrigation (Siamwalla, 1999). Furthermore, access to finance was assured through the Bank of Agriculture and Agricultural Cooperatives (BAAC) which offered subsidized credits for eligible borrowers. Hence, due to the low connectedness with international financial markets, the crisis had relatively small effects on agriculture. For example, Bresciani *et al.* (2002) found that agricultural export commodities benefited from more competitive prices due to the exchange rate devaluation, while on the other hand, small-scale farmers were burdened as they lost non-farm income sources.

The structure of agriculture did not change significantly during the 1990s, compared with the previous decades. Primary production for rice and field crops continued to be dominated by the small-scale farm sector while plantation crops, like sugarcane and rubber were in the hands of somewhat larger landowners. Total arable land increased only slightly by the end of the decade. However, farm sizes remained constant despite a continuous out-migration of labor, at an average rate of close to 2% per year. At the same time, agriculture became more mechanized as a labor shortage was setting in. Yet, still over 40% of the labor force remained in agriculture while the share of GDP went down to 12%. Labor productivity in agriculture was about seven times lower than in the industrial sector. As mentioned before, data on labor share and productivity could be flawed due the multi-functional nature of rural households. Yet, the agricultural sector modernized somewhat because of mechanical technologies but overall it had retained its traditional structure.

As far as significant events are concerned, the financial crisis negatively impacted on rural households by tightening credit constraints and through reduced off-farm income possibilities.

The performance of the agricultural sector somewhat followed the rapid growth of the economy. Rice yields increased moderately, compared with previous periods. However, they clearly remained behind other South East Asian rice producing countries. Growth in yields was due to the increased use of mineral fertilizer which was more widely available, before the financial crisis had hit. Livestock production continued its progressive growth path, which started in the mid-80s, until the early 1990s. However, the livestock production index leveled-off and even fell by the end of the decade. On the other hand, the crop production index continued a constant growth throughout the decade (Global Economic Indicators, 2022).

The Fifth Decade

Development policies during the first decade after the millennium were marked by a turbulent period in economic and political development. In 2001, Thaksin Shinawatra came to power with a landslide election victory. In part, this victory was driven by the highly negative impact of the 1997 Asian financial crisis on the poor, mostly rural population. Before Thaksin's election in 2001, the political elite largely overlooked the rural electorate (Keyes, 2014). During the first Thaksin administration, (2001–2005), the economy improved noticeably and a range of pro-poor social protection and economic stimulus policies were implemented. For example, a social security fund including the 30-baht universal health program was installed. Thaksin was the first prime minister in Thailand to complete an entire term in office and was overwhelmingly re-elected in 2005. However, his administration was ousted by a military coup in September

2006. While after the Asian financial crisis, the economy had recovered, following the 2006 coup, it slowed down again. Political instability with frequent changes of prime ministers followed. Differences in political views between the People's Alliance for Democracy (PAD) and the National United Front of Democracy Against Dictatorship (UDD) culminated in the capture of Bangkok's international airports in 2008 and in the "red shirt crisis" in 2010 with a series of bloody political protests.

Starting in 2007, until mid-2008, the global food price crisis and the global financial crisis hit the world economy. Inflation had increased significantly, prompting the Thai Government to deliver a $1.4 billion consumer-relief package which included subsidies on diesel, a lowering of the excise tax for gasohol, free electricity and water for poor households, and free rides on third class trains and buses in certain areas. Furthermore, publicly funded mega-projects in irrigation and hydropower electricity were implemented and bilateral, regional and global free trade agreements (FTAs) were reached. Also, Thailand joined the group of donor nations in development assistance.

Overall, in spite of the political instability, Thailand remained on its course as a free-market economy with a pro-poor, inclusive growth orientation. On the other hand, the country's economic development prospects were hindered by the persistence of what can be called "crony capitalism" (Bertelsmann Stiftung, 2009). Furthermore, there remained a large underground and informal economy (Phongpaichit, 1999; Roeksiripat, 2016) with many people in poor and insecure employment conditions. Problems posing continuous risks to Thailand's development included the unresolved rural-urban divide, the southern Malay-Muslim insurgency and the semi-dormant border conflict with Cambodia which ignited again in mid-2025.

Agricultural policies played an important role under the Thaksin administration. It was clearly paying attention to their electorate, i.e. farmers and the rural population. For example, the Thailand Village Fund ("one million Baht per village") was implemented as one of the largest of such funds in the world. Several studies found positive welfare impacts of the program. For example, Townsend (2011) concluded that village funds had increased total short-term credit, consumption, agricultural investment, income growth (from business and labor) and had a positive impact on rural wages, although they decreased overall asset growth (Kabowski and Townsend, 2012). Likewise, Kislat and Menkhoff (2013) found high adoption of the program by the village population. However, Boonperm *et al.* (2013) cautioned that borrowers were disproportionately poor and agricultural although he confirmed that the fund had a positive effect on income, expenditures and durable assets. Another rural development program that aimed to increase village-based non-farm income opportunities was "OTOP" (*One-Tambon-One-Product*), a community-based program.

Later in 2009, the government launched the "Price Insurance Scheme" (PIS) which replaced a traditional rice pledging (mortgage) scheme. As found by Wagener and Zenker (2021) the program was short-lived and was only carried out during the first year as announced. Delayed payments, regional inconsistencies, other administrative problems and high transaction costs for farmers caused the program to fail in spite of initially positive effects for rice production and income (Wagener and Zenker, 2021).

The structure of the agricultural sector followed the overall modernization of the Thai economy with the adoption and diffusion of mechanical technology, adapted to the needs of small-scale farmers. Small, tracked combine-harvester machines for rice became popular and widespread, driven also by an increasing labor shortage in rural areas. However, overall the labor shares in agriculture remained high and farm sizes remained stagnant or were even declining further (Yamauchi, 2021). With just about 4 ha on average, farms in Thailand are still much larger than in most other ASEAN countries but they are far too small to generate enough income for a family. Practically, most farms in Thailand became multiple-income-source, part-time farms, where a significant proportion of the family's labor force is engaged in off-farm wage-, or non-farm self-employment. Yet, as found by Chawanote and Barrett (2013, Unpublished manuscript), structural transformation through the creation of village-based, non-farm business were disappointing.

Generally, the agricultural sector continued to be characterized by the duality of a modern large-scale sector for some commodities and a traditional small-scale sector. Rubber, coffee, sugarcane, and many fruits were often produced by larger farmers or agribusinesses, while rice and field crops remained in the domain of the 5.9 million smallholder farmers.

The most significant events for agriculture during the first millennium decade, was the food price crisis in 2008. Although agricultural commodity prices had been gradually rising since 2005, the sudden surge in April 2008 still came as a shock to both producers and consumers. Most frequently mentioned reasons were the depreciation of the US dollar, low interest rates, reductions in grain stocks, biofuels and financial speculations. In addition, Thailand's trade policy measures, being the biggest rice exporter together with those of Vietnam and India played a significant role too in the surge of prices (Heady, 2011). For farmers in Thailand, high rice prices in principle were an opportunity, albeit a risky one. Because when prices were at their peak, wet season planting was yet to start and prices of fertilizer had increased too. Modeling decision-making of farmers in Northeast Thailand (Völker *et al.*, 2013) showed that high fertilizer prices at planting, combined with the uncertainty of output prices at harvest time kept farmers away from using higher rates of fertilizer.

Performance of the agricultural sector showed that agricultural value added increased remarkably during this decade from around US$10 billion to over US$40 billion, perhaps a result of the pro-farmer policies during the Thaksin period. In terms of agricultural intensity, fertilizer use, on average did not increase a lot. The same applies to cereal yields reaching just above 3 t/ha on average and around 2 t/ha for rice in the Northeast, the largest agricultural area in the country. Arable land remained rather constant, yet showing slight signs of decline towards the end of the decade. While well-being of farmers increased, income poverty was still high in Northeast Thailand with around 30–40% prior to the food price crisis based on a large panel data set from 2008 and 2009 (Hardeweg *et al.*, 2013), i.e. before the food price crisis.

The Sixth Decade

In development policy, the political turmoil of the previous years was carried over into the Second decade of the millennium (Sixth development decade). Although Thaksin was in exile, he continued pulling the strings of Thai politics from abroad and hereby influenced the country's development path. In August 2011, his youngest sister, Yingluck, was elected Thailand's first woman prime minister. Ultimately, this meant a continuation of the populist policies, largely directed at the rural- and low-income voters. For example, minimum wages were raised by 40% in 2012, with a plan for a nationwide uniform rate of Bt 300 per day. The country was faced with an extreme flood in 2011 that affected over 80% of the provinces in Thailand, including Bangkok. It was the worst flood during the last 50 years, causing extensive damage and requiring tremendous amounts of public and private investment for repairs, reconstruction and compensations. However, the disaster also served as a point of departure for gradually adopting a policy towards a more climate-resilient economy with more attention being paid to integrated water resources management, for example. In 2014, a military coup had ousted the Yingluck government and the coup leader and army chief General Prayuth, became prime minister. This event caused a significant shift in Thai political development, towards a more authoritarian government, with much repressed freedom of expression (Human Rights Watch, 2015). Eventually, a new constitution, as the basis for a 20-year national strategy that aimed to lift Thailand from an upper middle-income country to a high-income country (OECD, n.d.; ADB, 2021), was ratified. As a development policy, the concept of 'Thailand 4.0,' was promoted. It intended to transform the country into a value-based and innovation-driven economy (Yang, 2023).

A major event with implications for the country's development path was the passing of King Bhumibol Adulyadej on October 13, 2016, after his 70-year long reign as Thailand's widely respected head of state. While economic growth during the end of the decade was somewhat sluggish, with rates of economic growth falling behind other ASEAN countries, (e.g. Vietnam) the one sector that continued to grow, was tourism. It had reached a share of some 20% of GDP by 2019. Overall, while Thailand continued to grow and modernize during this decade, it could not solve its past social and political problems, while new ones came into play. These problems kept inhibiting the country to fully exploit its development potential.

In terms of Agricultural Policies, in 2011, the price insurance scheme (PIS) was phased out. The challenges faced by the PIS created an opportunity for the Yingluck Government to introduce the populist rice price-pledging scheme with significant effects for public finance, implemented immediately after the 2011 election. The scheme was estimated to cost around 3.5% of GDP each year (World Bank, 2012), causing a significant burden to taxpayers. While the idea was straightforward, it ultimately proved to be misguided. As the primary rice exporter globally, the Thai Government believed that by withholding large amounts of rice from export, the world market price of rice would exceed the floor price that farmers were guaranteed under the pledging scheme (Sriyakul and Jermsittiparsert, 2017). By removing price uncertainty farmers would increase rice output based on their marginal costs of production. As shown by the experience with the Common Agricultural Policy (CAP) of the European Union (EU), where floor prices had led to "*milk lakes*" and "*meat mountains*," the expectations were that farmers could enjoy significant gains in income. Furthermore, the country would benefit from rice exports due to higher world prices caused by the forced shortage in international rice supply. Hence, ideally, the scheme would be a "win-win" game. However, in view of numerous existing bilateral trade contracts, the world market of rice is a residual market. Thus, high market price fluctuations are common. Also, other rice producers, e.g. Vietnam and India, could quickly fill the gap caused by Thai rice, stored in warehouses in Thailand. So, in the end, the government had the choice between prolonging the storage period of the rice or selling it in the world market below the price guaranteed for farmers. Ultimately, this contributed to a huge public finance deficit and caused a fierce debate between rice traders and academicians on the one hand and farmer leaders and government officials on the other hand. For example, Dr. Nipon Poapongsakorn, distinguished fellow at the Thailand Development Research Institute (TDRI), observed that the program misled farmers to produce rice uneconomically, e.g. by planting inferior, short-lived varieties to achieve double- or triple cropping. On the other hand, Wichien Phuanglamjiak, president of a rice farmer group from Ayutthaya, blamed rice surveyors and warehouse operators for poor storage procedures and corruption. He considered the scheme to the best way to support farmers and bridge the rural-urban income gap (Arunmas and Chantanusornsiri, 2017). In a scientific paper Poapongsakorn and Pantakua (2014) offered an excellent analysis of the negative welfare implications of the program.

Following the departure of Yingluck, the Thai government revived the traditional haphazard farmer support schemes of the past, providing subsidies for seeds, fertilizers, irrigation and crop insurance, supporting research and training for good agricultural practices and also organic farming. However, adoption, scale and impact of such measures were largely unknown but are likely to be quite limited based on the experience of prior decades.

The Structure of Agriculture was marked by a decline of the GDP share of agriculture to less than 10% after 2010, although it leveled-off prior to Covid-19 in 2020. Labor in agriculture fell strongly between 2011 and 2014 but remained above 30% by 2020. Arable land, while declining slightly after 2011, picked up again towards the end of the decade to reach around 33%. This indicates that the transfer of agricultural land for urban and other uses has been modest. Partially the loss of agricultural land was compensated by deforestation which picked up again after 2014. As in the previous decades, average farm size did not show any upward trend and remained at around 4 ha per household, still above most other Asian neighbors, but too small for making a living from farming alone. A worrisome problem, building up gradually is household debt as evidenced by over-indebtedness and multiple borrowing (Chichaibelu and Waibel, 2017, 2018).

In terms of significant events, in 2011, caused by excessive and continuous rainfall from successive, powerful monsoons and subsequent, numerous dam breaches, Thailand suffered from severe floods. These inundated more than six million hectares of land in 66 of the country's 77 provinces, and affected more than 13 million people between July and December 2011. Some 66 provinces were affected by severe, record-high flooding, in particular affecting agricultural areas along the Chao Praya River, including Bangkok metropolis and its surrounding areas.

The total damage and losses from the 2011 floods in Thailand amounted to THB 1.43 trillion (US$46.5 billion). One also has to mention the change in agricultural policies away from a populist but yet pro-rural focus after the dramatic political events in 2014 (Arya, 2021). Apparently, the military-backed, rather authoritarian government, paid more attention to the interests of the urban population.

Regarding performance of agriculture, value added of the agricultural sector dropped markedly between 2011 and 2015. It recovered thereafter but it did not reach the level before when it had peaked in 2010. The FAO crop production index fell after 2011 as a result of the floods but recovered to reach the pre-disaster level of 2010. The same pattern can be observed for cereal yields. Fertilizer use dropped quite strongly around 2014 when the pro-farm policies were terminated. The decline continued until about the end of the decade. On the other hand, livestock continued to grow based on the FAO livestock production index. Using a long-term household panel data set Nantajit (2022) found that poverty headcount in three provinces in Northeast Thailand had increased from below 30% to close to 40% between 2010 and 2013 and fell thereafter to just above 10%. Note, these values are conservative since they are based on the $1.9 poverty line which is above the national Thai poverty line. In conclusion, while during this decade the agricultural sector continued its modernization path in terms of mechanical technology and had contributed to economic development of the country, its performance in terms of land and labor productivity did not change significantly.

After 2020

The Coronavirus year, 2020, marked a global slowdown of the world economy. Inevitably Thailand could not escape from the effects of the epidemic. For example, tourism came to a sudden halt and the Thai economy contracted by 6.1% in 2020 (BoT, 2020). Agriculture was negatively affected from export restrictions but overall the impact was modest and only temporary (Thammachote and Trochim, 2021). Although the Thai economy recovered quickly from Covid-19, to date, it has not reached the pre-Covid-19 levels. Growth in 2023 and future growth prospects are not encouraging. The political problems remain unresolved. In 2023, Srettha Thavisin replaced Prayut Chan-o-cha as Prime Minister of Thailand. The new government painted the vision of Thailand reaching the status of an industrial country by 2030. One of the images is the "Agriculture Food Hub." The popular metaphors "*Fish in the water, rice in the fields, money in the pocket*" and "*Thailand as the kitchen of the world*" were promoted. The government announced to triple farmer income within 4 years, ensure the fertility of land, water, crop varieties, livestock and fish species, more than double irrigation areas and promote precision agriculture and other advanced technologies. At best, these announcements are an indicator of the importance of the rural electorate as a political constituency, that politicians can no longer ignore (Ricks and Laiprakobsup, 2021). However, the targets are not realistic. This could also mean that populist policies with input subsidies in various forms will be pursued. Quite likely, also, that farmers will be pressured to modernize and digitalize agriculture, although the economic benefits of such new technologies are far from proven. Clearly, what is missing, is a well thought-out strategy to achieve agricultural transformation towards a progressive, productive and efficient agricultural sector that recognizes its multifunctionality.

In the next section, we exemplify the opportunities and constraints in a subsector of agriculture, i.e. vegetables. The case of vegetables is symptomatic for many of the features that were highlighted in the historical review. Despite significant advancements in the value chain, the production system of vegetables has not changed much, just as is the case for crop production in Thailand as a whole.

Case Study: "Vegetables"

Vegetables in Thailand are high value crops which are labor- and material-intensive in production and capital intensive in post-harvest processes. At the same time, there is a high potential for new technologies. On the consumer side, preferences for health and environmental issues are important. Thus, the vegetable sector is particularly suitable for studying structural change in agriculture because of dynamic changes on the supply as well as the demand side. On the producer side, the vegetable sector is dynamic, diverse, with high intensity, high productivity, high value and high profits. From the consumer side, the issues are price and income elasticity, substitutability, supplementary and complementary relationships, internal and external quality, environment, health and nutrition. In addition, vegetables have complex domestic and international value chains, including a range of processing and preservation technologies. In Thailand, a major issue has been the relocation of the traditional, Chinese-built ditch and dike vegetables systems in the vicinity of Bangkok to more market-distant rural environments. This process has been driven by rapid urban expansion.

The vegetable case study was carried out in early 2000 as a PhD research study (Hardeweg, 2008). A positive mathematical programming (PMP) model was developed that allows to depict national and regional vegetable supply and demand. The data used for the model were obtained from statistical sources, including agricultural censuses, own field surveys and expert workshops with vegetable producers and extension specialists. The model shows a detailed representation of the vegetable sector in Thailand as it covers 90% of national vegetable supply. It captures supply and demand of 23 vegetable crops in eight distinct vegetable producing regions. It includes the resource base of the agricultural sector, a description of vegetable production systems with its different technologies, levels of productivity and intensity. A linkage is established with external environmental and economic factors and the institutional setting and policy environment that preconditions the development of the vegetable sector. The analysis displays the situation of the Thai vegetable market during the early 2000s but also made predictions for the situation until around 2010. The model also provides input volumes for transportation, fertilizer and pesticide use and hence informs about the environmental implications. Changing policy parameters shows the impact on spatial distribution and the inter-regional dynamics of vegetable production.

The study found that domestic demand for vegetables was growing as incomes rise and urbanization was progressing. The share of vegetables in agricultural GDP has risen from 11% in 1980 to almost 19% in 2005. Production has expanded with marked regional differences, yet vegetables occupy only about 1% of the total agricultural land. Vegetable production in Bangkok and vicinity declined in favor of somewhat more distant production areas. The number of farms that included vegetables into their crop portfolio has grown but average vegetable area per farm decreased. The highest expansion took place in the northeastern region while most vegetables are grown in the Central and Northern regions. Also, Thailand has become a net exporter of vegetables, especially for frozen and processed products (e.g. canned baby corn).

The model results showed that some 60% of vegetables are consumed outside their production region. Hence, transportation is a major component of the value chain with an average distance of almost 300 km per transportation unit. This results in CO_2 emissions of 245,000 metric tons per year, equivalent to about 25% of the total CO_2 emissions of agriculture in Thailand. Although vegetables relocated away from peri-urban to the rural areas, input-intensity of chemical inputs did not change. For example, on average 11.8 kg active ingredients of chemical pesticides were used, a multiple of what farmers use in rice or field crops. Vegetables consume almost 4% of pesticide imports while only occupying 1% of arable land. Pesticide poisoning of farmers were frequent events as found by the study. Based on anecdotal reports some farmers carry with them atropine injections when applying pesticides in their vegetable fields.

Policy simulations using three strategy parameters, namely fuel costs, storage and transport technology and land zoning, suggest that the possibility to achieve improvements in health and the environment are extremely limited. Fuel cost increases will affect the far-away production locations. Zoning, although reducing vegetable area in urban and peri-urban areas will lead to a further intensification of material inputs in the remaining areas. Post-harvest loss reduction due to improved storage and transportation technology has a positive effect by reducing harvest and transportation quantities. The 2011 scenario calculations assumed mechanical-technical progress and predicted a decline in hired labor in the Central Region and a more efficient use of family labor. However, chemical input intensity will rise which keeps the negative externalities in terms of environment and human health effect unchanged. Simulation scenarios to curb excessive pesticide and fertilizer use by introducing a quota for chemicals and an area reduction target in the most intensive production locations shows that this would not lead to substantial reduction in externalities. However, a Pigouvian tax on chemicals would be economically efficient but for political reasons it is almost impossible to implement such policy.

To reflect the results of the vegetable case study from today's perspective, the study was correct in its conclusions and predictions. Improvements in the marketing, e.g. the increasing vegetable market share in supermarkets with potentially better monitoring systems may have improved quality but at the same time is producing more plastic waste. There are research efforts and pilot projects on programs such as good agricultural practices (GAP), protected cultivation, low and free chemical vegetables, and organic production (Poapongsakorn and Tey, 2016). These projects are positive signs. However, until now the total area of these is small and there is no indication of a rapid scale-out in the near future. There are private sector actors exporting premium products, e.g. Thai Chin Vegetables and Fruit Thailand Co. Ltd (https://www.thaichin.com/EN/about.html). Again, these do not go beyond pilot scale. An efficient and reliable labeling system is still to be implemented in order to increase the willingness of consumers to pay premium prices for healthier products. In summary, the fundamental problems of the vegetable sector, i.e. environment and health problems at the producer level are pretty much the same as 20–30 years ago. Thus, to conclude, the vegetable sector in Thailand has many modern elements but on a large scale, it has not changed. There is a need for both technical and institutional innovations if the sector will achieve modern standards in the near future.

Summary and Outlook

On the surface, the Thai agricultural sector has changed a lot over the six development decades, as described in this chapter. On the other hand, the sector faces challenges to further structural change as pointed out by Poapongsakorn and Pantakua (2021). When looking from below, farming in Thailand is still much the same. A dualistic structure, with a few large-scale and influential agribusiness companies, engaged in contract farming and marketing on the top. Below, there are some 6 million, dominantly traditional, semi-commercial, small-scale farm households, who must make their living from multiple sources, increasingly from non-agriculture. The average farm size has not changed noticeably during the past 50 years. Hence, the vast majority of Thai farmers cannot make a living from farming alone. Yet, they continue to do farming, together with other income-generating activities, rather than leaving agriculture altogether. So, what is the reason for this lack of structural transformation? Why do they stay on their farms and make less and less money from farming? For example, and just to paint a somewhat drastic scenario, why doesn't CP or any of the other agribusiness giants, not just take up all the land and create professionally and efficiently managed, large-scale farming firms or businesses? The answer to this admittedly hypothetical question can be found in economic theory and empirically by looking at the history of farming in Europe.

Development economic theory on the transformation of agriculture, has its foundations – among others – in Kuznets (1971) seminal work. Kuznets endorsed that marginal products of labor between agriculture and other sectors will equalize once sufficient labor has been transferred from agriculture to industry or service sectors. In his paper in the *American Economic Review*, "Sectoral Shares in Labor Force: A Different Explanation of the Kuznets' (I+S)/A Ratio," Schmitt (1989) pointed out that productivity comparisons between agriculture and other sectors are often wrong because in such comparisons, small-scale farms are treated as firms and not as what they really are, i.e. farm-household-systems, where production, investment and consumption decisions are interconnected. Hence, income-generating combinations with farming as one of several gainful activities, is simply the result of the households' utility maximization process. Thus efficient, intra- and inter-sectorial allocation of resources, can be found in a range of farming systems, regardless of the level of development of a country. Schmitt

(1989) also pointed out that transaction costs can exorbitantly rise with increasing farm sizes. While these costs certainly depend on the type of production system, it is easy to imagine what costs would emerge if, for example, the current 3–5 ha rice farms in Thailand would be remodeled into 1000 ha rice farms.

Let us undertake a simple "back-of-the-envelope" calculation to illustrate the problem of Thai agriculture in this regard. According to the Office of Agricultural Economics (OAE), in 2022 (see Table 4.2), the average net income per ha was 51,743 THB (about US$1437). Given an average farm size of 3.98 ha and the size of an agricultural household of 3.48 persons, to achieve an income equivalent of the average per capita income in Thailand in 2022 (around US$7000), an average Thai farm would have to be at least 100 rai (16 ha) in size. Ultimately, this would mean a reduction of currently 5.9 million farmers to about 1.5 million.

Needless to say that there is little prospect that this could happen within the next 10 or even 20 years, even when we consider that aging will make many rural people eventually give up farming. In fact, by the year 2033, Thailand is anticipated to transition into a super-aged society, characterized by a population where at least 20% are aged 65 and above (TGRI and IPSR, 2020). Also, a recent study submits that a majority of older farmers have plans to stop farming in the near future due to health concerns (Jansuwan and Zander, 2021). On the other hand, both economic theory and empirical evidence suggest that farming in Thailand will continue in different ways. Putting aside the agroindustry corporations that already exist, there are perhaps three farming models that could be relevant in the future. Firstly, full-time farming, where farms are businesses, conducted by agricultural entrepreneurs with advanced education levels, sufficient capital and a willingness to stay in rural villages. This will take a long time to come about. Secondly, part-time farming with a variety of types, depending on the labor hours spent for farming and other gainful activities (OGA). The latter traditionally includes off-farm wage employment, both on-site and off-site (migrants) and non-farm self-employment, by residents and migrant household members. This may include village-based services, services outsourced by urban firms and perhaps also "holidays-on-farms" or agri-tourism which has recently been promoted by the government. The distinction hereby is, whether OGA is the primary or secondary activity. As the EU example has shown, there could be transitions between both groups. In all likelihood, in Thai agriculture, this category will remain the major one for a long time to come. There is yet a third farming system group which should be recognized. This can be called "hobby-lifestyle-retirement" farming which may include extended home-gardening, small-scale livestock raising, mainly for home consumption but also for market sales as well as home-stay or holiday on farms.

While it is beyond the scope of this book chapter to come up with detailed policy recommendations, it is important to point out that policymakers must recognize the full variety of

Table 4.2. Average household income of agricultural households per year, average farm size and income share agriculture.

Year	Household Income (THB/year)	Income Share Agriculture (%)	Farm Size (rai)	Family Size No./HH
2013	247743	59.07	25.71	4.0
2014	268301	55.25	24.36	3.96
2015	283261	52.40	24.68	4.21
2016	300565	52.36	25.46	4.04
2017	309278	52.03	25.26	4.04
2018	370040	53.34	23.15	3.83
2019	394911	51.67	24.48	3.78
2020	390376	47.24	24.16	3.89
2021	397218	47.10	22.49	3.88
2022	420198	49.10	24.92	3.48

Source: Office of Agricultural Economics, different years.

farming systems. Therefore, policies must be tailored to these systems and their dynamics, instead of focusing on a fixed model. Generally, a comprehensive rural development strategy is needed which will help to convert the rural space into an attractive living and work environment. Useful suggestions on rural development have already been made earlier in a policy review of Thai agriculture, by Seidel and Adam (2013). The model of EU rural development policies may serve as some useful food for thought for a start (e.g. Augère-Granier, 2015; Bonfiglio *et al.*, 2017; Pelucha and Kveton, 2017).Empirically, the lessons learned from farming in Europe suggest that part-time farming, with non-farm income either as a major or minor share, has persisted for a long time. A recent study by Addo and Salhofer (2022) showed that by 2015, in the EU, there were still some 17% part-time farmers with farming as the major share of household income, down from 31.5%, 10 years earlier. Households engaged in part-time farming, where income from farming is a minor share, had increased, from almost 5% to over 7%, in the same period. Consequently, in 2015, about 75% of the 12.6 million EU farms, were full-time farmers. The total number of farms has declined by about 10% between the two periods. This means, there are fewer, but larger farms, while, at the same time, part-time farming is still around. Several authors have pointed out that even in advanced economies, farming takes place in different systems (e.g. Shahzad and Fischer, 2022), is multi-functional and is embedded in a "pluri-economic activity" model which includes part-time farming (Robson *et al.*, 1987; Fuller, 1990; Fuller, 2015).

So what does this mean for Thai agriculture? How realistic is it to have modern and economically efficient large-scale farming soon, as hoped by some political decision-makers and as proposed by some researchers (e.g. Otsuka *et al.*, 2016a, b; Yamauchi, 2021; Yamauchi *et al.*, 2021)?

References

Addo, F. and Salhofer, K. (2022) Part-time farming and scale efficiency. *German Journal of Agricultural Economics* 71(1), 16–35.

Akrasanee, N. and Ajanant, J. (1986). Thailand: manufacturing industry protection issues and empirical Studies. The political economy of manufacturing protection: Experiences of ASEAN and Australia, 78–96.

Arunmas, P. and Chantanusornsiri, W. (2017) How awful was rice pledging, really? Bangkok Post. Available at: https://www.bangkokpost.com/business/general/1312615/how-awful-was-rice-pledging-really (accessed 13 September 2025).

Arya, A.R. (2021) *Agricultural Transformation in Asia. Policy and Institutional Experiences*. Food and Agriculture Organization of the United Nations, Bangkok.

Asian Development Bank. (2021) Thailand Country Partnership Strategy 2021–2025: Prosperity and Sustainability Through Knowledge and Private-Sector-Led Growth. Asian Development Bank. Available at: https://www.adb.org/documents/thailand-country-partnership-strategy-2021-2025?utm_source=-chatgpt.com.

Augère-Granier, M.L. (2015) *EU Rural Development Policy*. Brussels, Belgium: European Parliamentary Research Service.

Bank of Thailand. (2020) Thailand's Economic Conditions in 2020 (Annual Report). Bank of Thailand. Available at: https://www.bot.or.th/content/dam/bot/documents/en/thai-economy/state-of-thai-economy/annual-report/annual-econ-report-en-2020.pdf.

Bertelsmann Stiftung (2009) *BTI 2010 – Thailand Country, Report*. Bertelsmann Stiftung, Gütersloh.

Booth, A. (2016) Falling behind, forging ahead and falling behind again: Thailand from 1870 to 2014. *Economies* 4(1), 1.

Bonfiglio, A., Camaioni, B., Coderoni, S., Esposti, R., Pagliacci, F. *et al.* (2017) Are rural regions prioritizing knowledge transfer and innovation? Evidence from rural development policy expenditure across the EU space. *Journal of Rural Studies* 53, 78–87.

Boonperm, J., Haughton, J. and Khandker, S.R. (2013) Does the Village Fund matter in Thailand? Evaluating the impact on incomes and spending. *Journal of Asian Economics* 25, 3–16.

Bresciani, F., Feder, G., Gilligan, D.O., Jacoby, H.G., Onchan, T. *et al.* (2002) Weathering the storm: the impact of the East Asian crisis on farm households in Indonesia and Thailand. *The World Bank Research Observer* 17(1), 1–20.

Benor, D., Harrison, J.Q. and Baxter, M. (1984) *Agricultural Extension. The Training and Visit System*. The World Bank, Washington, DC.

Chawanote, C. and Barrett, C. (2013) *Farm and Non-Farm Occupational and Earnings Dynamics in Rural Thailand*. Cornell University, Ithaca, NY.

Cernea, M. (1983) *Agricultural Extension by Training and Visit. The Asian Experience*. The World Bank, Washington, DC.

Chichaibelu, B.B. and Waibel H. (2017) Borrowing from "Pui" to pay "Pom": multiple borrowing and over-indebtedness in rural Thailand. *World Development* 98 (Oct.), 338–350.

Chichaibelu, B.B. and Waibel H. (2018) Over-indebtedness and its persistence among rural households in Thailand and Vietnam. *Journal of Asian Economics* 56 (June), 1–23.

Doner, R. (1974) The development of agribusiness in Thailand. *Bulletin of Concerned Asian Scholars* 6(1), 8–15.

ECAFE/FAO Agriculture Division. (1955) *Agricultural Development of Thailand*. ECAFE/FAO Agriculture Division, United Nations.

Fuller, A.M. (1990) From part-time farming to pluriactivity: a decade of change in rural Europe. *Journal of Rural Studies* 6(4), 361–373.

Fuller, A.T.M. (2015) From part-time farming to multifunctionality. In: Bakker, J.I. (ed.) *The Methodology of Political Economy: Studying the Global Rural–Urban Matrix*. Lexington Books, Lanham, MD.

Global Economic Indicators. (2022). Available at: Https://www.theglobaleconomy.com/Thailand/ (accessed 13 September 2025).

Goss, J. and Burch, D. (2001) From agricultural modernisation to agri-food globalisation: the waning of national development in Thailand. *Third World Quarterly* 22(6), 969–986.

Goss, J., Burch, D. and Rickson, R.E. (2000) Agri-food restructuring and third world transnationals: Thailand, the CP Group and the global shrimp industry. *World Development* 28(3), 513–530.

Hardeweg, B. (2008) *The Spatial Distribution and Inter-Regional Dynamics of Vegetable Production in Thailand*. Wissenschaftsverag Vauk, Kiel, 238 pp.

Hardeweg, B., Klasen, S. and Waibel, H. (2013) Establishing a database for vulnerability assessment. In: Klasen, S. and Waibel, H. (eds.) *Vulnerability to Poverty – Theory, Measurement, and Determinants*. Palgrave Macmillan, London, UK. pp. 50–79.

Headey, D. (2011) Rethinking the global food crisis: the role of trade shocks. *Food Policy* 36(2), 136–146.

Human Rights Watch (2015) *World Report 2015: Thailand*. Human Rights Watch, New York, NY.

Jansuwan, P. and Zander, K.K. (2021) What to do with the farmland? Coping with ageing in rural Thailand. *Journal of Rural Studies* 81, 37–46.

Kaboski, J.P. and Townsend, R.M. (2012) The impact of credit on village economies. *American Economic Journal: Applied Economics* 4(2), 98–133.

Kermel-Torrès, D. (2004) *Atlas of Thailand*. IRD Éditions, Marseille, France.

Keyes, C. (2014) *Finding their Voice: Northeastern Villagers and the Thai State*. Silkworm Books, Chiang Mai, Thailand.

Kislat, C. and Menkhoff, L. (2013) The village fund loan programme: Who gets it, keeps it and loses it? In: Klasen, S. and Waibel, H. (eds.) *Vulnerability to Poverty – Theory, Measurement, and Determinants*. Palgrave Macmillan, London, UK. pp. 283–304.

Kuznets, S. (1971) *Economic Growth of Nations: Total Output and Production Structure*. Harvard University Press, Cambridge, MA.

Mittal, S. (2021) Global review of agricultural policies. Non-peer reviewed manuscript. Available at: https://www.researchgate.net/publication/354461688_Global_Review_of_Agricultural_Policies (accessed 13 September 2025).

Nantajit, C. (2022) Essays on village development in rural Thailand. Doctoral dissertation, Leibniz Universität Hannover, Germany.

Nidhiprabha, B. (2019) Commodity price cycles, the agricultural trap, and Thailand's incessant subsidies. *Asian Economic Papers* 18(2), 49–69.

OECD. (2021) *OECD Investment Policy Reviews*. OECD Publishing, Thailand.

Office of Agricultural Economics. Available at: https://www.oae.go.th/view/1/Home/EN-US (accessed 13 September 2025).
Otsuka, K., Liu, Y. and Yamauchi, F. (2016) Growing advantage of large farms in Asia and its implications for global food security. *Global Food Security* 11, 5–10.
Otsuka, K., Liu, Y. and Yamauchi, F. (2016) The future of small farms in Asia. *Development Policy Review* 34(3), 441–461.
Pelucha, M. and Kveton, V. (2017) The role of EU rural development policy in the neo-productivist agricultural paradigm. *Regional Studies* 51(12), 1860–1870.
Phongpaichit, P. (1999) Guns, girls & gambling. In: *Thailand's Illegal Economy and Public Policy*. Seminar Paper Delivered at the Centre of Southeast Asian Studies, Kyoto University, Japan.
Poapongsakorn, N., Anuchitworawong, C. and Mathrsuraruk, S. (2006) *The Decline and Recovery of Thai Agriculture: Causes, Responses, Prospects and Challenges*. Food and Agriculture Organization of the United Nations, Rome, Italy.
Poapongsakorn, N., Ruhs, M. and Tangjitwisuth, S. (1998) Problems and outlook of agriculture in Thailand. *TDRI Quarterly Review* 13(2), 3–14.
Poapongsakorn, N. and Tey, Y.S. (2017). Institutions, governance and transformation in Southeast Asian agriculture. In: *Proceedings of the Regional Forum: Farms, Food, and Futures: Toward Inclusive and Sustainable Agricultural and Rural Development in Southeast Asia*. Chapter 3: SEARCA, Los Banos, Philippines.
Poapongsakorn, N. and Pantakua, K. (2014) Assessing the Thai paddy pledging policy: its performance and social costs. *World Food Policy* 1(2), 2–29.
Poapongsakorn, N. and Pantakua, K. (2021) Agricultural transformation in the Kingdom of Thailand. In: Arya, A.R. (ed.) *Agricultural Transformation in Asia – Policy and Institutional Experiences*. FAO, Bangkok.
Ricks, J.I. and Laiprakobsup, T. (2021) Becoming citizens: policy feedback and the transformation of the Thai rice farmer. *Journal of Rural Studies* 81, 139–147.
Rieber, L. (2013) Making the rice bowl bigger: agricultural output and rural poverty in Thailand. *Global Majority E-Journal* 4(1), 4–14.
Robinson, D., Teja, R.S., Byeon, Y. and Tseng, W.S. (1991) *Thailand: Adjusting to Success: Current Policy Issues*. International Monetary Fund, Washington, DC, USA.
Roeksiripat, K. (2016) Informal economy: case study of street vendors in Bangkok. *International Journal of Business and Administrative Studies* 2(2), 50.
Robson, N., Gasson, R. and Hill, B. (1987) Part time farming – implications for farm family income. *Journal of Agricultural Economics* 38(2), 167–192.
Schmitt, G. (1989) "Simon Kuznets' Sectoral shares in labor force": A different explanation of his (I+ S)/A ratio. *The American Economic Review* 79, 1262–1276.
Seidel, J. and Adam, R. (2013) Thailand: Agrarpolitik in der Sackgasse (in German). In: Thailand Rundschau der Deutsch-Thailändischen Gesellschaft e.V. Jg 26 (2), 52–65. Köln.
Shahzad, M.A. and Fischer, C. (2022) The decline of part-time farming in Europe: an empirical analysis of trends and determinants based on Eurostat panel data. *Applied Economics* 54(42), 4812–4824.
Shintani, M. (2003) *The Process of Agricultural Growth in Thailand*. Kyushu University Press, Fukuoka.
Siamwalla, A. (1995) Land-abundant agricultural growth and some of its consequences: the case of Thailand. In: Mellor, J.W. (eds.) *Agriculture on the Road to Industrialization*. Johns Hopkins University Press, Baltimore, MD, USA, pp. 150–174.
Siamwalla, A. (1999) The impact of the bubble and crisis on Thai agriculture. *Quarterly Journal of International Agriculture* 38(4), 359–374.
Singhapreecha, C. (2014) *Economy and Agriculture in Thailand*. Food Fertilizer and Technology Center for the Asian and Pacific Region, Taipei, Taiwan, 2134592583–1590046580.
Sriyakul, T. and Jermsittiparsert, K. (2017) Politicization of rice price: Who gain and who lose from the populist policies to intervene rice price in Thailand? *Asian Political Science Review* 1(1), 19–31.
Suphannachart, W. and Warr, P. (2011) Research and productivity in Thai agriculture. *Australian Journal of Agricultural and Resource Economics* 55(1), 35–52.
TGRI (Foundation of Thai Gerontology Research and Development Institute) and IPSR (Institute for Population and Social Research, Mahidol University) (2020) *Situation of the Thai Eldery 2019*. Printery Co., Ltd., Nakhon Pathom. Available at: https://thaitgri.org/?wpdmpro=situation-of-the-thai-eldery-2019 (accessed 28 April 2024).
Thammachote, P. and Trochim, J.I. (2021) The Impact of the COVID-19 Pandemic on Thailand's Agricultural Export Flows (PRCI Research Paper No. 4). Food Security Group, Department of Agricultural, Food and Resource Economics, Michigan State University. https://doi.org/10.22004/ag.econ.320709.

Thaxton, R. (1973) Modernization and counter-revolution in Thailand. *Bulletin of Concerned Asian Scholars* 5(4), 28–40.

Townsend, R.M. (2011) Financial systems in developing economies: Growth, inequality and policy evaluation in Thailand. Oxford University Press.

Viksnins, G.J. (1973) United States Military Spending and the Economy of Thailand, 1967–1972. *Asian Survey* 13(5), 441–457.

Völker, M., Tongruksawattana, S., Schmidt, E. and Waibel, H. (2013) Impact of food price shocks on vulnerability to poverty: a mathematical programming approach. In: Klasen, S. and Waibel, H. (eds.) *Vulnerability to Poverty – Theory, Measurement, and Determinants*. Palgrave Macmillan, London, UK, pp. 163–188.

Wagener, A. and Zenker, J. (2021) Decoupled but not neutral: the effects of counter-cyclical cash transfers on investment and incomes in rural Thailand. *American Journal of Agricultural Economics* 103(5), 1637–1660.

Waibel, H. (1991) Pesticide subsidies in Southeast Asia. *FAO Plant Protection Bulletin* 38(2), 111–120.

Warr, P.G. (1999) What happened to Thailand? *The World Economy* 22(5), 631–650.

World Bank (2012) *Thai Flood 2011: Rapid Assessment for Resilient Recovery and Reconstruction Planning*. World Bank, Washington, DC.

Yamauchi, F. (2021) *Changing Farm Size and Agricultural Productivity in Asia*. International Food Policy Research Institute, Washington, DC, USA.

Yamauchi, F., Huang, J. and Otsuka, K. (2021) Changing farm size and agricultural development in East Asia. In: *Agricultural Development: New Perspectives in a Changing World*. International Food Policy Research Institute (IFPRI), Washington, DC. pp. 79–110.

Yang, Y. (2023) Thailand. Country Report. In: Anheier, H.K. (ed.). *The External Cultural Policy Monitor*. IFA, Stuttgart.

Zhang, Y., Qu, H. and Tavitiyaman, P. (2009) The determinants of the travel demand on international tourist arrivals to Thailand. *Asia Pacific Journal of Tourism Research* 14(1), 7–92.

5 Agricultural Development and Policy in Vietnam: Also a Success Story?

Jonathan Pincus and Hermann Waibel

Abstract

In this chapter on Vietnam, the review of economic and agricultural policies starts with the Vietnam War period, covers the post-war phase, the economic reform period and subsequent development decades. The connection between general development policies, agricultural policies, agricultural transformation and the performance of the agriculture is discussed in detail. A case study of a rural household drawing from a unique panel data base complements the analysis and a summary of the major issues in relation to agricultural and rural transformation is added. Essential statistics are included to support the conclusion and outlook for Vietnam's agriculture in the near future.

Introduction

Vietnam's economic development is widely regarded as a success story in Southeast Asia. At the time of reunification in 1975, Vietnam was one of the poorest countries in the world, burdened by the legacy of two major wars, economically and diplomatically isolated and heavily dependent on foreign aid from the socialist bloc. Following a short-lived attempt to implement Soviet-style central planning, the government launched reforms in stages to legalize private farming and business activity, impose financial discipline on state firms and encourage foreign trade and investment. As the Cold War wound down, Vietnam was slowly reintegrated into the world community. The European Community restored diplomatic relations in 1990, followed by the USA in 1994. Vietnam acceded to the World Trade Organization in 2007.

The results of the reforms, or *doi moi* (renovation) as they came to be known, far exceeded expectations. Real GDP growth has averaged 6.7% per annum since 1990, outperforming Indonesia (4.6%) and Thailand (3.6%). Vietnam's growth has been export-led, in the early years propelled by agriculture and light manufacturing, and increasingly dominated by electronics, machinery and metals. Consistently high levels of foreign direct investment have deepened the country's integration into East Asian production systems, a process that has accelerated with US trade restrictions on China. The percentage of the population living in poverty was reduced from 76% in 1992 to 4% in 2022, using the World Bank's $3.65 per day poverty line.

In this chapter, we review agricultural policy and outcomes within the context of Vietnam's development strategy starting from

 Asian Agriculture in a Climate Change-driven and Risky World (H. Waibel, S. Liebenehm and Ernah eds)
DOI: 10.1079/9781800627338.0005

the wartime era in 1975 and bringing the story up to the post Covid-19 period. Agricultural development has contributed significantly to the country's economic success, lifting rural incomes and living standards, stabilizing national food prices and earning foreign exchange through exports. Once a food-deficit country, Vietnam is now a major exporter of rice, seafood, coffee, pepper, fruits and nuts, rubber and cut flowers. The government has prioritized rural development, especially in remote regions, and while disparities in living standards have persisted, Vietnam has largely avoided the emergence of extreme levels of inequality of income and outcomes like Thailand, for example (see: www.wid.world). To sustain these gains, Vietnamese agriculture will need to graduate from its present reliance on bulk commodities to specialize in higher value-added products, while at the same time adopting sustainable, climate-resilient practices. Fragmentation of landholdings and small average farm sizes remain an obstacle to mechanization, and therefore land consolidation will be necessary to sustain productivity growth, especially as wages rise and the share of the labor force in agriculture contracts (Nguyen and Warr, 2020).

Development Phases

Table 5.1 summarizes trends in economic and agricultural development policy from the pre-1975 period until the present. The table is not an exhaustive list of agricultural policies, but is intended as a guide to shifts in policy in response to changes in the internal and external context.

Pre-1975 Period

From the 1954 Geneva Accords until April 30, 1975, Vietnam was divided at the 17th parallel into the Democratic Republic of Vietnam in the north and the Republic of Vietnam in the south. The Communist north carried out a land reform program from 1953–1956 to redistribute land from landlords and rich peasants to land-poor peasants, a process that even the Communist Party later admitted had been marred by excessive violence and brutality. This was followed by collectivization, in stages, in which farmers were organized into agricultural cooperatives, pooling their land and tools and assigned to production brigades. Payment was made on the basis of work points earned, which rewarded quantity rather than quality of labor and encouraged shirking (Pingali and Xuan, 1992). Productivity declined on cooperatively farmed land, especially as households redirected their efforts to private land – known as "5 percent plots" – that they were allowed to cultivate from 1959 (Kirkvliet, 1995).

In the south, tenancy was the dominant form of land tenure during and after the colonial period. The Viet Minh redistributed land from landlords and large farmers to tenants and imposed reductions in areas under their control during and after the French War. President Ngo Dinh Diem carried out his own half-hearted land reform in the 1950s, but the program actually strengthened the grip of landlords on tenants (Prosterman, 1970). No further action was taken until President Nguyen Van Thieu's Land to the Tiller program was launched in 1970 with US financial backing. The program redistributed one million hectares of land to about 800,000 tenant farmers over three years but came too late to reverse the declining fortunes of the southern regime (Gawthorpe, 2022).

Reunification Period

After reunification in 1975, a new round of land redistribution was ordered in the southern provinces, but the program achieved only modest results. To accelerate socialist transformation of agriculture, the government decided in 1979 to move directly from land reform to collective farming on the northern model, pooling land, livestock and farm machinery and forming production teams to work the land. The policy was fiercely opposed by southern farmers, many of whom slaughtered livestock and abandoned land rather than join collectives. More than two-thirds of production units had collapsed by the end of 1980, by which time collectives covered only 6% of land in the Mekong River Delta. Rice output in the south initially declined before regaining pre-unification levels in 1980. At the national level, per capita output level fell to historically low levels, and Vietnam remained dependent on food imports (Dang, 2018).

Table 5.1. Overview of development by time periods and development decades. (Authors' own table).

#	Period	General Development Policy	Agricultural Policies	Structure of Agriculture	Significant Events	Performance of Agriculture
1	Pre- 1975	North Vietnam: Central planning; South: Market economy during wartime (1955–1975);	Tenant system in the South; collectivized farming in North	Land reform followed by collectivization in the North; land reform in the South carried out by nationalists before American intervention	End of War	Low productivity, but effective management of war economy; higher productivity in the South but tenancy remains dominant
2	1975–1985	National reunification; collective agriculture imposed on the southern provinces but fails due to farmer resistance	Collectivized farming followed by first experiments with household contract system (Directive 100)	Food deficits; some exports of coffee, tea, cashew nuts; return to household-based farming systems	Storms "Cecil" and "Wayne"	Rice production growth still below population growth
3	1986–1990 (Third Decade)	*Doi moi* economic reforms to create a "socialist-oriented market economy"; gradual elimination of restrictions on private activity	Resolution 10 (land use rights), some for individuals but ownership remains with the state	Collectivization effectively reversed by 1990; liberalization of food markets and trade	High rainfall in Central Highlands	Growth in food production exceeds population growth; Vietnam becomes a rice exporter; labor productivity increased
4	1991–2000 (Fourth Decade)	Abolishment of State Monopoly in Foreign Trade; Member of ASEAN/AFTA/ASEM/APEC	Irrigation investments	Privatization of input markets, land tax (in-kind) GDP share ~30 %; labor share still >70%; Agricultural surplus invested in industry and service sector	Several tropical storms; no extreme storm	Yields increased
5	2001–2010 (Fifth Decade)	WTO Membership, strong increases in FDI, ODA, remittances and capital funds	GDP share of AG-R&D, ~0.2% only	Labor transfer out of agriculture increased; GDP 20%; labor ~50%	Heatwave due to El Niño in 2010; TVSEP and VAHRS panel	

6	2011–2020 (Sixth Decade	Significant progress in labor market reforms and education; climate change policy	New Rural Development Program; master plan on agricultural production; agriculture as a contributor to growth	GDP 15%; Labor ~35%; rice still major crop (50%), followed by maize, pepper and coffee; 70% of farms are less than 1 ha; >90% are less than 3 ha; ~8% more than 3 ha	Covid-19	Major exporter of cashew, pepper, coffee, cassava, rice
7	After 2020	Rigid Covid-19 measures; economic growth quickly regained	Continue to promote large-scale farms model (LSFM)	Important player in agricultural markets; average farm size remains small (~1 ha); marketing small-scale	Severe flooding in Central Vietnam, 10–11/2020	No major change

In response, the Central Committee of the Communist Party issued Directive 100 in January 1981, abandoning the work point system in favor of product contracts. Individual households carried out the bulk of farming operations on land allocated to them, contributing funds to the cooperative for farm inputs and centralized services such as ploughing and irrigation. The contract system incentivized farmers to cultivate land assigned to them more intensively, and yields did rise in the aftermath of the decree. But product contracts did not represent a long-term solution. Because the cooperative retained the authority to redistribute land parcels, farmers had no reason to invest in the farm, improving irrigation or conserving soil. Centralized provision of farm inputs and procurement of output was inefficient and perceived as inequitable, especially as cooperatives raised production quotas as yields recovered (Pingali and Xuan, 1992; Tran, 1998).

Doi Moi

In response to slow economic growth, repeated bouts of hyperinflation, goods shortages and growing concerns relating to economic and political instability in the Soviet bloc, a significant change in economic policy was announced in December 1986. The new policy, popularly known as "*doi moi*" or renovation, effectively abandoned central planning and sanctioned economic activity outside of state control including fewer restrictions on international trade and inward investment. A hard budget constraint was imposed on state-owned enterprises to control credit expansion, and the first steps were taken toward creating a modern banking system.

In agriculture, bad weather and shortages of farm inputs in 1987 led to a decline in food output and local grain shortages, especially in the northern provinces. By 1985, the gains from Directive 100 had been exhausted, and the perilous state of road transport, combined with resistance from local officials, complicated efforts to move rice from south to north. The severity of the crisis sparked another round of reforms. In April 1988, collectivization was reversed and long-term land use rights allocated to households, while reducing the role of the cooperative in cultivation decisions and the provision of services. Farmers were free to sell their surplus on the private market after fulfilling tax and quota obligations. Agricultural output rebounded and the country made a rapid transition from food importer to exporter in the following years (Barnett and Wailes, 2011; Beresford, 1990).

Opening up to the World

The *doi moi* reforms were consolidated in the 1990s. The 1993 Land Law extended household tenure to 20 years for annual crops and 50 years for perennial crops, and more importantly legalized the transfer of land use rights, an important step toward the creation of a land market (Bellemaire *et al.*, 2020). Rice purchases through cooperatives and state agencies were abolished in 1989 and input markets were privatized. The Vietnam Bank for Agriculture was established in 1990 to provide credit directly to farmers, although land was not accepted as collateral until 1994, limiting the reach of formal sector lending.

Market incentives were important, but sustained growth required investment in public goods, especially in irrigation and drainage, electrification and transport infrastructure. Household farming lifts productivity by linking effort to income, but agriculture exhibits increasing returns to scale in infrastructure, especially water control. The Mekong Delta had emerged from the war with an archaic and severely damaged irrigation system in desperate need of modernization. Improved water control in the Mekong River Delta made possible a shift from one to two or three rice crops per year. Public investment crowded in private spending on pumps for irrigation and drainage, which increased productivity and promoted crop diversification.

Formal diplomatic relations with the USA were restored in 1994, opening the path to international economic integration. Vietnam joined the Association of Southeast Asian Nations (ASEAN) and the ASEAN Free Trade Area (AFTA) in 1995. A Framework Cooperation Agreement with the European Community was signed in 1995 and entered into force in 1996, and a bilateral trade agreement with the USA was finalized in 2000. Agricultural output soared, growing at an average rate of 15% from 2001 to 2012. Accession to the World Trade

Organization (WTO) in 2007 forced the sector to become more competitive, reducing import barriers and export subsidies, and introducing international Sanitary and Phytosanitary (SPS) standards. Industrialization continued to progress as agricultural surplus was transferred to other sectors of the economy. Consequently, the GDP share of agriculture fell to around 30%, but the labor share was still over 70%. Also, the agricultural sector became more resilient. For example, despite several extreme weather conditions during this decade, average rice yields had increased to over 4 t/ha by the year 2000.

Exports of fish and shellfish, mostly from the Mekong River Delta, expanded rapidly in the decade from 1995. Exports reached $1 billion in 2007 and $2 billion in 2018. Intensive shrimp farming spread rapidly as the government relaxed rules against converting rice land to other uses in the early 2000s. Extensive, low-input shrimp production expanded in coastal regions of the Mekong River Delta, but capital-intensive methods using aerators, pumps, reservoirs and mechanized feeding have come to dominate the sector. Higher intensity brings greater risk of disease and tremendous environmental costs. Expansion of shrimp ponds led to the destruction of half of the Mekong Delta's mangrove forests. In 2022, Vietnam was the largest shrimp exporter in the world with exports worth $4.3 billion.

Meanwhile Vietnam also became an active partner in the international agricultural and development research community. This had resulted in at least two long-term empirical research projects in agriculture and rural development. One was the "Viet Nam Access to Resources Household Survey" (VARHS), Between 2006 and 2014, five panel waves of some 2100 rural households (<200 households per province) were carried out (Tarp, 2017). The VARHS provided a good basis for a range of relevant research topics. These include, for example, *Vietnam war and agricultural productivity* (Appau *et al.*, 2021), *rental markets, gender and land certificates* (Ayala-Cantu and Morando, 2020), *land use rights* (Markusen *et al.*, 2011), *climate change and migration* (Baronchelli and Ricciuti, 2018), *self-employment and subjective well-being* (Markussen *et al.*, 2018), *the economics of communist party membership* (Markussen *et al.*, 2019) and *tenure security and agricultural investment* (Bellemaire *et al.*, 2020).

Another research panel data project was the Thailand Vietnam Socioeconomic Panel (TVSEP), financed by the German Research Foundation which started in 2007. The aim of the project, which is still ongoing, was to address the problems of rural development in Central Vietnam, in the context of a long-term research project on vulnerability to poverty in Vietnam (and Thailand). The panel of 2200 rural households in 110 communes was carried out in the provinces of Ha Tinh, Thua Thien Hue and Dak Lak and was first administered in 2007. With some 750 households per province, the sampling ratio is higher than that of the VARHS. TVSEP data is very comprehensive, including the usual living standards measures but also capturing shocks and risks as well as personality traits and other behavioral questions relevant for development and poverty reduction. Special measures were initiated to ensure data quality (e.g. Phung *et al.*, 2015; Brooks *et al.*, 2020). Research based on the TVSEP panel data dealt with a range of topics related to agricultural development. A major theme was *rural urban migration* (e.g. Nguyen *et al.*, 2015; Amare and Hohfeld, 2016; Gröger and Zylberberg, 2016), *shocks, coping and household welfare* (e.g. Phung and Waibel, 2010; Völker and Kanchanitra, 2021), *climate change* (Liebenehm *et al.*, 2023a, b), *natural resource extraction* (e.g. Völker and Waibel, 2010; Bierkamp *et al.*, 2024) and *land rental markets* (e.g. Nguyen *et al.*, 2021).

New Rural Development Strategy

The government unveiled a "New Rural Development Strategy" in 2010 to "build a new countryside," modernizing agriculture through market integration, consolidation of small landholdings and technological change, and developing supporting industries and services. Explicitly modeled on South Korea's New Village Movement in the 1970s, the program has been implemented in 9000 communes across the country. Activities include development of transport infrastructure, irrigation and drainage systems, and the electrical grid, health and education services and the provision of cultural amenities such as sports facilities and village centers.

The central government contributed $15 billion in funding over the period 2010–2019, supplemented by international donors, local government and domestic commercial banks. The government claims that in its first decade of implementation the program increased non-agricultural incomes and reduced poverty in participating villages, although it is difficult to know how much of these gains can be directly attributed to the program (Nguyen *et al.*, 2021).

The "One Commune One Product" program adopts a similar approach, this time adapted from a model originally from Japan designed to reverse rural depopulation. The objective is to mobilize local resources and capabilities to build the value chain, apply advanced quality standards and build brand awareness. Products are assessed based on marketing potential, product quality and community attributes and awarded zero to five stars. Eighty percent of products rated three stars or higher are from the food group.

Vietnam achieved lower middle-income country status in 2011, a milestone widely celebrated at home and among the development organizations. Attention quickly turned to new benchmarks, especially prospects for breaking into the high-income group by the time of the country's centennial in 2045. The question of the middle-income trap, and whether Vietnam could escape it, loomed large. In the context of agriculture, avoiding the trap meant accelerating the transition from production of bulk commodities – cheap rice and coffee – to higher value-added, specialized products. The problems facing Vietnam in the agricultural sector are similar to those in manufacturing. Vietnam has achieved three decades of export-led growth as a competitive, low-cost producer. In manufacturing this has meant labor-intensive assembly operations, creating millions of steady jobs for low-skilled workers and earning billions in foreign exchange. Analogously, Vietnamese agriculture has specialized in labor-intensive activities on small farms, mostly producing low- to medium-grade bulk commodities for export such as rice, robusta coffee and raw black pepper. In both manufacturing and agriculture, this amounts to an extensive, "vent-for-surplus" strategy to mobilize underutilized land and labor to produce competitive exports. The strategy has been successful, transforming Vietnam from a poverty-stricken country to a thriving middle-income economy.

In both manufacturing and agriculture, the challenge facing Vietnam – similar to many other middle-income countries – is to find a path from the present model of vent-for-surplus, extensive growth to production of higher-value added exports using technology- and innovation-intensive methods. The history of the high-income countries shows us that the process of moving from extensive to intensive growth is unbalanced and asynchronous, so there is evidence to support the views of optimists and pessimists (Hirschman, 1958). In agriculture, optimists can point to the spectacular pace of change in the fish and shellfish industries, and, on a smaller scale, the rise of specialty coffee producers. The slowdown in productivity growth in recent decades seems to corroborate the pessimists' view, which holds that small, part-time farmers lack the capital and knowledge needed to move into higher value-added products (Gray and Jones, 2022).

The effects of climate change make the transition all the more urgent (Ho and Shimada, 2021). The Mekong River Delta is one of the three delta regions most vulnerable to climate change and sea level rise due to its low elevation, population density and reliance on agriculture. Coastal provinces suffer saline intrusion during the dry season, forcing farmers to adjust cropping schedules or convert to aquaculture. The frequency and intensity of El Niño effects, especially drought, have increased in recent years. The negative effects of climate change are compounded by excessive groundwater extraction, sand mining and upstream dams, which reduce sediment flow and thereby accelerate erosion, subsidence and sea level rise (Triet *et al.*, 2020). As farming and fisheries have become less viable in some locations, migration from the delta to other regions, especially Ho Chi Minh City, has accelerated.

The government's response to the impact of climate change on the nation's agriculture is indicative of both the strengths and weaknesses of policymaking in Vietnam. Public agencies in Vietnam have shown a strong capacity for mobilization, as reflected in the country's performance delivering basic health, education, electricity, sanitation and water supply to its citizenry, even in rural and remote locations. The government has been less effective when top-town directives require local or sectoral consultation and adaptation. As evidence of the impact of climate change began to mount, the government acted swiftly, launching a National Target Program on Responding to Climate Change in 2007. Among the strategies and plans released since then, Resolution 120, issued in 2017, aims to reorient agriculture in the Mekong Delta away from rice monoculture and towards nature-based approaches. However, implementation of strategies and plans has been hampered by fragmentation of public institutions, both horizontally between ministries and provincial level departments, and vertically between central and local government (Pincus, 2023). The government embarked on a restructuring program in 2025, merging ministries and provinces in an attempt to overcome fragmentation and coordination problems.

Covid-19 and Beyond

The response to Covid-19 was a demonstration of the Vietnamese government's authoritarian mobilization capacity. Vietnam was among the first countries to mandate flight suspensions, medical declarations for international arrivals, mask wearing and contact tracing. Free Covid-19 testing was introduced and quarantine centers established, controlled by the military. These measures helped to contain the disease in 2020, and Vietnam was one of the few economies to achieve positive growth (2.9%) for the year. Although the vaccination program was delayed by difficulties obtaining materials, once underway in 2021 vaccinations were administered free of charge to all residents, achieving 70% coverage by December 2021. The descent into crisis in the second half of 2021 highlights the government's limitations. The government failed to adapt flexibly to the emergence of more contagious variants, and showed reluctance to impose new restrictions on movement until the situation was out of control (Hoang, 2022). Also, a shadow was cast over the government's management of the pandemic by the Viet A scandal, in which a private company bribed officials to sell Covid-19 RT-PCR tests at a heavy mark-up to provincial health departments and hospitals.

Current Major Issues

Fragmentation of Land Holdings

Agricultural development aims to promote the use of advanced and climate-resilient technologies, to encourage downstream processing and diversify rural incomes. One of the legacies of the relative equality of land redistribution after collectivization is small average farm sizes and fragmentation of holdings. This is especially true in the Red River Delta region. In a labor surplus economy, small farms achieve high yields per hectare as marginal labor costs are low. However, as labor is drawn off into higher productivity pursuits small, fragmented holdings can inhibit labor productivity growth as they are ill suited to mechanization. As wages rise, labor intensive cultivation methods become less viable and farmers apply less labor or convert farms to lower value crops. In the Vietnamese context, Nguyen and Warr (2020) find that land consolidation in rice farming reduces labor use, increases use of machinery and raises farm income.

The Land Law was revised in 2013, to lengthen land tenure to between 20 and 50 years for farm households and 70 years for organizations. However, the maximum household land under annual crops is limited to 3 hectares in the Mekong River Delta and the Southeast and 2 hectares elsewhere. The 2024 revision to the land law maintained the area limits but allows households and individuals not involved in rice cultivation to receive transfers of land use rights if they establish an organization and have a plan to use rice land.

In rice, different models of large-scale field farming (LSFM) have been implemented on pilot scale. According to Dao *et al.* (2013), the area under LSFM had reached more than 80,000 ha in three provinces in the Mekong River Delta. However, LSFM was less successful in the north and in particular in the Red River Delta (Pham and La, 2014). By and large, the success of LSFM is doubtful as found in a CGIAR study (Thang *et al.*, 2017). Also, the data on farm size distribution from the most recent agricultural census does not support the notion of progress in this regard. By 2020, 85% of all farm households were below 2 ha with 70% of them below 0.5 ha. Remarkably there hasn't been much change since 2011 (see Figure 5.1). Farm households in the category below 0.2 had even increased, perhaps due to the aging farm population and as a result of the Covid-19 crisis. By 2023, the vast majority of farms in Vietnam were mini- to small-scale part-time farmers. This is evidenced by the growing share of wage employment, as shown in Figure 5.2. While in 2011, barely 30% of rural households listed as wage workers in the labor force census, by 2020 the corresponding figure was close to 50%, with moderate gender differences (see Figure 5.2). There are marked differences across Vietnam's major regions. For example, in the Red River Delta wage employment was 60% of the labor force in 2023 (Agricultural Census) which indicates that rural industrialization is progressing within the region. In the Mekong Delta, only 43% of rural households were listed as wage employees as many workers migrate to industrial areas in the Southeast region.

Rural households engage in many income earning pursuits including agriculture, which elevates the share of the labor force in agriculture above that of GDP. In some areas, for example in northern Vietnam and the Central Highlands, the gap between rural and urban areas has widened. Likewise, the gap between ethnic minorities and the Kinh majority continues to exist.

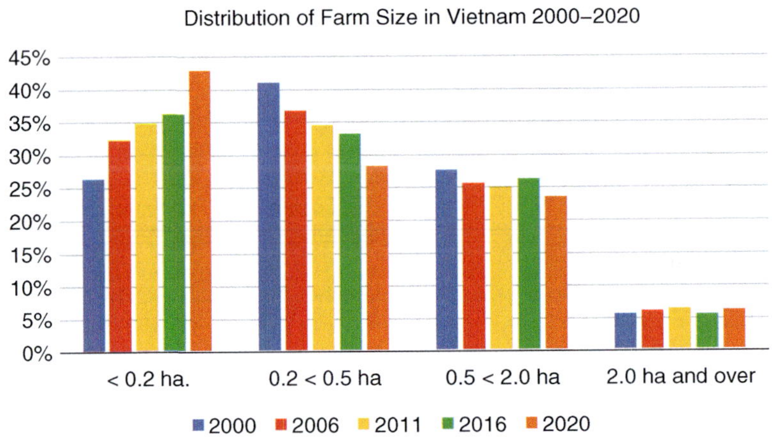

Figure 5.1. Farm size distribution. (Author's own image)

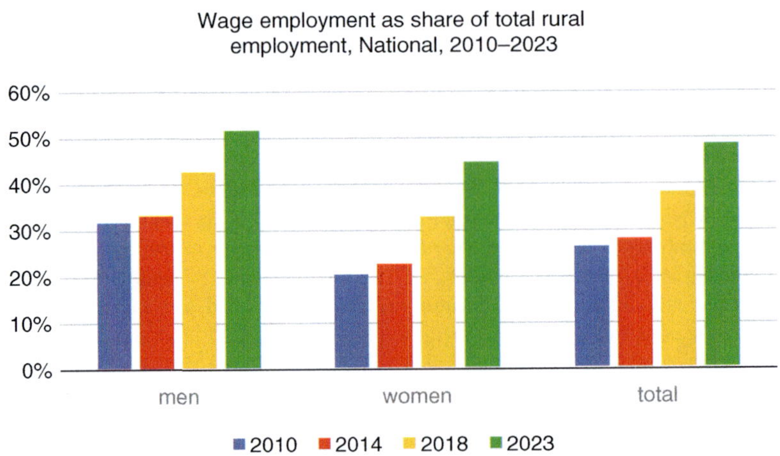

Figure 5.2. Wage employment of rural household members by gender, 2011–2023. (Author's own image)

Mekong Delta and Climate Change

Climate change will significantly impact agriculture in the Mekong River Delta, the country's main source of exports and rice for domestic consumption. The region is vulnerable to sea level rise and salinity intrusion (Dang *et al.*, 2020; Yuen *et al.*, 2021). Increased incidence of droughts and floods will negatively affect soil quality and productivity, and damage infrastructure (Dung and Phuc, 2012; Kontgis *et al.*, 2019; Phuong *et al.*, 2024). These effects are compounded by reduced sedimentation flow due to upstream dams, groundwater extraction and sand mining.

The government proposed a fundamental restructuring of agriculture in the Mekong Delta away from rice monoculture and toward more climate-resilient crops. A master plan for the Mekong River Delta region signaled a shift from triple cropping of rice to "living with floods" and sustainable farming practices through diversification and local adaptation to ecosystem changes. However, the implementation of the plan is progressing slowly. There are gaps in enforcement, monitoring, scaling up and the practical application of concepts such as "nature-based solutions" and "active adaptation." Resource allocation still heavily favors rice production and large-scale irrigation infrastructure, rather than promoting more diverse, climate-resilient systems.

In connection with Vietnam's Nationally Determined Contribution (NDC) under the Paris Agreement, the government approved a program entitled "Sustainable development of one million ha of high-quality and low-carbon rice together with green growth in Mekong Delta up to 2030" to reduce emissions from agriculture and increase farmer incomes. The technology involves reduced seed volumes (from around 200 kg to below 70 kg per ha), reduction of chemical fertilizers and pesticides by 30%, and water use reduction by 20%. Pilot projects in the Delta region have demonstrated that these practices can reduce input costs and increase profits. However, there is a lack of recognized standards for "high quality" and "low carbon" rice. There are also gaps in project feasibility evaluation, particularly in identifying suitable land sites for large-scale rice farming and assessing market demand for high-quality, low-carbon rice. There is a lack of demand and market price protection for high-quality, low-carbon rice. With a large production of one million hectares of rice, excessive supply might occur, so higher rice prices might not be guaranteed. Therefore, financial incentives associated with the new farming models may not materialize.

Agricultural Diversification

The Mekong River Delta region produces 50% of national rice production and 90% of Vietnam's rice exports. Given the perceived stability and lower risk associated with rice farming, many farmers are reluctant to adopt new systems like aquaculture or fruit farming, which are promoted as more adaptable to the region's evolving environmental conditions. While the government introduced policies to promote diversified farming systems, such as the integration of rice with shrimp farming or horticulture, progress has been slow in large part because existing infrastructure has been developed for rice farming (Tarp *et al.*, 2003). Irrigation systems are optimized for providing fresh water to rice paddies and are less suited to other agricultural activities such as shrimp farming or fruit cultivation. Thus, the lack of targeted support for farmers to move away from rice farming is one of the major barriers to scaling climate-resilient agricultural practices in Vietnam's Mekong River Delta. While livelihood diversification through activities such as aquaculture and fruit farming is being promoted, the region's infrastructure, policies and market systems are still heavily skewed toward rice production.

Using the TVSEP panel data, we present a case study describing the change in the life of a rural household in Ha Tinh province. The purpose of presenting this story is to supplement the analysis of rural agricultural transformation by means of a concrete case derived from primary empirical data.

Case Study: A Rural Development Story from Ha Tinh Province

This is a case study of a household from the TVSEP panel in the province of Ha Tinh (Waibel, 2023).

The household head, let's call him Mr. Nguyen (his real name is different), when we first interviewed him in 2007, was living alone because his parents had recently passed away. At that time, he had 0.35 ha of land, growing upland rice, sweet potato and peanuts. He had one cow and some poultry. He did not have a motorbike. His per capita income was just about $2 per day, almost 80% coming from farming.

When looking at his records 10 years later, the situation had changed significantly. He meanwhile got married, had two children and found a job in a local construction company. His wife took care of the farm which had enlarged to 1.17 ha by renting in land. Mr. Nyugen, aside from being a construction worker, also helped his wife with the farm. In 2017, he had a motorbike and his per capita income went up to $7 per day with still almost 50% coming from farming. So, Mr. Nguyen found his way out of poverty as a local wage labourer, combined with farm investment – instead of migrating to Hanoi, Da Nang or Ho Chi Minh City. On the downside, Mr. Nguyen's health had severely deteriorated during this period, due to occupational health exposure, in the absence of worker safety measures. What does this case tell us about rural development in Vietnam? First, the way out of poverty was income diversification, not migration with wage employment, more often than not, ending up in poor employment conditions. In the case of Mr. Ngyuen's household, he stayed in his village, and combined farming with wage employment as a part-time farmer. Clearly, during his lifetime, Mr. Nguyen won't become a large-scale, full-time and highly commercialized farmer, as policy-makers may envision and as many researchers suggest. Second, we must ask, what could the government do for a family like the one of Mr. Nguyen? Much of the family's future will depend on Mr. Nguyen's health and on the prospects for his children. Will there be good education for them in the village or the commune? Will there be adult education options and health enhancement programs for Mr. Nguyen (and his wife) to enable him to change from his construction worker job to a service type of job, when his health inhibits hard physical labor? Will there be access to smart and sustainable farming technologies at affordable prices? Will there be good infrastructure in his village, both physical and social, including good roads, high standards in sanitation, fast internet and social activities that would make the rural space a more attractive work and living place?

Summary and Outlook

This brief run-through the history of economic and agricultural development in Vietnam shows a diverse picture of agriculture. Perhaps it can be summarized by the metaphor of the *bountiful rice bowls held by a thin bamboo stick and carried by a poor person*.

The "rice bowls" stand for the wealth of the urban centers such as Hanoi, Ho Chi Minh City and Da Nang, which emerged in an unbelievably short time. The "bamboo stick," represents the rural areas with the millions of extremely small farms who are unable to make end meet with farming alone. While market conditions have been introduced in the rural communes and villages, the social organization and the viewpoint of people is still dominated by the socialist school of thought. For example, as observed by Tria Kerkvliet (2006), farming families are often ambiguous on the one hand between free land markets with the possibility of renting in and out as well as buy and sale of land use rights and, on the other hand, the need for redistribution of land according to the wants of people in order to maintain social cohesion in the village. This may be one of the constraints towards more rapid structural change in agriculture with much larger average farm sizes.

Similar to the other Asian countries (see the example of Thailand in Chapter 4) average farm size did not increase and the share of farms with less than 2 ha of size was almost 95% in 2023 (see Figure 5.1). In fact, the proportion of farms above 2 ha has dropped significantly between 2020 and 2023 which could be a post-Covid-19 effect, when poor people had not many other choices than revert to subsistence farming. The overwhelming majority of farm households in Vietnam to date, are part-time farmers with a growing share of non-agricultural income.

While the vast share of agricultural goods is still produced by small-scale farmers, the government has promoted agri-food companies (Junaid and Dinh, 2017), often organized as cooperatives. According to the Vietnam General Statistics Office (GSO, 2013, 2014), there are about 6610 companies in the agri-food sector. According to a survey of 200 agricultural companies carried out by the Institute for Policy and Strategy for Agriculture and Rural Development (IPSARD) in 2014, the technological and managerial capacity of these firms is limited, and they face constraints on access to land (Shaikh and Dinh, 2017). Even in poultry production, small-scale enterprises have been dominant, as reported by Burgos *et al.* (2007). For example, large-scale, indoor chicken farms, mostly located in the outskirts of major cities produce about 10% of Vietnam's chickens.

Although the government has repeatedly stated its intention to promote a restructuring of agriculture, the politics of land remains a major obstacle. While successive land laws have strengthened user rights to land, government agencies still have considerable discretion in individual cases (Nguyen *et al.*, 2018). Lack of clarity and security of tenure has held back the development of agricultural land markets and is a source of corruption and local land disputes.

As long as labor costs remain low, especially in comparison with other ASEAN countries, Vietnam will continue to attract investment in labor-intensive manufacturing. We can expect migration from rural to urban areas, and from farming to manufacturing and services, to continue for some years to come. Eventually, if current trends continue, surplus or underemployed rural labor will be exhausted, and urban wages will rise. If productivity and wage growth outside of agriculture are sustained, rural households will find it less attractive to hold on to their land and operate it as part-time farmers. At that point, pressure for land consolidation would increase as labor-intensive farming would no longer be economically viable without government support. However, if job creation outside of agriculture slows, it is likely that households would seek to hold on to their smallholdings. It is important to remember in this context that as more than half of the labor force is already dependent on wage employment, the persistence of household farming would be relevant to a shrinking minority of rural dwellers. Most would be cast into the marginalized, "footloose" population of semi-urbanized wage workers (Breman, 2012).

Who will be Vietnam's farmers of the future? As in other Asian countries, farmers are aging and their descendants are seeking employment opportunities outside agriculture. Already in the medium term, this will have consequences for technical progress in agriculture, aggregate agricultural output and national food security. Government policy envisions an agricultural system consisting of a high share of full-time farmers, who are entrepreneurs residing in the rural areas, using climate-smart technology and high-tech equipment, such as GPS-supported crop management tools. But there are still substantial hurdles to be overcome, including investment in the requisite infrastructure, education and research, more secure land title for farmers and agricultural corporations, and greater consistency in policy implementation. In the absence of these changes, the agricultural sector will remain a safety net rather than an engine of growth.

References

Amare, M. and L. Hohfeld (2016) Poverty transition in rural Vietnam: the role of migration and remittances. *Journal of Development Studies* 52 (10), 1463–1478.

Appau, S., Churchill, S.A., Smyth, R. and Trinh, T.A. (2021) The long-term impact of the Vietnam War on agricultural productivity. *World Development* 146, 105613.

Ayala-Cantu, L. and Morando, B. (2020) Rental markets, gender, and land certificates: evidence from Vietnam. *Food Policy* 94, 101842.

Barnett, J.H. and Wailes, E. (2011) Vietnamese agricultural policies during Doi Moi: 1986–2002, an analysis. *Journal of the Southwest Conference on Asian Studies* 7, 3–25.

Baronchelli, A. and Ricciuti, R. (2018) *Climate Change, Rice Production, and Migration in Vietnamese Households* (No. 2018/86). WIDER Working Paper, Helsinki, Finland.

Bellemare, M.F., Chua, K., Santamaria, J. and Vu, K. (2020) Tenurial security and agricultural investment: evidence from Vietnam. *Food Policy* 94, 101839.

Beresford, M. (1990) Vietnam: socialist agriculture in transition. *Journal of Contemporary Asia* 20(4), 466–486.

Bierkamp, S., Trung, T.N. and Ulrike, G. (2024) Farming efficiency and environmental resource dependence: evidence from panel data for rural Central Vietnam. *Australian Journal of Agricultural and Resource Economics* 68, 146–167.

Breman, J. (2012) *Footloose Labor*. Cambridge University Press, Cambridge, UK.

Brooks, M., Lippe, R.S. and Waibel, H. (2020) *Comprehensive Data Quality Studies as a Component of Poverty Analysis*. TVSEP Working Paper No. 19, Hannover. Available at: https://www.tvsep.de/en/publications-1#c69283 (accessed 26 February 2026).

Burgos, S., Hanh, P.H., Roland-Holst, D. and Burgos, S.A. (2007) Characterization of poultry production systems in Vietnam. *International Journal of Poultry Science* 6, 709–712.

Dao, M.H.T., Nguyen, A.V. and Vo, L.M.T. (2013) Linkages in rice production and consumption in Vietnam. VIE Report. Vietnam Institute of Economics, Hanoi, Mimeo.

Dang, A., Kumar, L. and Reid, M. (2020) Modelling the potential impacts of climate change on rice cultivation in Mekong Delta, Vietnam. *Sustainability* 12, 9608. DOI: 10.3390/su12229608.

Dang, T. (2018) Vietnam's post-1975 agrarian reforms: how local politics derailed socialist agriculture in Southern Vietnam. ANU Press, Canberra.

Dung, N. and Phuc, L. (2012) How severe is the impact of climate change on crop production in the Mekong Delta-Vietnam? *Journal of International Business Research* 11, 97.

Gawthorpe, A.J. (2022) Modernization, agricultural economics, and US policy towards land reform in South Vietnam. *The International History Review* 44(2), 282–299.

General Statistics Office (2013) Results of the 2012 Establishments Census. Available at: https://www.nso.gov.vn/wp-content/uploads/2019/03/KQTDTCSKT2012-1.pdf (accessed 26 February 2026).

Gray, E. and Jones, D. (2022) Innovation, agricultural productivity and sustainability in Viet Nam. *OECD Food, Agriculture and Fisheries Papers* 181. Available at: https://doi.org/10.1787/9cc1f47a-en (accessed 26 February 2026).

Gröger, A. and Zylberberg, Y. (2016) Internal Labor Migration as a Shock Coping Strategy: Evidence from a Typhoon. *American Economic Journal: Applied Economics* 8(2), 123–153.

Hirschman, A.O. (1958) *The Strategy of Economic Development*. Yale University Press, New Haven.

Ho, T. and Shimada, K. (2021) The effects of multiple climate change responses on economic performance of rice farms: evidence from the Mekong Delta of Vietnam. *Journal of Cleaner Production* 315, 128129. DOI: 10.1016/J.JCLEPRO.2021.128129.

Hoang, V. (2022) The COVID-19 pandemic in Vietnam – success, crisis, and endemic: Key thresholds and lessons. *Journal of Global Health* 12, 03065.

Junaid, M.S. and Dinh, T.B.L. (2017) Main policies affecting corporate performance of agri-food companies Vietnam. *Academy of Accounting and Financial Studies Journal* 21(2), 1–17.

Kang, H., Sridhar, V., Mainuddin, M. and Trung, L. (2021) Future rice farming threatened by drought in the Lower Mekong Basin. *Scientific Reports* 11, 9383. DOI: 10.1038/s41598-021-88405-2.

Kerkvliet, B.J.T. (1995) Village-state relations in Vietnam: the effect of everyday politics on decollectivization. *Journal of Asian Studies* 54(2), 396–418.

Kerkvliet, B.J.T. (2021) Everyday politics in agrarian societies. In: *Handbook of Critical Agrarian Studies*. Edward Elgar Publishing, Cheltenham, UK, pp. 463–468.

Kerkvliet, B.J.T. (2006) Agricultural land in Vietnam: markets tempered by family, community and socialist practices. *Journal of Agrarian Change* 6(3), 285–305.

Klingler-Vidra, R. and Wade, R. (2020) Science and technology policies and the middle-income trap: lessons from Vietnam. *Journal of Development Studies* 56(4), 717–731.

Kontgis, C., Schneider, A., Ozdogan, M., Kucharik, C., Tri, V. *et al.* (2019) Climate change impacts on rice productivity in the Mekong River Delta. *Applied Geography* 102, 71–83. DOI: 10.1016/J.APGEOG.2018.12.004.

Lee, S. (2020) Implication of the Korean agricultural technology group activities during the Vietnam War. *Journal of Rural Development/Nongchon-Gyeongje* 43(4), 115–139.

Li, Z., Saito, Y., Dang, P.X., Matsumoto, E. and Vu, Q.L. (2009) Warfare rather than agriculture as a critical influence on fires in the late Holocene, inferred from northern Vietnam. *Proceedings of the National Academy of Sciences USA* 106(28), 11490–11495.

Liebenehm, S. Jaretzky, H. and Waibel, H. (2023) Extreme weather and agricultural management decisions among smallholder farmers in rural Thailand and Vietnam. *Agricultural Economics* 54(5), 593–622.

Liebenehm, S., Schumacher, I. and Strobl, E. (2023) Rainfall shocks and risk aversion: evidence from Southeast Asia. *American Journal of Agricultural Economics* 106, 12403.

Lipton, M. (1984) Urban bias revisited. *Journal of Development Studies* 20(3), 139–166.

London, J.D. (2022) How to study contemporary Vietnam? In *Routledge Handbook of Contemporary Vietnam*. Routledge, London, UK, pp. 1–17.

London, J.D. and Pincus, J. (2022) Welfare, inequality, and opportunity in contemporary Vietnam. In: *Routledge Handbook of Contemporary Vietnam*. Routledge, London, UK, pp. 273–290.

Majumdar, S., Mani, A. and Mukand, S.W. (2004) Politics, information and the urban bias. *Journal of Development Economics* 75(1), 137–165.

Marcus, R.A. and Greene, G. (2006) Exploring the Vietnam war. *Education about Asia* 11(1).

Markussen, T. and Ngo, Q.T. (2019) Economic and non-economic returns to communist party membership in Vietnam. *World Development* 122, 370–384.

Markussen, T. and Tarp, F. (2014) Political connections and land-related investment in rural Vietnam. *Journal of Development Economics* 110, 291–302.

Markussen, T., Fibæk, M., Tarp, F. and Tuan, N.D.A. (2018) The happy farmer: self-employment and subjective well-being in rural Vietnam. *Journal of Happiness Studies* 19, 1613–1636.

Markussen, T., Tarp, F. and Van Den Broeck, K. (2011) The forgotten property rights: evidence on land use rights in Vietnam. *World Development* 39(5), 839–850.

Markussen, T., Tarp, F., Thiep, D.H. and Tuan, N.D.A. (2016) *Inter-and Intra-Farm Land Fragmentation in Vietnam* (No. 2016/11). WIDER Working Paper, Helsinki, Finland.

Markussen, T., Smriti, S., Saurabh S. and Finn, T. (2019) Inequality, corruption and cooperation: Evidence from Vietnam. Available at: https://www.academia.edu/download/81830081/vncsaesmritisharma.pdf (accessed 26 February 2026).

Masina, P. (2006) *Vietnam's Development Strategies*. Routledge, London, UK.

McCargo, D. (Ed.). (2004) *Rethinking Vietnam*. Routledge, London.

Nguyen, H.Q., and Warr, P. (2020) Land consolidation as technical change: economic impacts in rural Vietnam. *World Development* 127, 104750.

Nguyen, H. and Grote, U. (2004) *Agricultural Policies in Vietnam: Producer Support Estimates, 1986–2002*. ZEF Discussion paper, Berlin, Germany 93.

Nguyen, Hoang Linh, Jin Duan and Guo Qin Zhang (2018) Land politics under market socialism: the state, land policies, and rural–urban land conversion in China and Vietnam, *Land* 7, 2.

Nguyen, L.D., Katharina, R. and Ulrike, G. (2015) Rural-urban migration, household vulnerability, and welfare in Vietnam. *World Development* 71(July), 79–93.

Nguyen, T.T., Viet, T.T., Thanh-Tung, N. and Ulrike, G. (2021) Farming efficiency, cropland rental market and income effect: evidence from panel data for rural Central Vietnam. *European Review of Agricultural Economics* 48(1), 207–224.

Nguyen-Viet, T.A. and Imai, M. (2018) The effects of ethnic Chinese minority on Vietnam's regional economic development in the post-Vietnam War period. *Journal of Development Studies* 54(9), 1680–1697.

Pham, N.Q. and La, A.H. (2014) Household welfare and pricing of rice: does the large-scale field model matter for Vietnam?. *Trade Policies, Household Welfare and Poverty Alleviation*. Case Studies from the Virtual Institute Acadmic Network. Geneva: UNCTAD. 301–335.

Phung, T.D., Hardeweg, B., Praneetvatakul, S. and Waibel, H. (2015) Non-sampling error and data quality: what can we learn from surveys to collect data for vulnerability measurements? *World Development* 71(July), 25–35.

Phung, T.C. and Waibel, H. (2010) Diversification in land and labor allocation in response to shocks among small-scale farmers in Central Vietnam. *Schriften der Gesellschaft fur Wirtschafts- und Sozialwissenschaften des Landbaues e.V.* 45, 91–111.

Phuong, T., Vien, T., Son, C., Thuy, D. and Greiving, S. (2024) Impact of climate change on agricultural production and food security: a case study in the Mekong River Delta of Vietnam. *Sustainability* 16(17), 7776. DOI: 10.3390/su16177776.

Pincus, J. (2016) In search of a new growth model. *Southeast Asian Affairs* 2016, 379–397.

Pincus, J. (2023) Institutional Bottlenecks and Vietnam's Prospects for Growth and Development, UNDP Policy Brief, Available at: https://www.undp.org/vietnam/publications/institutional-bottlenecks-and-viet-nams-prospects-growth-and-development (accessed 26 February 2026).

Pingali, P.L. and Xuan, V.T. (1992) Vietnam: decollectivization and rice productivity growth. *Economic Development and Cultural Change* 40(4), 697–718.

Prosterman, R.L. (1970) Land-to-the-tiller in South Vietnam: the tables turn. *Asian Survey* 10(8), 751–764.

Shaikh, J. and Dinh, T.B.L. (2017) Main Policies Affecting Corporate Performance of Agri-foodCompanies In Vietnam. *Academy of Accounting and Financial Studies Journal* 21(2). Availbale at: http://ur.aeu.edu.my/484/1/Main%20Policies%20Affecting%20Corporate%20Performance%20of%20Agri-Food%20Companies%20in%20Vietnam.pdf (accessed 26 February 2026).

Shifa, A.B. (2013) The dual policy in the dual economy – the political economy of urban bias in dictatorial regimes. *Journal of Development Economics* 105, 77–85.

Sok, S., Chhinh, N., Hor, S. and Nguonphan, P. (2021) Climate change impacts on rice cultivation: a comparative study of the Tonle Sap and Mekong river. *Sustainability* 13(16), 8979. DOI: 10.3390/su13168979.

Szalontai, B. (2008) The diplomacy of economic reform in Vietnam: the genesis of Doi Moi, 1986–1989. *아세아연구* 51(2), 199–252.

Takeshima, H., Liu, Y., Cuong, N.V. and Masias, I. (2018) *Evolution of Agricultural Mechanization in Vietnam: Insights from a Literature Review and Multiple Rounds of a Farm Household Survey*. International Food Policy Research Institute, Washington, DC, USA.

Takeshima, H., Liu, Y., Cuong, N.V. and Masias, I. (2020) Evolution of agricultural mechanization in Vietnam. In: *An Evolving Paradigm of Agricultural Mechanization Development: How much can Africa Learn from Asia*. The International Food Policy Research Institute, 203–231.

Tarp, F. (2017) *Growth, Structural Transformation, and Rural Change in Viet Nam: A Rising Dragon on the Move*. Oxford University Press, Oxford, UK.

Tarp, F., Roland-Holst, D. and Rand, J. (2003) Economic structure and development in an emergent Asian economy: evidence from a social accounting matrix for Vietnam. *Journal of Asian Economics* 13(6), 847–871.

Thang, T.C., Khoi, D.K., Thiep, D.H., Lan, V.T., Tinh, T.V., Pede, V.O. (2017) Assessing the Potential of Climate Smart Agriculture in Large Rice Field Models in Vietnam. CCAFS Working Paper No. 211. Wageningen, the Netherlands: *CGIAR Research Program on Climate Change, Agriculture and Food Security (CCAFS)*. Copenhagen, Denmark. Available online at https://cgspace.cgiar.org/server/api/core/bitstreams/34be7c38-b1ba-49ed-b79d-01849458a9dc/content

Tran, T.Q. (1998) *Vietnam's Agriculture: Challenges and Achievements*. Institute of Southeast Asian Studies, Singapore.

Triet, N., Dung, N., Hoang, L., Duy, N., Tran, D., *et al.* (2020) Future projections of flood dynamics in the Vietnamese Mekong Delta. *Science of the Total Environment* 742, 140596. DOI: 10.1016/j.scitotenv.2020.140596.

Völker, M. and Kanchanachitra, M. (2021) Long-term effects of health shocks and insurance on rural households' consumption and wealth in Vietnam: evidence from panel data. *Journal of Public Health and Development* 19(3/Sept.–Dec.), 1–16.

Völker, M. and Waibel, H. (2010) Do rural households extract more forest products in times of crisis? Evidence from the mountainous uplands of Vietnam. *Forest Policy and Economics* 12(6), 407–414.

Yuen, K., Hanh, T., Quynh, V., Switzer, A., Teng, P. *et al.* (2021) Interacting effects of land-use change and natural hazards on rice agriculture in the Mekong and Red River deltas in Vietnam. *Natural Hazards and Earth System Sciences* 21, 1473–1493. DOI: 10.5194/NHESS-21-1473-2021.

6 The Philippines – Increasing Productivity in a Risky Environment

Hermann Waibel, Gesa Horstkotte-Wesseler and Justus Wesseler

Abstract

The Philippines is a densely populated country whose population had grown rapidly during the past six decades. While in the 1960s the population was about 30 million, there were about 117 million people in 2024, living on just over 300,000 km^2. Economic growth in The Philippines since the 1960s had its ups and downs which seems to be related to the respective government administrations. However, since around 2010 growth was more stable, also due to a more stable political environment. The chapter reviews the different economic and agricultural policies by government administrations since independence. Conclusions are drawn as regards the performance of the agricultural sector. In addition, two case studies are incorporated to highlight special issues of agricultural development. The chapter ends with an outlook on Philippine agriculture.

Introduction

The Philippines is a country of more than 7600 islands, of which only around 2000 are inhabited. Geographically, it consists of two major islands, Luzon in the north and Mindanao in the south. The Visayas in the center are a group of smaller and larger islands such as Cebu and Panay. In terms of climate, it is one of the most exposed countries to tropical storms, with flooding a common occurrence from June to September. The annual damage to agriculture from typhoons, droughts and floods is estimated to have reached an equivalent of 3% of total agricultural output in the late 2000s (World Bank, 2013). Climate change is likely to increase the frequency and severity of tropical storms (Overland and Vakulchuk, 2017). The climate risk index (German Watch, 2021) includes The Philippines in the group of countries most affected by recurring extreme events, based on the long-term index between 1993–2022. Earthquakes and volcano eruptions are other risk factors, since the country is located along the Pacific Ring of Fire, a seismically active belt of volcanoes and tectonic plate boundaries. There are more than 20 active volcanoes in The Philippines, with significant eruptions causing major damage in 1991 (Mount Pinatubo) and 2023 (Mount Mayon). Major earthquakes had occurred in Northern Luzon in 1990, the Central Visayas in 2013 and recently in Mindanao in 2023, all with a magnitude above 7 on the Richter scale. For the third straight year, in 2024, the country registered the highest world risk index (WRI) in the World Risk Report (WRR, 2024).

Population density in The Philippines is very high, with some 117 million people living on just over 300,000 km^2. Over 20 million people live in or around the national capital, Manila, where

 Asian Agriculture in a Climate Change-driven and Risky World (H. Waibel, S. Liebenehm and Ernah eds)
DOI: 10.1079/9781800627338.0006

population density reached 21,765 people/km^2 in 2024 – more than in Macau (21,000) or Monaco (18,000). On the other hand, in the mountain areas of Northern Luzon, 91 people share one square kilometer in 2020.

Population growth rate in The Philippines was over 3% in the 1960s, still 1.5% by 2015, and only recently has fallen below 1% (World Bank, 2024). Consequently, the Philippine population is very young, about one-third are under 15 years of age and only 5% older than 65. The median age in 2020 was 25.3 years (PSA, 2022) giving the country a strong population dividend. In this regard, a special characteristic of The Philippines is its high share of international migrants who send substantial amounts of remittances, significantly contributing to the country's GNP.

Economic growth in The Philippines since the 1960s had its ups and downs which seems to be closely related to the respective government administrations (see Figure 6.1). During the over 20-year rule of President Marcos, between 1964 and 1986, per capita growth was mostly between 3–6%, except towards the end of his autocratic regime, when a popular revolution had ousted him. Generally, growth rates were higher in subsequent democratic governments, although negative growth was experienced in 1991, 1992, 1998, 2009 and of course during the Covid-19 pandemic in 2020 (Figure 6.1). The frequent natural calamities and economic shocks, such as the 1990 earthquake, the Mount Pinatubo volcano eruption (1991), the Asian financial crisis, the food price crisis, and the Covid-19 pandemic which started in 2019, are possible explanations for the negative growth rates.

As in most other Asian countries, poverty in The Philippines has declined albeit more slowly than elsewhere, for example in Thailand. In 2023 the headcount ratio was still at a level of 15.5%, using the national poverty line (PSA, 2025). Also, inequality is high with a Gini coefficient of 39.3 in 2023 but has been falling steadily since 2012 (World Bank, 2024). Furthermore, while prevalence of undernourishment was reduced, down from 13.6% in 2011 to 5.9% in 2022, malnutrition remains high. The rate of stunting among children below 5 years old was still 28.8% in 2019 with very little change during the past 15 years (Global Nutrition Report, 2024).

Agriculture in The Philippines is marked by a scarcity of land due to high population density. Shortage of agricultural land is further aggravated by continuing demand for residential, commercial and industrial space. There is about 5.6 million ha arable land in The Philippines (World Bank, 2024), of which about 65% is irrigated (National Irrigation Administration, 2022). While there is potential for expanding land under irrigation, the maintenance of existing irrigation systems is a problem (Inocencio and Baker, 2018). Deforestation and unsustainable farming practices contribute to environmental degradation, further limiting the productivity of land. Thus, to meet the growing demand for food, land

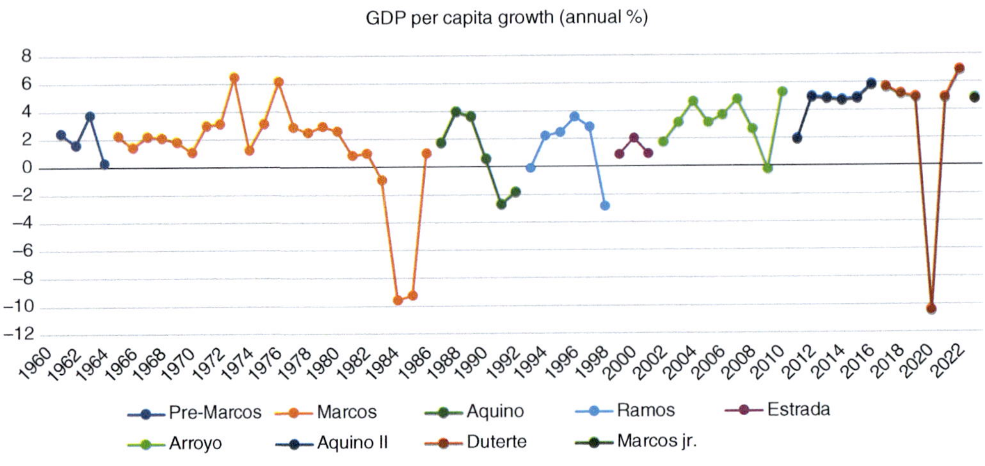

Figure 6.1. GDP per capita growth (annual %).

productivity must be increased by means of good natural resources management if rising food imports are to be avoided. For example, while conservation agriculture is technically feasible, often, economic incentives and effective extension services limit its adoption (Lapar and Pandey, 1999).

As reported in the agricultural statistics (PSA, 2024), over half of the gross value added of the agricultural sector comes from crop production, some 36% from livestock, including fish, while the remainder is attributable to agricultural support activities by the government. Among crops, rice is by far the dominant crop with 35% of farm land or some 4.8 million ha. Rice is grown under a wide range of ecosystems, all over the country. For example, in the mountain region, the famous Banaue Rice Terraces are on the UNESCO World Heritage list. However, most rice production takes place in the lowlands, which are predominantly wet rice under irrigated conditions. While average rice yields had a nearly constant growth rate during the past 60 years, they are still at a modest level of 4.15 t/ha. Rice production regularly falls short of domestic consumption. For example, in 2023, The Philippines imported 3.6 million tons of rice, making it the second-biggest rice importer globally.

Maize is the second largest annual crop with about 2.5 million ha. It is mostly grown under rainfed conditions with an average yield of 3.26 t/ha in 2022 (PSA, 2024). Among the perennial crops, coconut is dominant with about 3.6 million ha, followed by banana and sugarcane with less than half a million ha each (see Case Study below). Farm sizes in general are mini- to small-scale except for banana and pineapple where plantations in Mindanao are managed by international corporations. In livestock, although the sector is transforming towards larger units, so far, only poultry is a dominantly large-scale, commercial type of business while cattle and hog production are still dominantly small-scale.

The contribution of agriculture, forestry and fishing to GDP is gradually declining, in line with developments in other Asian countries. In 2023, it stood at 9.4%, down from 12.5% a decade earlier and from about 25% in the 1960s, indicating resource transfer to the non-agricultural sectors as is common during the course of development. On the other hand, labor shares declined less, i.e. from 31% to 22% in the same period (see Figure 6.2), indicating again that labor productivity in agriculture is much lower than in other sectors.

More than half of The Philippines' inhabitants live in rural areas, with agriculture as their primary source of income. However, rural households increasingly diversify into non-farm income sources, as income from farming is not enough to meet the needs of an average family of 4 persons (PSA, 2020). Overall, however, there has been little structural change in Philippine agriculture, an issue that will be highlighted in the last section of the chapter.

Following this brief introduction to The Philippines and its agriculture, the chapter continues with a discussion of central economic and agricultural policies since the 1960s, as well as significant events and changes relevant for agricultural development. A summary of key developments, using keywords, is presented

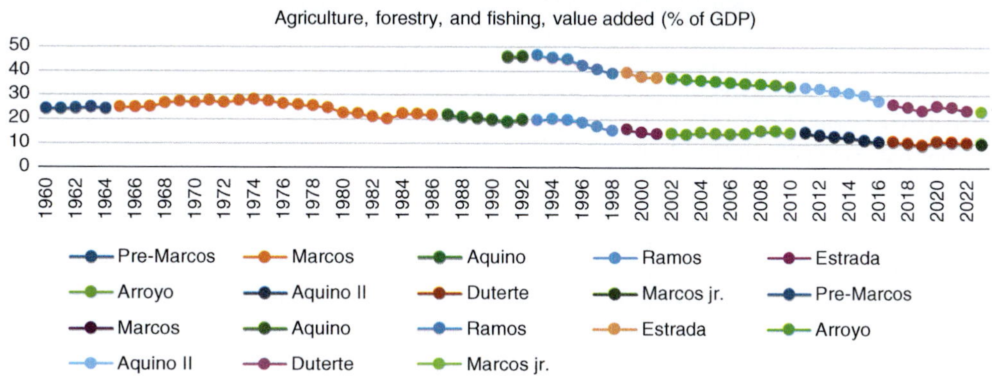

Figure 6.2. Agriculture, forestry and fishing, value added (% of GDP).

in Table 6.1, with more detailed explanations thereafter. Following the policy history, two case studies are presented. These aim to highlight specific agricultural policy issues that have either been the subject of our research in The Philippines or that further emphasize particular important policy issues. The chapter concludes with an outlook on the future of agricultural development in the country.

Policy Developments since World War II

The following description of policy developments is divided according to the government periods of successive presidents of The Philippines, each of whom defined new policy priorities and enacted laws that shaped agricultural development in the country. Table 6.1 outlines for each of the periods, the general development policy, the agricultural policies, significant events that affected the course of general and agricultural development and drawing conclusions on the performance of the agricultural sector.

We start out with the period after World War II with five different government administrations, Roxas, Quirino, Magsaysay, Garcia and Macapagal. The 21-year rule of Ferdinand Marcos is split into an early and later phase. The subsequent administrations of Corazon Aquino and Fidel V. Ramos are joined into one period, while the five succeeding administrations are discussed separately.

From Independence to Marcos 1946–1965

The Philippines gained independence from the US colonial power with the Treaty of Manila on July 4, 1946. During World War II, Japan occupied the country, challenging US military presence. After Japan's surrender in the aftermath of the Hiroshima and Nagasaki atomic bombs, The Philippines was liberated by the American army under General Douglas MacArthur. With its first president, Manuel Roxas, a constitution similar to that of the USA was adopted. In return for a 99-year lease agreement to allow armed forces to be stationed in US bases, the USA provided substantial aid for reconstruction.

It is important for a good understanding of Philippine politics to date, to know that political power is exercised by elite groups, clans and families, largely determined by the evolution of the country's political structure since Spanish and American colonial times (Lande, 1968). The colonial period laid the foundation for the large estates (haciendas) of the post-independence period. initially owned by Spanish and later American owners. Hence large tracts of land were owned by a small number of landlords, while most farmers worked as tenants or laborers. Neither during US colonial rule, nor soon after independence, were any effective land reform programs undertaken. The US supported the existing landowning class, which served its interests, especially in export crops. This resulted in widespread poverty among the peasantry, even before independence (Pomeroy, 1978).

Not surprisingly, during the early years of independence, insurgencies, particularly by various peasant movements took place. The most notable one was the "Huk" rebellion, a tenant grassroots movement that emerged in response to exploitative rents demanded by the landlord class. Although the Huk's influence waned in the early 1950s, it regained momentum with the rise of the New People's Army (NPA) movement. Both the Huk and the NPA withdrew to rural areas, with the NPA retreating particularly to the Cordillera Central in Northern Luzon (Wurfel, 1983) but also later in Mindano.

A good example of agricultural commercialization is the coconut industry which transformed from a domestic activity into a major export industry, during US colonial times and persisted after independence. Similarly, sugarcane has been grown in The Philippines for centuries, long before Spanish colonization. During the second half of the 19th century and well into the colonial period, it became an export industry dominated by large plantations run by wealthy and influential families.

While initiatives for land reform had been started in the pre-independence period, it was only in 1963 when the Republic Act No. 3844, the Agricultural Land Reform Code of 1963 was established (see Table 6.1). This code aimed to abolish tenancy and introduce a leasehold system. The implementation had some initial success with income gains for those farmers who shifted from

Table 6.1. Development periods by governments. (Authors' own table).

Period	General Development Policy	Agricultural Policies	Significant Events	Ag. Sector Performance
After World War II until Marcos: 1946–1965	Open economy policy, foreign direct investment, dominated by USA; dependency on US investments and development assistance	Initial land reform programs; Republic Act No. 3844, the Agricultural Land Reform Code, support rice production for self-sufficiency goal	Construction of first large hydroelectric power plant (Ambuklao); establishment of International Rice Research Institute (IRRI) in Los Banos, 1960.Huk rebellion mainly in Central Luzon	Increase in agricultural productivity, including rice; moderate GDP growth (1.9%)
Ferdinand Marcos – 1965–1972(Phase I)	Monopolizing key industries, infrastructure development by foreign debt	Republic Act No. 6389 (Code of Agrarian Reform) and RA No. 6390; promote green revolution technologies; strong support for the sugarcane industry	Establishment of the Department of Agrarian Reform and the Agrarian Reform Special Account Fund; rise of rural insurgency ("New People's Army" since 1969)	Green Revolution Technology adoption; growth in agriculture: >5% p.a.Rice self-sufficiency in 1968; poverty rate about 49%
Ferdinand Marcos 1972–1986(Phase II)	Imposition of Martial Law 1972–1986; international donor community started to withdraw from supporting the country	Masagana 99, launched in 1973 to increase rice self-sufficiency, World Bank, Training and Visit (T&V) extension project	Farmer-Scientist Partnership for Development (MASIPAG) 1985; founding of Philippine Rice Research Institute (PhilRice) in 1985; oil price shock	Average growth rate declined to 1.06% p.a.; rice self-sufficiency uncertain; increase of rural poverty to almost 60%
Corazon Aquino 1986–1992; Fidel V. Ramos 1992–1998	Constitution, democratization, privatization and deregulation of critical industries, social reform agenda for poverty reduction	Comprehensive Agrarian Reform Program (CARP) in 1988; measures to increase rice yields	EDSA revolution in 1986; Luzon earthquake 1990; eruption of Mount Pinatubo 1991; deadly storm "Thelma"	Growth rates in rice productivity and rice production declined initially; later increased. Poverty down to 36%

Continued

Table 6.1. Continued.

Period	General Development Policy	Agricultural Policies	Significant Events	Ag. Sector Performance
Joseph 'Erap' Estrada, 1998–2001	Starting large-scale infrastructure projects	MAGKASAKA (Shoulder-to-shoulder for Agrarian Development) Program for rice productivity; continuation of CARP	Second EDSA Revolution, January 2001; Asian financial crisis	Economic downturn caused by the Asian financial crisis. The average growth rate declined to 1.39% per year. Poverty up again in almost 40%
Gloria Macapagal Arroyo 2001–2010	Continuation of infrastructure investments. Economic openness with emphasis on good relationship with USA; privatization of state enterprises; poverty reduction program (Pantawid Pamilyang Pilipino Program) (4Ps), peace actions	Agrarian Reform Program Extension with Reforms (CARPER); goal to modernize agriculture	Extreme El Niño events in 2009 and 2010, causing severe drought	Average growth in agriculture 2.9%, overall economic growth 3.20%. Poverty rate down to below 30%
Benigno Aquino III, 2010–2016	Further opening of the economy. High priority infrastructure projects; expansion of 4Ps	Continuation of CARPER; Agri-Pinoy Framework Sugarcane Industry Development Act	Typhoon Haiyan in the Visayas; Agri-Pinoy Trading Centre in Benguet	Political stability continued. Increase in economic growth with an average growth rate of 4.83%; poverty rate of 23.5% on average
Rodriguo Duterte, 2016–2022	Initially restrictive import and export policies for rice. War on drugs	Rice Tariffication Law (Republic Act No. 11203) in 2019; continuation of CARP	Social Amelioration Program (SAP) to cope with Covid-19 for farmers and rural communities	Agriculture productivity decreased. Actual distribution of land to 175,000 farmers; poverty rate 18.1% in 2022
Ferdinand "Bongbong" Romualdez Marcos Jr. since 2022	Recovery started. Poverty reduction and food security program (Walang Gutom Program)	Masaganang Ani 200 program to increase rice productivity; Coconut policy	Key initiative is to support the coconut palm sector	Recovery expected. Strong economic growth is forecasted

share tenancy to leasehold tenancy, because they benefited more from the introduction of modern rice varieties (e.g. Ostsuka, 1991). In the longer term, however, the program was marked with difficulties due to insufficient funding and political resistance (Jandina *et al.*, 2025).

A significant event for rice production was the foundation of the International Rice Research Institute in Los Banos under the Rockefeller Foundation giving The Philippines access to modern rice breeding technology with high-yielding, fertilizer-responsive varieties.

Overall, the period immediately after independence is characterized by modest growth in both the economy and agriculture.

Ferdinand Marcos 1965–1986

During Ferdinand Marcos's 21-year rule, from his election in 1965 until his removal in 1986, The Philippines experienced a blend of economic prosperity and challenges. Generally, developments were positive during the first Marcos phase, i.e. 1965–1972 but deteriorated constantly during phase II after the declaration of martial law in 1972 until 1986. This era saw significant investments in infrastructure projects funded by foreign debt and substantial support for rice irrigation initiatives in Luzon, alongside the founding of the Philippine Rice Research Institute (PhilRice). The country's Gross Domestic Product (GDP) saw significant growth, increasing from $5.27 billion in 1964 to $37.14 billion by 1982.

In the initial years, the rice intensification program, Masagana 99, yielded positive results. However, by the early 1980s, environmental sustainability concerns grew, sparking a countermovement opposing the intensification strategy. The MASIPAG movement, a grassroots organization, emerged as the most successful opposition, actively resisting technological changes and opposing biofortified rice, such as vitamin A-enriched "Golden Rice" (see Case Study below).

In the 1980s, economic initiatives such as the Movement for Livelihood and Progress, sought to attract more foreign direct investment. However, the creditworthiness of the government had dramatically deteriorated. For example, documents from the Presidential Commission on Good Government (PCGG) alleged that the Marcos family embezzled between $5 billion and $10 billion from the Central Bank of The Philippines.

During Ferdinand Marcos's administration in the 1970s, sugar became one of The Philippines' primary export commodities. As international sugar prices surged in the early 1980s, Marcos brought domestic and international sugar trading under government control by creating the Philippine Sugar Commission (PHILSUCOM). This enabled him and his cronies to reap profits from a substantial dividend between low prices paid to producers and high world market prices (Billig, 1993). Besides, sugar production was highly concentrated in the Visayas Region. On Negros Island, sugarcane became virtually the only agricultural crop, with most agricultural land devoted to its production. When sugar prices crashed in the early 1980s, the region's overreliance on sugarcane together with the government monopoly, led to the bankruptcy of many sugar planters and the loss of livelihood for over 190,000 sugar workers with severe consequences for malnutrition, especially among children. This event became widely known as the Negros Famine which peaked in 1985 and was a key trigger for the downfall of the Marcos administration.

The Marcos regime was confronted with strong opposition from farmer groups because of the failure to achieve sustainable rural development. One example was the establishment of MASIPAG, an NGO-driven initiative, aiming at farmer control of genetic and biological resources and against technocratic and top-down development. Finally, with the assassination of Marcos' political opponent, Benigno (Ninoy) Aquino, Jr. on the tarmac of Manila International Airport, upon his return from his US exile, in August 1983, political instability heightened and ultimately led to a severe recession, reducing the country's GDP to $30.7 billion by 1985.

In conclusion, the 21-year Marcos rule failed to make the agricultural sector a driver of overall economic growth, mainly because technology access and thus income gains among the rural population were far from equal and inclusive. Given the generally small farm-size structure, where income from farming is insufficient to make a decent living, a rural industrialization policy with the generation of non-farm employment for the rapidly growing labor force was missing. Instead, in the export crop sector, large-scale plantation production and processing remained under the control of multinational companies with little benefit for the rural population. Furthermore, the resulting political and economic crisis increased the pressure for domestic and international migration.

The EDSA Revolution and the Corazon Aquino and Fidel V. Ramos Administrations, 1986–1998

In 1986, Ferdinand Marcos was overthrown by a popular uprising outside two military camps along Epifanio de los Santos Avenue (EDSA), led by Army General Fidel V. Ramos and former Senator Juan Ponce Enrile (see Table 6.1). This uprising, called the EDSA Revolution, followed the "snap election" that Marcos had called for as a last attempt to stay in power. However, instead, Corazon Aquino, the widow of Benigno Aquino, Jr., was installed as President. Under the newly ratified Constitution, Aquino could only serve a 6-year term and was succeeded by General Ramos in 1992. During this period, significant support and investment in agriculture came from organizations such as the World Bank, FAO and IMF. The Philippines emerged as a pioneer in utilizing Build-Operate-Transfer (BOT) schemes to drive development. Under this model, private investors are invited to construct specific government projects – such as tollways, power plants, and railways – operate them for a predetermined period, and eventually transfer ownership to the government, addressing the limited financial capacity of the state.

The era was marked by two major environmental catastrophic events. First, in 1990, an earthquake in Northern Luzon severely impacted the mountain region of Northern Luzon, especially Baguio City, with severe loss of infrastructure. The following year, the eruption of Mount Pinatubo further shocked the nation, severely affecting rice and sugarcane production in surrounding provinces, causing significant infrastructure disruptions in Central Luzon.

The international donor community supported both administrations during this period. Agriculture and regional development initiatives, such as the Highland Agriculture Development Project, emphasized sustainability (Wesseler, 1996). However, issues related to pesticide use in agricultural production became evident, with pesticide resistance posing serious challenges, particularly in vegetable production. Despite international donor support for alternative control strategies, success was limited (Amend and Basedow, 1997). At the peak of the issue, farmers resorted to using cyanide to control the diamondback moth, a major pest in cabbage cultivation. This cyanide, obtained through exchanges with miners who used it for gold extraction, led to a dramatic drop in vegetable prices from the Cordilleras (Wesseler, 1996).

Major reforms in agriculture were the Comprehensive Agrarian Reform Program (CARP) through Republic Act No. 6657, in 1998, to promote equitable land ownership and strengthen agricultural support services. About 75% of all agricultural lands over 5 ha were to be distributed to landless farmers and regular farmworkers. The Department of Agrarian Reform (DAR) was tasked with implementing CARP, which included both land distribution and provision of support services such as credit, infrastructure and training. However, CARP faced numerous challenges, including resistance from landowners, legal loopholes, lack of support services and insufficient funding. For example, large landowners distributed land among their children and other relatives to avoid the loss of land.

In sugarcane, policies were introduced to promote agricultural diversification on the "sugar island" of Negros, implementing a "60-30-10" plan, i.e. 60% of landholdings were to remain in sugar cultivation, 30% was designated for other high-value export crops and the remaining 10% was set aside for home gardens and contract farming. By 2016, 85% of sugarcane was cultivated by farmers with 5 ha or less, only 1% had more than 50 ha.

In summary, GDP growth during the Aquino/Ramos period fluctuated as a result of natural disasters and the Asian financial crisis, albeit to a lesser degree than other Asian countries. Agricultural growth only picked up towards the end of the Aquino/Ramos phase.

Joseph 'Erap' Estrada, 1998–2001

In 1998 Joseph Estrada, a former movie actor and major of Manila, was elected as president. He emphasized poverty reduction through comprehensive rural development, boosting agricultural production, an affordable housing program, enhancing infrastructure, pursuing economic reforms, and adopting an aggressive stance against the Moro Islamic Liberation Front. When Estrada assumed office in 1998, The Philippines faced significant challenges, including the Asian Financial Crisis and severe weather conditions that distressed the agricultural sector. Consequently, economic growth plummeted to −0.6% in 1998, a sharp decline from the 5.2% growth rate in 1997. However, the economy started to rebound in 1999, achieving a 3.4% growth rate, followed by 4% in 2000. During this brief period, the Estrada administration recorded an annual average of 1.025 million metric tons of imported rice.

Estrada's administration expanded the Comprehensive Agrarian Reform Program (CARP) to benefit landless peasants, distributing over 266,000 hectares of land to 175,000 farmers, including areas owned by the traditional rural elite. In his second year, the total distribution reached 523,000 hectares to 305,000 farmers. In September 1999, Estrada issued Executive Order 151, known as the Farmer's Trust Fund, to facilitate the voluntary consolidation of small farm operations into medium- and large-scale enterprises capable of accessing long-term capital.

In 1999, Estrada introduced the Magkabalikat Para sa Kaunlarang Agraryo (MAGKASAKA) program, aiming to foster investments in rural areas and boost farmers' income through joint ventures and contract growing schemes. This initiative resulted in a 6% output growth, a record high at the time, and reduced the inflation rate from 11% in January 1999 to just over 3% by November. However, these short-term gains did not yield sustained improvements, especially after a period of low rice yields.

Despite having a capable economic team, Estrada's administration was criticized for cronyism, incompetence and corruption, diminishing foreign investors' trust. Investor confidence further eroded when President Estrada was accused of interfering in an investigation involving a friend's alleged stock market manipulation during his second year in office. The administration also faced social unrest, exacerbated by bomb threats, actual bombings, kidnappings and other violent incidents. Compounding these issues were the negative impacts of extreme weather conditions due to climatic disturbances on the nation's economy. On January 20, 2001, after a popular uprising, Estrada resigned amid charges of plunder and corruption. Note that by the end of the Estrada administration poverty headcount was up to a level of 40% again.

Gloria Macapagal Arroyo Government, 2001–2010

Following Estrada's resignation, Gloria Macapagal Arroyo, his vice-president took over. In 2004, she was then elected for a full six-year term. During her presidency, The Philippines experienced relatively robust economic progress, with GDP growth averaging around 4.6%, despite global economic challenges. Her administration launched several infrastructure projects, focusing on transportation, including roads, airports and bridges. Arroyo took a firm stand against terrorism, particularly targeting the Abu Sayyaf group and other insurgent organizations. In 2007, she enacted the Human Security Act, equipping the government with additional tools to combat terrorism, though critics expressed concerns over potential human rights violations and power abuse.

Arroyo maintained a close relationship with the USA, particularly in the war on terror, positioning The Philippines as a key US ally. She permitted American forces to conduct joint operations with Philippine troops. Her presidency, however, faced multiple corruption allegations, including the NBN-ZTE broadband deal scandal, leading to an impeachment attempt which however failed.

Under Arroyo, the Comprehensive Agrarian Reform Program (CARP) continued, although success was still hindered by land distribution issues and opposition from large landowners (Table 6.1). Her administration pushed for the liberalization of the mining industry to generate jobs and attract foreign investment, a policy that received criticism due to environmental impact concerns. Additionally, Arroyo continued privatizing state-owned enterprises, including the National Power Corporation (NAPOCOR) and other utilities, which critics argued resulted in higher costs for consumers.

One of the central goals of Arroyo's administration was to modernize the agricultural sector to improve productivity, reduce poverty and enhance competitiveness. Modernization efforts included adopting new technologies, improving irrigation systems and refining farming techniques. While rice self-sufficiency by 2013 was a significant goal, it was not fully achieved. Her administration sought to boost national rice production through programs such as the National Rice Program, which provided subsidies, new seeds, fertilizers and enhanced irrigation for farmers. Between 2001 and 2010, the average annual agricultural growth rate was 2.9%. However, drought caused by extreme El Niño events in 2009 and 2010 resulted in a roughly 1% contraction in output. Initiatives such as the Ginintuang Masaganang Ani (GMA-CARES) program and the Agriculture and Fisheries Modernization Act (AFMA) were aimed to modernize agriculture, for example, by promoting hybrid rice. However, success was modest. In fact, in 2007, The Philippines recorded the highest volume of imported rice, amounting to 2.5 million metric tons, almost 15% of domestic requirements.

In 2009, the Comprehensive Agrarian Reform Program Extension with Reforms (CARPER) was enacted through the Republic Act No. 9700 to extend and improve the original program. CARPER aimed at accelerating land acquisition and distribution and enhancing the provision of support services to agrarian reform beneficiaries. However, as pointed out by Elvinia (2011), while not a complete failure, its multiple objectives and biased exemptions and exclusions had led to unsatisfactory outcomes.

In summary, the Arroyo period was one of relative success with an average annual economic growth of over 3%, agricultural growth of close to 3% and poverty down to 30% at the end of her term.

Benigno Aquino III Government, 2010–2016

Benigno Aquino III, the son of former president Corazon Aquino, was elected president in 2010. During his administration economic growth and political stability further strengthened. At the same time, agriculture remained a crucial component of the Philippine economy, contributing significantly to the country's GDP and providing rural employment. However, industry and service sectors experienced faster growth and therefore agriculture's GDP share decreased more than in previous periods. Average annual agricultural growth rate was recorded at 1.2%. Despite efforts to achieve rice self-sufficiency, also driven by the food price crisis in 2008/2009, rice imports still stood at an average of 1.37 million metric tons per year, almost 10% below domestic requirements.

In 2011, the "Agri-Pinoy" framework was established to optimize natural and human resource development in agriculture and fisheries. The Comprehensive Agrarian Reform Program now concentrated on completing land distribution under CARPER, extending its mandate until 2014. The Agricultural Competitiveness Enhancement Fund (ACEF) with low-interest loans to farmers and fishers was launched and a Farmers' Trust Fund to improve access to finance was established. Additionally, social protection programs such as crop insurance and disaster relief for farmers and fishermen to aid recovery from natural disasters was introduced.

Despite efforts to bolster the agricultural sector, Aquino's policies faced numerous challenges. Natural disasters, such as Typhoon Haiyan (Yolanda) (see Table 6.1), severely impacted agricultural production, particularly in the Visayas region. Furthermore, ongoing issues such as land reform implementation and the need for greater support for small-scale farmers persisted throughout his administration. Further challenges were experienced by the coconut industry. Despite steady export growth, it struggled with declining yields, owing to Typhoon Haiyan damaging or destroying around 44 million trees, affecting about 1 million coconut farmers. Additional challenges such as aging trees, low investment, limited research and development, and inadequate infrastructure continue to keep many coconut farmers in poverty.

In the sugarcane industry block farming was introduced to increase total productivity including crop yields. Block farming aims to consolidate the operations of several small farms to enable economies of scale, while remaining individual land ownership. Furthermore, the Sugarcane Industry Development Act (Republic Act No. 10659) was enacted in 2015. The aim was to increase productivity, diversify products, generate jobs and improve the efficiency of sugar mills, ultimately raising farmers' and farm workers' incomes. It mandates government agencies to provide essential services such as ploughing, harrowing, weeding, fertilization, harvesting and other mechanized farm operations that small farmers cannot afford or access through loans.

To sum up, the Benigno Aquino III phase was marked by economic and social progress and political stability. He also succeeded to bring poverty down to close to 20% at the end of his period. Yet, some of the challenges in agriculture such as land reform and slow structural transformation remained.

Rodrigo Duterte Government, 2016–2022

Economic progress achieved during the previous administration continued during the Rodrigo Duterte administration, albeit at slightly lower growth rates until the economic decline owing to Covid-19. However, the political and social climate in the country deteriorated due to Duterte's brutal war against drugs, which recently has brought him to the International Court of Justice in The Hague on charges of crimes against humanity.

Agricultural policy concentrated on key areas aimed at enhancing the agricultural sector and addressing challenges faced by farmers. The focus was on boosting productivity, ensuring food security and fostering rural development. Infrastructure improvements, particularly farm-to-market roads under the "Build-Build-Build" program, were prioritized. This included the improvement of several highways accessing the Cordillera Central, for better access to the mountainous areas growing temperate vegetables.

The Covid-19 pandemic underscored the critical importance of food security and self-reliance. A major policy reform during this period was the passage of the Rice Tariffication Law (Republic Act No. 11203) in 2019, which replaced rice import quotas with tariffs. As a result, rice imports became easier and prices were lower, hereby ensuring a more stable rice supply. The downside, of course, were lower prices for domestic rice producers, lowering the profits of Filipino rice farmers. On the other hand, initiatives to introduce genetically modified, vitamin A enriched rice (Golden Rice, see Case Study) gained renewed momentum.

The Duterte administration launched various support programs to assist local farmers, particularly in the rice, coconut and sugarcane sectors, offering cash assistance, training programs and financial support. The Farmers and Fisherfolk Productivity Enhancement Program was designed to help smallholder farmers improve their production techniques. During the Covid-19 pandemic, the government provided financial support to farmers through the Social Amelioration Program (SAP) to alleviate the economic pressures caused by the pandemic on farmers and rural communities.

Aside from his negative human rights records, Duterte's government continued the positive trends in economic transformation. Also, the economy quickly recovered from the Covid-19 shock prior to his departure in 2022. Poverty headcount was down to 18%.

Ferdinand "Bongbong" Romualdez Marcos Jr. since 2022

Ferdinand Romualdez Marcos Jr., called Bongbong, is the son of President Ferdinand Marcos, ousted in 1986. Marcos Jr. was elected in democratic elections in 2022. It is too early to assess his administration; however, some initial observations can be made. Economic and agricultural policy generally follows the path of previous administrations. Generally, while adopting a slightly more liberal US-oriented economic approach, he maintains considerable government interference in key sectors such as rice and coconuts.

However, he pushed for rice sector reforms, the CARP, infrastructure as well as direct production support in agriculture, for example, fertilizer and crop subsidies. The highly ambitious program to increase rice productivity, i.e. the Masaganang Ani 200, with the aim to increase rice yield to 200 cavans per ha (about 8.8 t) was adopted from the previous administration and further promoted. His "Walang Gutom" (zero hunger) poverty-reduction and food security program is focused on low-income families whereby beneficiaries receive monetary assistance in the form of an Electronic Benefit Transfer (EBT) card, amounting to PhP 3,000 per month, to purchase a selected list of food commodities from accredited stores. While the Marcos Jr. period is only about mid-term, the prospects are positive. Economic growth during 2024 was over 5% and poverty was 15.5% in 2023, projected to decline further.

Summary

In summary, the development of agricultural policy in The Philippines since the 1960s can be divided into three significant stages:

First the 1960s until towards the end of the Marcos era in the mid-80s. This period was characterized by heavy government intervention in agricultural markets, with a government monopoly over the trade of rice, sugar and maize. During this period, high-yielding rice varieties were introduced. Public spending, particularly on irrigation increased, financed through taxes on major agricultural exports and foreign loans. Between 1978 and 1983, The Philippines achieved rice self-sufficiency and even became a net rice exporter.

The second phase until about 2000 can be labeled as partial liberalization. The private sector became more prominent in agricultural credit policy, and market interventions were reduced. Despite this, the provision of input subsidies remained the primary agricultural policy tool. In 1997, the government enacted the Agriculture and Fisheries Modernization Act (AFMA) to help farmers adjust to trade policy changes following The Philippines' accession to the World Trade Organization (WTO). While rice self-sufficiency remained an official policy, public expenditures on agriculture decreased in the late 1990s due to tight fiscal measures, following the 1997 Asian Financial Crisis. Thus, rice imports became necessary again.

The third phase is from 2000 to the present. This period is marked by a renewed focus on rice and significant increases in public spending for agriculture. Following the global food price crisis of 2008, the various governments concentrated on intensifying rice production programs, enhancing public spending on irrigation and providing input subsidies to achieve self-sufficiency. The National Development Plan during the second Aquino government addressed key challenges in agriculture, such as high farming input costs, inefficient supply chains, inadequate irrigation infrastructure, low technology adoption rates and limited access to formal credit. Launched in 2011, the Food Staples Sufficiency Program focused on rice and other selected staples. Still, it shifted emphasis from input subsidies to providing agricultural public services, such as extension services and infrastructure improvements like farm-to-market roads.

In summary, it can be concluded that while the different administrations never neglected agriculture in the country's economic development, the overall outcome was not impressive. For example, Briones (2021) found weak growth in total factor productivity due to inefficient public spending and lack of coordination among government agencies among others. It remains an open question how well agriculture in The Philippines will be able to compete with its ASEAN counterparts.

Case Study: Big Crops Big Challenges – Protecting the Banana Industry. Source: Author's Compilation

The Philippines has a rich history of fruit production, with bananas, mangoes and pineapples commonly grown across its islands. Major multinational companies such as Chiquita, Del Monte and Dole are involved in their cultivation through a combination of plantation agriculture with wage labor and contract farming. In the Cordilleras of Northern Luzon, the German government has supported fruit tree cultivation as a means of poverty alleviation and environmental protection. Fruit production is a vital income source for farm households in this region (Wesseler, 1997), including crops like bananas, mangoes and citrus fruits such as calamansi, mandarins, oranges, lychees and longans. Major fruits such as bananas, calamansi, mangoes and pineapples make up 68.2% of this production, while other fruits, including citrus varieties like oranges, account for the remaining 31.8%. Benguet was the leading fruit producer in 2023, with 8,694.5 metric tons, constituting 24.2% of the region's total output. Other key contributors included Abra (23.5%), Apayao (21.9%), Ifugao (14.3%) and Mountain Province (8.3%) (Philippine Statistical Authority, 2023).

Bananas are a staple food crop in The Philippines and a major export crop as well. While for domestic consumption, Lakatan and Saba are the most popular varieties, for exports it is Cavendish banana which were introduced to The Philippines in the early 20th century by international fruit companies. To date, The Philippines is the third largest banana exporter, with Japan as the major market. Banana for export is mainly produced by agro-industry companies such as Dole, Tadeco and Lapanday Foods Corporation, in large-scale plantations in the Visayas and Mindanao. Recently, also, contract production by small-scale producers has emerged (Yamasaki, 2022), among other grower-buyer arrangements (Loquias *et al.*, 2022). Banana cultivation in The Philippines faces challenges due to the spread of the so-called Panama Disease (Ploetz, 2015). Hence, the development of disease-resistant varieties is crucial for the survival of the banana industry in The Philippines. The Department of Agriculture and the Bureau of Plant Industry facilitate access to these newly developed resistant varieties. Earlier research (Wesseler, 1997) has highlighted that fruit production in The Philippines has good potential to diversify household income and to enhance food security, emphasizing the importance of adopting resistant varieties.

Case Study: Food Security and the 'Golden Rice' Debate. Source: Author's Compilation

Golden Rice was first introduced to the scientific community and the wider public 25 years ago. The second generation of Golden Rice, incorporating genes from maize and a common soil bacterium, can produce 20–30 μg of β-carotene per gram of edible rice (Palmer, 2025). The Golden Rice technology was transferred to public sector breeding programs in low- and middle-income countries, prioritizing Bangladesh and The Philippines. Although it is the most widely recognized biofortified crop and has the potential to prevent micronutrients deficiency, especially among children, the technology has faced strong opposition from consumer advocacy groups in The Philippines. A recent discussion paper of the Philippine Institute of Development Studies, i.e. Briones *et al.* (2023), using scenario analyses, concluded that most Filipino households will be able to afford adequate levels of energy and protein by 2030, but not for vitamin A. Micronutrient deficiencies among children, particularly in vitamin A, iron and zinc are prevalent. Approximately 17% of children aged 6–59 months experienced vitamin A deficiency in 2018, with the highest prevalence among children aged 12–24 months (22%) in 2019, followed by those aged 6–12 months (18%) (World Bank, 2021). In response, the Philippine government has been collaborating with the International Rice Research Institute (IRRI), which began introducing vitamin A-enriched rice, known as Golden Rice, in 2004. Research indicates that Golden Rice could significantly reduce vitamin A deficiency (Stein *et al.*, 2008; Wesseler and Zilberman, 2014; Wu *et al.*, 2021). Wu *et al.* (2021) estimate that delayed introduction results in annual welfare losses of about 1.5 billion Euros. Under the Duterte government policies, The Philippines became more open to the use of modern biotechnology in agriculture. Field trials for Golden Rice began in Ilocos Norte in 2022. However, its further deployment has faced opposition from environmental and other advocacy groups, leading to a legal case against its rollout (see e.g. https://masipag.org/sgrn/). Also, nutritionists found that the prevalence of vitamin A in storage declines (Schaub *et al.*, 2017). Such information may prompt additional investigations and contribute to further delays in roll-out programs.

The legal battle over the introduction of Golden Rice is ongoing in The Philippines. Possibly, Golden Rice will eventually be introduced (Malusog Rice E-Newsletter, November–December 2024, Issues 11–12), however, in any case, with significant delays.

Agriculture Development Outlook

While there is always uncertainty in forecasting development policies, for The Philippines it is expected that for the foreseeable future priorities may remain unchanged. Investment in rural infrastructure, rice and maize production, and diversifying agricultural output will continue to be key focuses. Coconut and sugarcane production will remain significant, while the development of the banana industry will hinge on effective disease management. Overall, the outlook is positive, bolstered by two decades of relative political stability and market liberalization, which have fueled continuous growth. Future governments will probably maintain this trajectory. Modest population growth in the country and region will drive increased food demand, benefiting the agricultural sector.

The country must remain open to the application of modern technologies in agriculture. This will be crucial for boosting agricultural productivity and reducing pressure on natural resources. A major issue will be the effects of climate change. There could be even more frequent extreme weather events such as storms, floods and droughts than the country has experienced in the past. This will make investment in agriculture riskier. Especially, tree crops such as coconut could suffer significantly. Climate smart adaption measures will be necessary yet regional variation will increase with the typhoon susceptible areas such as Bicol and Leyte falling behind further.

A major issue in the future development of the agricultural sector will be the size of farms. As shown in Table 6.2 since the 1960s, farms are constantly getting smaller. In 2022, average farm size was 0.83 ha, over four times less than their size 60 years ago. The vast majority, some 85% of particularly rice farms operate on less than 2 ha and 95% were less than 3 ha in size. This pattern basically did not change since 1996 (PhilRice, 2025). At the same time, there are over three times more farmers, reflecting the growth of the country's population but also the declining share of the farm population. The land reform program has been stated as one of the reasons why farms are getting smaller (Adamopoulus and Restuccia, 2020), while the aging farmer population, land fragmentation and lack of rural employment opportunities are other possible factors leading to rural households holding on to their land. Ultimately, smaller farms mean that, in order to gain a sufficient income, rural households have to engage in off-farm labor markets. This is also important for women whose incomes directly benefit children's health and education (Salazar and Quisumbing, 2009).

There seem to be two larger questions for the future of the rural areas in The Philippines, which are quite similar to the other Asian countries covered in this book. The first question is on the structure of agriculture. Given that the small-farmer world will continue to exist for quite some time more, how can modern technologies such as improved germplasm, advanced mechanization and information and communication technologies for increasing productivity be introduced? As the older people will eventually give up their farms, where will the farmer entrepreneurs in the rural villages who are willing to rent-in land from exiting farmers and operate larger, more efficient farms, come from? Second, in view of the considerable urban-rural income gap (Chua *et al.*, 2015) will the government have the courage and political will to implement the necessary reforms such as in land markets (re-zoning of agricultural land to commercial use) and provide incentives for more investment in rural non-farm businesses? A comprehensive, forward-looking rural development concept will be needed to facilitate and support the transformation process in order to make the rural villages and small towns attractive work and living places, in which opportunities will be needed.

Table 6.2. Population and farm size in The Philippines, 1960–2022 in ha.

	1960	1970	1980	1991	2002	2012	2022
Population (millions)	26.3	36.7	48.1	66.6	76.5	92.3	117.3
No. of Farms (millions)	2.17	2.35	3.42	4.61	4.82	5.56	7.43
Farm Area (million ha)	7.77	8.49	9.73	9.97	9.67	7.27	6.16
Average Area per Farm (ha)	3.59	3.61	2.84	2.16	2.01	1.31	0.83

Source: Philippine Statistical Authority, 2022; Agricultural and Fisheries Census Report & Census of Population and Housing, different years.

References

Adamopoulos, T. and Restuccia, D. (2020) Land reform and productivity: A quantitative analysis with micro data. *American Economic Journal: Macroeconomics* 12(3), 1–39.

Amend, J. and T. Basedow (1997) Combining release/establishment of *Diadegma semiclausum* (Hellen) (Hym., Ichneumonidae) and *Bacillus thuringiensis* Berl. for control of *Plutella xylustella* (L.) (Lep., Yponomeutidae) and other lepidopteran pests in the Cordillera Region of Luzon (Philippines). *Journal of Applied Entomology* 121, 117–342.

Billig, M.S. (1993) "Syrup in the wheels of progress": The inefficient organization of the Philippine sugar industry. *Journal of Southeast Asian Studies* 24(1), 122–147.

Briones, R.M. (2021) *Philippine Agriculture: Current State, Challenges, and Ways Forward*. Policy Notes. Philippine Institute of Development Studies, Manila, Philippine, 8 pp.

Briones, R.M., Bouis, H.E., Angeles-Agdeppa, I., Espineli, I.B. and Maniego, M.L.V. (2023) *Will Food Be Affordable to Filipinos by 2030? Alternative Expenditure Policies toward Ending Hunger by 2030 (No. 2023-23)*. PIDS Discussion Paper Series, Manila, Philippines.

Chua, K.K., Limkin, L., Nye, J., & Williamson, J. (2015) Urban-rural income and wage gaps in the Philippines: measurement error, unequal endowments, or factor market failure? *Philippine Review of Economics* 52(2), 1–21.

Elvinia, J. (2011) *Is Land Reform a Failure in the Philippines? An Assessment on CARP*. *Limits of Good Governance in Developing Countries*, pp. 333–362.

German Watch (2021) Available at: https://www.germanwatch.org/sites/default/files/Global%20Climate%20Risk%20Index%202021_2.pdf (accessed 13 September 2025).

Global Nutrition Report (2024) Available at: https://globalnutritionreport.org/resources/nutrition-profiles/asia/south-eastern-asia/philippines (accessed 18 June 2025).

Inocencio, A. and Barker, R. (2018) Current challenges in agricultural water resource development and management in the Philippines. *DLSU Business & Economics Review* 28(1), 17.

Jadina, M.C., Nissen, S. and Rennie, H. (2025) The unfinished implications of 'finished' land reform: Local experiences of the comprehensive agrarian reform program in Leyte, Philippines. *Land Use Policy* 157, 107658.

Lande, C.H. (1968) Parties and politics in the Philippines. *Asian Survey* 8(9), 725–747.

Lapar, M.L.A. and Pandey, S. (1999) Adoption of soil conservation: The case of the Philippine uplands. *Agricultural Economics* 21(3), 241–256.

Loquias, M.P., Digal, L.N., Placencia, S.G., Astronomo, I.J.T., Orbeta, M.L.G. *et al.* (2022) Factors affecting participation in contract farming of smallholder Cavendish banana farmers in the Philippines. *Agricultural Research* 11(1), 146–154.

Malusog Rice E-Newsletter, November–December 2024, Issues 11–12. Philippine Rice Research Institute.

National Irrigation Administration (2022). Available at: https://www.nia.gov.ph/monthly-archives/202206

Overland, I. and Vakulchuk, R. (2017) *Impact of Climate Change on ASEAN International Affairs: Risk and Opportunity Multiplier*. Norwegian Institute of International Affairs and Myanmar Institute of International and Strategic Studies. Available at: https://www.researchgate.net/publication/320622312_Impact_of_Climate_Change_on_ASEAN_International_Affairs_Risk_and_Opportunity_MultiplierPublished by the Norwegian Institute of International Affairs and Myanmar Institute of Strategic and International Studies ISSN: 1894-650X (accessed 04 March 2026).

Palmer, A.C. (2025) Golden rice: A quarter-century of innovation, challenges, and the promise of better nutrition. *Journal of Nutrition* 155, 2846–2853.

Philippine Statistical Authority (2022) *Population Projection Statistics*. Availbale at: https://psa.gov.ph/statistics/census/projected-population (accessed 04 March 2026).

Philippine Statistical Authority (2025) *Poverty Statistics*. Available at: https://psa.gov.ph/statistics/poverty (accessed 04 March 2026).

PhilRice (2025). https://ricelytics.philrice.gov.ph/.

Ploetz, R.C. (2015). Fusarium wilt of banana. *Phytopathology* 105(12), 1512–1521.

Pomeroy, W. (1978) The Philippine peasantry and the Huk revolt. *Journal of Peasant Studies* 5(4), 497–517.

Salazar, L. and Quisumbing, A. (2009). Assessing the impact of gendered labor markets in the rural Philippines. In *Paper for the FAO–IFAD–ILO Workshop on Gaps, Trends and Current Research in Gender Dimensions of Agricultural and Rural Employment: Differentiated Pathways out of Poverty*. (Vol. 31). Rome, Italy.

Schaub, P., Wüst, F., Koschmieder, J., Yu, Q., Virk, P. *et al.* (2017). Nonenzymatic β-carotene degradation in provitamin A-biofortified crop plants. *Journal of Agricultural and Food Chemistry* 65(31), 6588–6598.

Stein, A.J., Sachdev, H.P.S. and Qaim, M. (2008). Genetic engineering for the poor: Golden Rice and public health in India. *World Development* 36(1), 144–158.

Wesseler, J. and Zilberman, D. (2014). The economic power of the Golden Rice opposition. *Environment and Development Economics* 19(6), 724–742.

Wesseler, J. (1996) *Die Ökonomik der Einführung von Obstkulturen in der Cordillera Central von Nordluzon, Philippinen.* Shaker-Verlag, Aachen (in German).

World Bank (2013) Getting a grip on climate change in the Philippines. https://documents1.worldbank.org/curated/en/285281468296160884/pdf/787980WP0P13010e0Change0Philippines.pdf (accessed 13 September 2025).

World Bank (2024) Open Data. Available at: https://data.worldbank.org/indicator/SP.POP.GROW?locations=PH (accessed 13 September 2025).

World Risk Report (2024) Available at: https://www.ifhv.de/news/world-risk-report-2024 (accessed 04 March 2025).

Wu, F., Wesseler, J., Zilberman, D., Russell, R.M., Chen, C. *et al.* (2021). Allow golden rice to save lives. *Proceedings of the National Academy of Sciences USA* 118(51), e2120901118.

Wurfel, D. (1983). The development of post-war Philippine land reform: political and sociological explanations. In: Perla Makil, Q. and Virginia Miralao, A. (eds) *Second View from the Paddy, Antonio Ledesma*. Institute of Philippine Culture, Ateneo de Manila, Philippines.

Yamasaki, Y. (2022) The impact of commercial crop expansion on the local people: The Cavendish banana industry in the Philippines. In: *Paper prepared for presentation at the Agricultural & Applied Economics Association Annual Meeting, Anaheim, CA; July 31-August 2022.*

7 Agricultural Development and Policy in China – with Two Case Studies from Southwest China

Dan Huang, Shi Min, Shaoze Jin and Haowen Zhuang

Abstract

This chapter reviews the evolution of China's agricultural institutions and policies over four distinct periods: (1) privatization to agricultural cooperatives (1949–1956), (2) cooperatives to people's communes (1957–1977), (3) communes to the household contract responsibility system (1978–2012), and (4) the new era of agricultural modernization and rural revitalization (2013–present). Each period is marked by key policy changes and corresponding impacts on agricultural productivity, rural economies, and farmers' livelihoods. Furthermore, a case study from Xishuangbanna, Southwest China illustrates these dynamics, namely, the carbon balances of rubber plantations, highlighting ecological trade-offs between economic benefits and sustainability; and the development of off-farm labor markets. The chapter concludes that ongoing policy efforts should focus on agricultural modernization, sustainable land use, and rural infrastructure development to address remaining challenges and achieve long-term agricultural sustainability and rural prosperity.

Introduction

Since 1949, China's agriculture has undergone significant transformations and achieved remarkable progress. In the early years, land reform redistributed land to peasants, establishing public land ownership and laying the foundation for agricultural development. The household responsibility system introduced in the late 1970s further motivated farmers by linking their efforts to income, leading to a substantial increase in grain production. Mechanization and technological advancements have also played crucial roles in boosting agricultural productivity. Farmland has been better managed through policies such as the "red line" for arable land protection, ensuring stable grain production. Rural labor has shifted from traditional farming to more diversified employment, including non-farm activities, contributing to rural economic growth. Grain production reached a record high of 706.5 million tons in 2024, marking the first time it surpassed 700 million tons which matched over 80% of domestic consumption. Food security has been a top priority, with measures such as minimum purchase price policies and subsidies to support farmers. Farmers' incomes have steadily increased, with the per capita disposable income of rural residents reaching 23,119 yuan (about US$3190) in 2024. Rural development has seen significant improvements in infrastructure, environmental protection, and public services, with over 90% of villages now having access to 5G internet. Overall, these

DOI: 10.1079/9781800627338.0007

achievements reflect the combined effects of institutional reforms, policy support, and technological innovation in China's agricultural sector.

China's agricultural development still faces several key challenges, including water and land resource constraints, climate change impacts, an aging rural population, and the need to balance self-sufficiency with global market integration. Specifically, China has less than 10% of the world's arable land and only 6% of its freshwater resources, yet must feed nearly 20% of the global population. Additionally, climate change exacerbates food production uncertainties, while urbanization and industrialization continue to encroach on arable land. To address these challenges, China has implemented a range of strategies. For instance, the government has emphasized protecting farmland through policies such as the "red line" for arable land and promoting sustainable agricultural practices, including climate-smart agriculture and precision farming. Technological advancements such as high-yield crop varieties, digital agricultural platforms, and water-saving irrigation systems are being adopted to boost productivity. Policies support smallholder farmers through subsidies, low-interest loans, and training programs to enhance their resilience and integration into modern farming. China is balancing domestic production with strategic imports and participating in global food security initiatives to ensure stable supplies. For instance, through the Belt and Road Initiative, China collaborates with countries along the Belt and Road routes to develop agricultural infrastructure and improve food production capabilities. These integrated strategies aim to achieve sustainable agricultural development, enhance food security, and improve rural livelihoods while addressing environmental and demographic pressures.

This chapter will provide a detailed description of the evolution of China's agricultural institutions and policies and their corresponding effects. Since 1949, China's land use system has undergone several stages: "land to the tiller," agricultural cooperation, collective and unified management, the household contract responsibility system, and the separation of rural land ownership rights, contract rights, and management rights. Each significant adjustment to the land system has been accompanied by profound changes in agricultural production, efficiency, and rural economic structures. We will organize the historical context of agricultural development across five categories (Table 7.1): (1) general development policies, (2) agricultural policies, (3) agricultural structures, (4) major events, and (5) agricultural performance. We will show how agricultural and land policies have promoted long-term, rapid economic growth and agricultural development in China. Moreover, we will provide two case studies on land use change and rural labor transfer in Southern China. These case studies involve the carbon balances of rubber land use systems and the development of off-farm labor markets and their implications for small-scale farming, respectively.

From Private Farming to Agricultural Cooperatives (1949–1956)

The development policies in the early days of the People's Republic of China were tailored to China's specific political system at that time, involving the integrated advancement of confiscating bureaucratic capital and restoring and developing the national economy on the basis of land reform in urban areas and suburbs (Lian and Shen, 2024). Such adjustments rapidly restored production, leading to significant growth in both agriculture and industry. Compared with 1949, national grain output increased by 46.1%, and total agricultural output value grew by 48.47% in 1952; over 70% of rural households experienced noticeable improvements in production conditions and living standards (Statistics Bureau of China, 2020). After three years of national economic recovery, China's development policies shifted towards socialist industrialization and the three great transformations. During this period, the government leveraged national strength and fiscal investments to develop large-scale enterprises, promote economic construction, and vigorously support industrialization. Simultaneously, the Communist Party of China organized the socialist transformation of agriculture through the agricultural cooperative movement.

Table 7.1. Development overview. (Authors' own table)

Period	Development Policy	Agricultural Policy	Structure of Agriculture	Significant Events	Agricultural Performance
1949–1956: From farmers' privatization to agricultural cooperatives	Confiscation of capital, establishment of state-economy to stabilize prices; adjustment of industry and commerce; three great reconstructions; socialist industrialization	Land reform, the abolition of feudal land ownership, to achieve private ownership of farmers' land; guiding farmers towards collectivization and common prosperity	Single structure, dominated by crop farming and grain production	Implementation of a unified grain purchase and sale system; putting forward comprehensive high-yield requirements for agricultural production	Substantial increases in grain production under agricultural cooperative economy
1957–1977: From Agricultural Cooperatives to People's Communes	Continued industrialization under planned economy formation	Means of production were owned by three levels of organizations: communes, production brigades, and production teams, with production team collective ownership as the foundation	Proportion of crop farming declined slightly, especially in grain crops, while non-grain crops gradually increased	Great Leap Forward, People's Commune Movement, Cultural Revolution	Fluctuations in agricultural production, improvements in rural infrastructure
1978–2012: From People's Communes to Household Contract Responsibility System	Household contract responsibility system; reform and opening up; special economic zones construction, socialist market economy system	Gradually abolished the unified purchase and sale system, introduced market mechanisms; implementation of rural tax and fee reforms, abolition of agricultural taxes; strengthening of rural infrastructure	Transformation of agriculture from grain-based to diversified; from low value to high value-added enterprises	No 1 Central Document market-oriented reform	National grain production significantly increased; family farms, side-by-side with cooperatives; rural non-farm employment increased; farmers' income levels substantially increased
2013–Present: The New Era	Belt and Road Initiative; innovation-driven development strategy; Rural Revitalization Strategy; High-quality development strategy	Agricultural subsidy policy; Rural infrastructure improvement and protection of farmland; pilot projects on collectively owned construction land; land and homestead system reform	Diversification and regional specialization, promote green sustainability	Completion of rural land registration and certification process, amendment of land management law; rural revitalization law implemented; priority given to grain production	Land productivity and agricultural area increased, modernization of farming methods; further trends towards non-agricultural employment

In terms of agricultural policy, land reform facilitated a smooth transition in government and political change, breaking the barrier between the state and farmers. In the 1950s, legislation ensured farmers' land ownership, completely overthrowing the traditional rural order in China and reshaping grassroots organizations and power structures (Du, 1995). However, as society developed, the ownership of agricultural land by individual farmers contradicted the "inherent requirement of adhering to Marxism and socialism," which stipulates that land ownership must be public. This implied that private land ownership would not last long and signaled the end of farmers' private ownership (Zhong *et al.*, 2020). From 1951 to 1952, the Central Committee of the Communist Party of China (hereafter, CPC) issued the "Resolution on Agricultural Production Mutual Cooperation" and the "Resolution on the Development of Agricultural Production Cooperatives," stating that "the Party's task in rural areas is to lead farmers on the path of organization, mutual cooperation, collective progress, and common prosperity." The transition from primary to advanced agricultural cooperatives marked a significant transformation, characterized by the transfer of farmers' private land and other means of production, without any form of compensation, to a collective ownership structure of the advanced cooperatives. Under this newly adopted system, land management became centralized and unified, with land allocation contingent upon the work performed by the farmers. This marked the replacement of farmers' land private ownership with rural land collective ownership (Liu and Cheng, 2007).

Regarding the agricultural structure, China was predominantly reliant on planting and grain production, with grain crops occupying an absolutely dominant position. In the 1950s, given the weak overall national strength and low agricultural productivity, food and clothing security for the majority of the population faced challenges. Farmers actively engaged in agricultural production. In terms of output value composition, agriculture accounted for 85.9% of the total output value of agriculture, forestry, animal husbandry, and fishery in 1952, occupying an absolutely dominant position, while forestry, animal husbandry, and fishery accounted for 1.6%, 11.2%, and 1.3%, respectively (Rural Department of the National Bureau of Statistics, 2019).

In terms of significant events, domestic food shortages and inadequate supply of various resources emerged. The government implemented a grain purchase and sales system and put forward the requirement for comprehensive high yields in agricultural production. The land revolution granted land ownership to farmers, which unleashed unprecedented enthusiasm for production among farmers (Gu, 2016).

In September 1951, China adopted the "Resolution on Agricultural Production Mutual Cooperation (Draft)," which stated "two types of enthusiasm" of farmers for individual economies and labor mutual aid. It required the extensive development of labor mutual aid groups based on the needs and possibilities of production. In areas where conditions were conducive, it called for the focused development of primary agricultural production cooperatives based on land shares. By 1956, the advanced collectivization of agriculture was completed, with each cooperative averaging around 200 households. The agricultural cooperative movement promoted the transformation of the individual agricultural economy to a cooperative economy, initially resulting in substantial increases in grain output.

From Agricultural Cooperatives to People's Communes (1957–1977)

The development policies during this period focused on industrialization and the implementation of a planned economy. After 1957, China gradually entered the primary stage of socialism, with industrialization becoming a critical task. During this time, China continued to advance industrial construction centered on heavy industry, aiming to establish a solid foundation for socialist industrialization (Yan, 1989). At that time, China's productivity level was low, the national economy was weak, and there was a shortage of supplies for essential livelihood goods.

Following the completion of socialist transformation within the country, the people's participation was still high, with the hope of quickly transforming China from a backward agricultural country into an advanced industrial country. By the end of 1956, agricultural collectivization was essentially completed, with 96.3% of rural households joining agricultural cooperatives. However, some smaller cooperatives could not meet the goals of agricultural productivity development. The government tended to merge smaller-scale cooperatives into larger ones. In 1958, documents such as the "Opinions of the CPC (Communist Party of China) Central Committee on Appropriately Merging Smaller Agricultural Cooperatives into Larger Ones" and the "Resolution of the CPC Central Committee on Establishing People's Communes in Rural Areas" were successively issued, and people's communes were widely established across the country, promoting collectivization policies in agricultural production. In the early stages of the collectivization movement, the forms of mutual aid teams, primary cooperatives, and advanced cooperatives were used to boost agricultural productivity and significantly increase grain output. As advanced cooperatives transitioned into people's communes, the People's Commune Movement was launched nationwide in 1958, forming a collective management system characterized by "three-level ownership with the production team as the base." However, due to rigid management and insufficient incentive mechanisms, under a centralized resource system, farmers' enthusiasm for production declined and agricultural productivity dropped.

In 1962, production materials were owned by three levels of organizations: communes, production brigades, and production teams, with the collective ownership of production teams as the foundation (Zhu *et al.*, 2021). People's communes and production brigades became the "ownership" entities, while production teams became the de facto "management right" entities. Agricultural land ownership shifted from private ownership by farmers to collective ownership, forming a "separation of two rights" within the collective.

The "Great Leap Forward," a campaign initiated by China in 1958, intended to propel agricultural growth at a rapid pace. The central government had set excessively high industrial and agricultural production targets, aiming for quick success and rapid structural transformation (Chen, 2011). The commune movement attempted to accelerate the socialist paradigm by changing the rural grassroots organizational structure. However, many problems arose during the actual implementation of this movement, such as being detached from the actual level of rural productivity and violating the principle of farmers' voluntariness (Yang, 2013). This led to confusion in the management system of agricultural production and to a decline in productivity, severely impacting agricultural production. As a result of these policies and the shocks of natural disasters, China experienced three years of severe food shortages, commonly referred to as the Great Chinese Famine, from 1958 to 1961. Consequently, the period from 1958 to 1978, was one of the slowest in China's agricultural development. Per capita grain output increased from 310 kg in 1956 to only 318.5 kg in 1978, with an average annual increase of less than 0.4 kg. Growth in agricultural labor productivity was a negative with −0.2% per year. The "integration of government and commune" model of the people's commune system deprived collective economic organizations of their autonomy, and serious issues with the conduct of cadres further exacerbated the difficulties in agricultural production.

Despite the overall slow development of agriculture, some achievements were made during this period. The structure of agricultural production gradually improved, with the area of non-grain (economic) crops (such as cotton, oilseeds, and sugarcane) increasing. Meanwhile, the state made large-scale investments in farmland water conservancy and agricultural mechanization, leading to some improvements in agricultural infrastructure. For example, by 1978, the total power of agricultural machinery nationwide reached 117.5 million kilowatts. However, due to the excessive extraction of rural resources by the state, the progress of agricultural modernization and mechanization remained slow.

During this period, the rural economic structure was singular, with farmers' income primarily dependent on agricultural production. Due to the shortage of agricultural products, the basic living needs of urban and rural residents were hardly met, and the per capita output of agricultural products was low. The daily caloric intake of urban and rural residents was below the global average level. Rural infrastructure and public service levels were insufficient, and the living conditions of farmers were generally poor.

Although some infrastructure development took place with the People's Commune Movement, due to the rigidity of the system and policy mistakes, agricultural productivity remained low, motivation of farmers' enthusiasm was poor, and consequently rural economic development stagnated. The lessons learned from this period provided important references for rural reforms following the reform and opening up.

Transformation from the People's Commune to the Household Contract Responsibility System (1978–2012)

From 1978–2012, China's agricultural economy underwent significant transformations, driven by a series of policy reforms and market-oriented adjustments. This period can be divided into four distinct stages, each marked by key policy changes and their corresponding impacts on rural land management, labor, income, and overall agricultural performance.

From 1978–1984, the basic rural management system was established. During this period, the introduction of the household responsibility system was a pivotal reform. It allowed farmers to contract land from collective organizations and manage it independently, significantly boosting agricultural productivity. In December 1978, following the Third Plenary Session of the Eleventh CPC Central Committee, China embarked on a policy of internal reform and opening up to the outside world. The reform initiated in rural areas, marking the beginning of China's internal reform through measures such as the implementation of the household contract responsibility system (Sun and Zhang, 2024). The government began to support farmers' spontaneous actions to separate collective land ownership rights from land use rights and delegated farmland contracting and management rights, ultimately evolving into the household contract responsibility system, a unique Chinese rural land arrangement (Li *et al.*, 2020). The separation of collective land ownership rights from land use rights significantly boosted farmers' production enthusiasm and increased agricultural output (Cai and Li, 2021). As a result, agricultural output grew at an average annual rate of 7.7%, and the total agricultural output value increased by 42.23% by 1984 compared with 1978. In 1984, national agricultural output reached its highest level since the establishment of the People's Republic of China, with the household contract responsibility system contributing up to 42% of this growth (Lin, 1994). This reform also led to a substantial reduction in rural poverty, with the number of poor rural people decreasing from 250 million to 125 million, and the incidence of poverty in rural areas dropping from 30.7% to 14.8% (Yan *et al.*, 2019).

From 1985 to 1992, the agricultural product supply system was reformed. In this stage, the state reformed the unified purchase system for agricultural products, allowing farmers to sell produce in open markets. In 1985, China's No. 1 Central Document abolished the unified purchase and distribution system for agricultural products, allowing market-based pricing of agricultural products. This reform measure further promoted the marketization of the rural economy, enhancing its vitality and competitiveness. It improved the efficiency of agricultural production and led to a diversification of rural economic activities. During the same period, the rise of township and village enterprises (TVEs) provided alternative employment opportunities for rural labor, contributing to a 9.0% annual growth rate in rural per capita income in the early 1990s. The reform of the agricultural product supply system enhanced the role of market mechanisms in resource allocation and laid the groundwork for further rural economic development.

From 1993–2002, agricultural and rural economic structures experienced strategic adjustments, including optimizing the agricultural production structure and promoting the diversification of the rural economy. This period focused on encouraging the development of cash crops, livestock, and fisheries. Policies were introduced to support agricultural modernization, including the use of advanced technologies and mechanization. The agricultural structure shifted from a grain-dominated model to a more diversified model and from low added value to high added value enterprises (Du, 2012), with the area of cash crops increasing significantly.

The production of livestock and fisheries also expanded, contributing to a stabilization of grain production and a substantial improvement in the quality and variety of agricultural products.

From 2003–2012, the integrated development of urban-rural social and economic system was actively pursued and implemented. This stage emphasized narrowing the urban-rural income gap and promoting integrated development. Increased investment in rural infrastructure, support for rural education and healthcare, and measures to encourage rural labor transfer to non-agricultural sectors were key policies during this period. The proportion of non-agricultural employment among farmers continued to grow, and rural incomes increased significantly. The overall living standards of rural residents improved, and the rural economy became more integrated with the national economy. These efforts aimed to address long-standing disparities and foster more balanced national development.

Over the 1978–2012 period, China's agricultural economy achieved notable progress. The proportion of cash crops increased from 11.5% in 1949 to over 30% by 2012, contributing to a more balanced and diverse agricultural structure. Grain yield per unit has been increasing, whilst grain production remained stable (Figure 7.1), supported by technological advancements and policy initiatives. The household responsibility system enabled farmers to manage land more efficiently, leading to higher productivity and income (Figures 7.2 and 7.4). The transfer of rural labor to non-agricultural sectors accelerated, with a significant increase in the share of off-farm employment and the number of rural workers employed in manufacturing and services (Figure 7.3). Rural per capita income grew significantly (Figure 7.4), wherein off-farm employment played an increasing role (Figure 7.5). Despite these achievements, the period also highlighted the need for further agricultural modernization, sustainable development, and continued efforts to bridge the urban-rural divide.

The New Era (2013–Present)

Since 2013, China has made remarkable progress in agricultural and rural development through comprehensive policy reforms, technological advancements, and strategic initiatives.

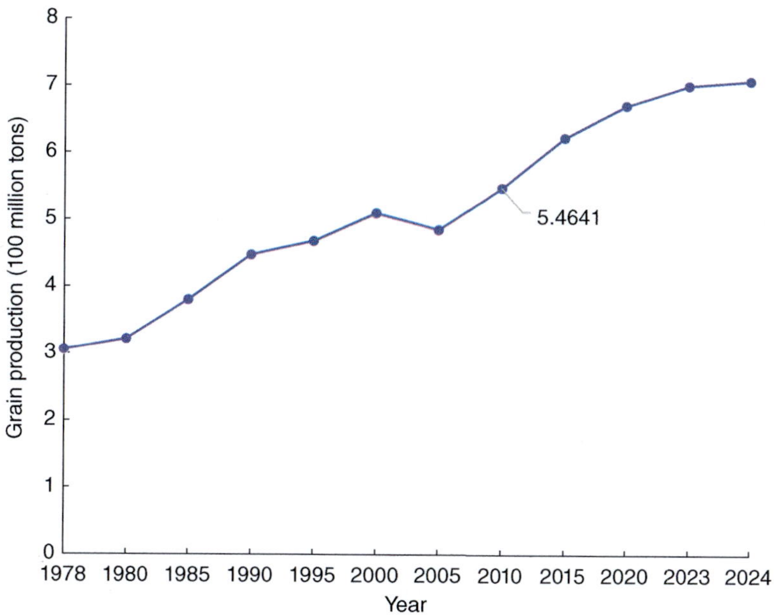

Figure 7.1. Grain Production. Data Source: National Bureau of Statistics of China.

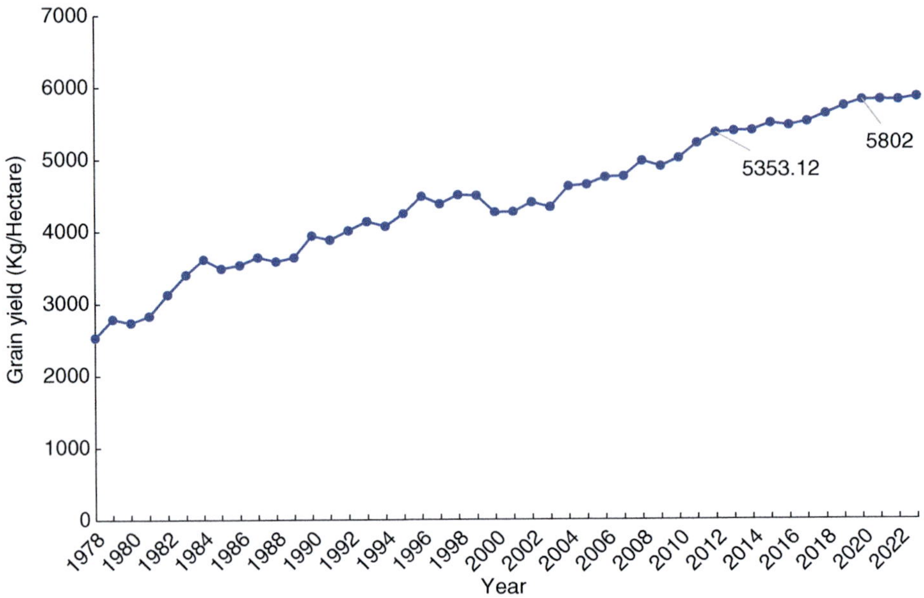

Figure 7.2. Grain yield per hectare. Data Source: National Bureau of Statistics of China.

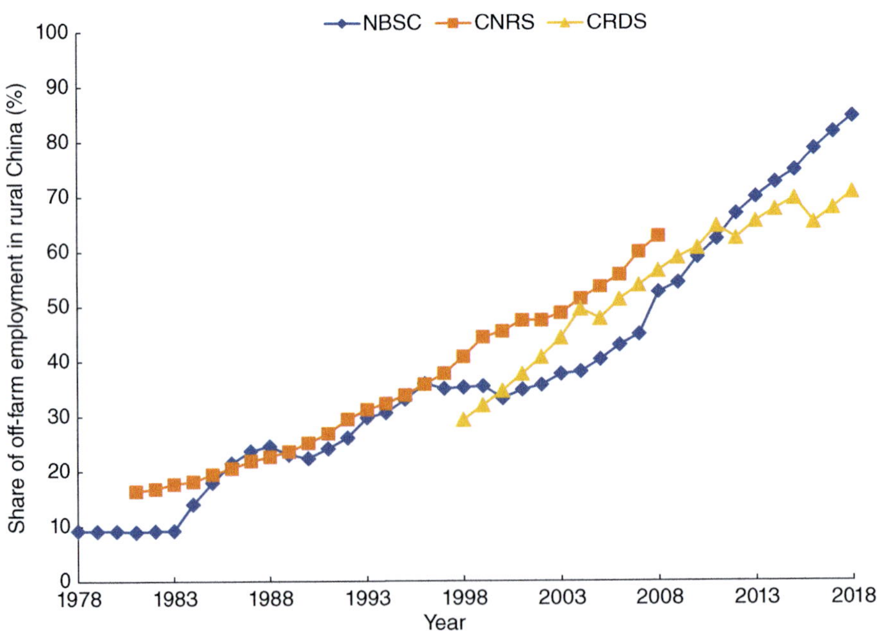

Figure 7.3. The share of off-farm employment in rural China. Data source: Li *et al*. (2021) collected from National Bureau of Statistics of China (NBSC), China National Rural Survey (CNRS), and China Rural Development Survey (CRDS).

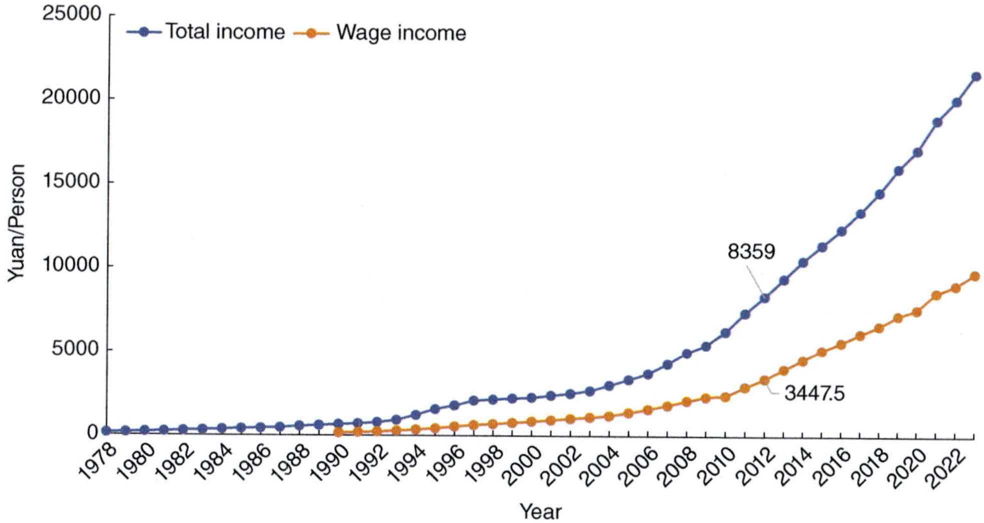

Figure 7.4. Household income of rural residents (Yuan/person). Data Source: National Bureau of Statistics of China.

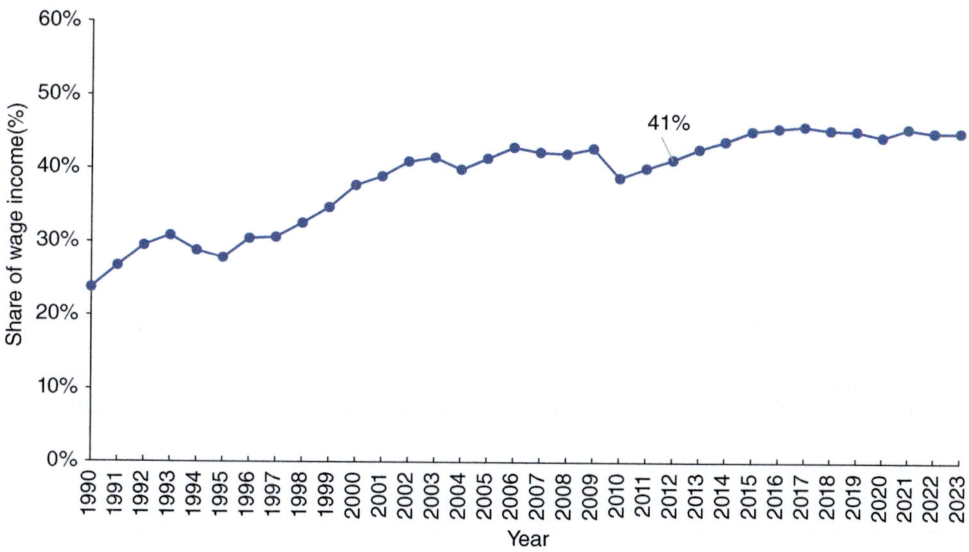

Figure 7.5. Share of wage income in total household income (%). Data Source: National Bureau of Statistics of China.

These efforts have primarily focused on addressing the "three rural issues" – agriculture, rural areas, and farmers – and promoting rural revitalization. This chapter reviews the key achievements in these areas and identifies the ongoing challenges that require further attention.

Institutional and policy reforms have played a crucial role in driving agricultural modernization and rural development. The introduction of the "three-rights separation" system for rural land – distinguishing ownership, contract rights, and management rights – has facilitated the more efficient use of land resources and has paved the way for the establishment of large-scale farming operations (Kong and Mu, 2018). This system, first proposed in 2014, has guided the orderly circulation of land management rights, supporting the development of modern agriculture by improving productivity and encouraging land circulation (Qu *et al.*, 2021; Chen *et al.*, 2023). However, despite promotion of rural land transfer and moderate-scale management (Zuo *et al.*, 2015), and the development of new types of agricultural management entities such as family and corporate farms, as shown in Table 7.2, the situation, where small-scale farmers remain the main body of land management has not changed (Ji *et al.*, 2016). The percent of farms with more than 2 ha is still below 5% and those above 10 ha had not yet reached 1% by 2022 but are gradually increasing.

From 2013 to 2020, China's policy priorities centered on eradicating absolute poverty, addressing key challenges in agriculture and rural areas, and promoting high-quality agricultural development. Through a series of effective policy measures, China achieved comprehensive poverty eradication in rural areas, contributing to the building of a moderately prosperous society (Zheng and Zheng, 2021). In 2017, China put forward a rural revitalization strategy which is a multi-faceted approach that seeks to transform rural areas into modern, prosperous, and sustainable communities. The strategy includes measures to support rural development through financial, technological, and institutional innovations. In the 2020s, the focus shifted to advancing rural revitalization and accelerating agricultural and rural modernization. Key policy areas included ensuring national food security, preventing large-scale relapses into poverty, enhancing rural industries, improving rural infrastructure, and strengthening rural governance. The 2024 No. 1 Central Document further emphasized learning from successful projects such as the "Thousand Villages Demonstration, Ten Thousand Villages Renovation" to drive comprehensive rural revitalization (Zhang, 2021).

Summary

China's agricultural policies have consistently prioritized food security, modernization, and income growth for farmers (Zhu *et al.*, 2021; Hu *et al.*, 2023). The government has implemented various subsidy policies, such as direct grain subsidies and comprehensive agricultural material subsidies, to incentivize grain production while strengthening farmland protection and promoting high-standard farmland construction. In addition, policies have encouraged agricultural technological innovation (e.g. funding supports research and development of smart agriculture and other technologies), enhanced agricultural equipment levels (e.g. subsidies for the purchase and application of agricultural machinery), and promoted green development (e.g. promoting techniques such as soil testing, formula fertilization and increased application of organic fertilizer to reduce the use of chemical fertilizers and pesticides).

China has achieved significant milestones in agricultural production and food security (see

Table 7.2. Farm size distribution by category over 2009–2022.

	Farmers operating farmland with an area of 0–2 hectare (%)	Farmers operating farmland with an area of 2–3.33 hectare (%)	Farmers operating farmland with an area of 3.33–6.7 hectare (%)	Farmers operating farmland with an area of 6.7–13.33 hectare (%)	Farmers operating farmland with an area of over 13.33 hectare (%)
2009	96.22	2.57	0.84	0.26	0.11
2013	95.82	2.55	0.86	0.25	0.11
2016	95.99	2.61	0.94	0.33	0.13
2019	96.22	2.54	1.02	0.38	0.17
2022	95.88	2.53	1.06	0.40	0.19

Data source: China Rural Operation Management Statistical Annual Report, Annual Statistical Report of China's Rural Policy and Reform.

Figure 7.1). Grain output has reached record highs, with the yield per unit reaching to 5802 kg/ha and the total production hitting 706.5 million tons in 2024, marking the first time the country has surpassed the 700 million-ton mark. In the same year, China imported 158 million tons of grain, down 2.3% year-on-year, and exported 2.25 million tons of grain, down 13.7% year-on-year. Grain output contributes over 80% of domestic grain consumption. This success is attributed to robust support for agricultural water conservancy infrastructure and breakthroughs in agricultural production technology such as biological breeding, UAV plant protection, and better irrigation systems. The soybean planting area expanded to 10.3 million hectares in 2024, and soybean production exceeded 20 million tonnes in 2024. However, soybeans production was still far below domestic consumption and imports reached 105 million tons. Winter rapeseed output has also seen annual growth for eight consecutive years. These achievements reflect China's commitment to ensuring food security through technological innovation and policy support.

Rural labor and farmer income have shown positive trends (see Figures 7.3 and 7.4). Rural residents' income has continued to rise steadily, with a nominal increase of 7.7% in 2023 and a 6.3% increase in the first nine months of 2024. The number of migrant workers increased by 2.7% in 2023, with a growing preference for employment in central and western regions. Rural industries such as logistics, e-commerce, and leisure tourism have created new employment opportunities, contributing to the diversification of income sources for rural households.

China's agricultural structure has undergone significant adjustments to meet evolving economic and social demands. The sector has shifted from a traditional, grain-focused model to a more diversified and comprehensive direction, with coordinated growth across agriculture, forestry, animal husbandry, and fisheries (Chen *et al.*, 2018). While the proportion of agricultural output value has gradually declined, the shares of animal husbandry, fisheries, and other industries have increased. This diversification reflects China's efforts to meet diverse food demands. Agricultural modernization has accelerated through technological innovation and smart agriculture, improving production efficiency and crop quality. The level of agricultural product processing and value-added conversion rates have also increased, while the ecological, social, and cultural functions of agriculture have been further explored.

Rural revitalization has been a key focus, with significant improvements in infrastructure and living conditions. Over 90% of villages now have access to 5G internet, and paved roads are more prevalent in rural areas. Basic public services, including education, healthcare, and elderly care, have continuously improved. Environmental conditions have also seen progress, with approximately 75% of rural areas having access to sanitary toilets and over 90% of villages implementing proper solid waste management. Over 45% of rural domestic sewage is now treated and controlled.

Several landmark events have profoundly impacted agricultural modernization and sustainable development since 2013. First, the completion of rural land confirmation, registration, and certification clarified land property rights and safeguarded farmers' interests (Wei and Kong, 2024). Second, significant breakthroughs in rural land system reform were achieved with the 2019 amendments to the Land Administration Law, which removed legal obstacles for rural collectively operated construction land to enter the market and strengthened the protection of rural homestead rights (Ji, 2017). Third, the implementation of the Rural Revitalization Promotion Law in 2021 provided a legal basis for rural development. Finally, measures to ensure stable grain production, such as increasing grain production capacity and providing agricultural insurance, have maintained grain output at over 1.3 trillion kilograms.

Overall, China's agricultural and rural development from 2013 to 2024 has been marked by significant achievements in policy reforms, increased grain production, enhanced food security, improved farmer incomes, and better rural living standards. However, the ongoing challenges emphasize the importance of maintaining a balanced approach between economic growth, social equity, and environmental sustainability. These challenges highlight the need for continued innovation and policy support to ensure sustainable agricultural development and rural prosperity. Future efforts should focus on leveraging technological advancements, strengthening policy coherence, and fostering rural-urban integration to achieve long-term agricultural and rural development goals.

Case Study: Rubber. Source: Author's Compilation

The case study on small-scale rubber cultivation is mainly based on three PhD theses by Chinese scholars, namely Min (2016), Jin (2020), and Zhuang (2022). The studies draw upon a panel data set of some 600 rubber farmers in Xishuangbanna, Southwest China.

Since the rural reforms in the late 1970s, Southern China has experienced widespread and dramatic land use changes such as deforestation, agricultural expansion and conversion of secondary vegetation into monocultures, in particular rubber plantations (Ahrends *et al.*, 2015). Driven by the relatively higher profits of rubber farming than the profits previously obtained by cultivating other crops (Liu et al., 2006; Maes, 2012; Ahrends *et al.*, 2015), natural rubber plantations have been expanding rapidly in Xishuangbanna Dai Autonomous prefecture (XSBN) in the Southern Yunnan province, China, which is one of China's few tropical rainforest areas.

A number of socioeconomic studies suggest that natural rubber presents a profitable opportunity for smallholders, and is hence a potentially effective proposition in moving households and communities out of poverty (Fox and Castella, 2013; Fox *et al.*, 2014; Liu *et al.*, 2016). However, the transition from traditional agriculture and virgin forest land to rubber plantations has led to an imbalance in the local ecological system and has caused severe ecological degradation (Xu *et al.*, 2005; Zhang *et al.*, 2007; Fu *et al.*, 2009; Qiu, 2009; Ahlheim *et al.*, 2015), and thereby has resulted in largely environmental pressures and policy challenges toward sustainable land use in XSBN. On the other hand, the altering of traditional land use patterns and labor allocation towards specialized rubber plantation also implies a higher liability to climatic and economic risks. The significant land use change with rubber expansion in XSBN has also affected the landscape carbon sequestration in local forests. Although the cultivation of rubber trees on non-forested land could act as a carbon sink by sequestering carbon in biomass and indirectly in soils (Gnanavelrajah *et al.*, 2008; Nizami *et al.*, 2014; Wauters *et al.*, 2008), the carbon sequestration of rubber trees is much lower than that of natural forest (de Blécourt *et al.*, 2013; Yang et al., 2016; Jiang *et al.*, 2017). The research by Min *et al.* (2019) shows the carbon stocks of rubber plantations of the 612 sample smallholder rubber farmers were about 173,934 Mg by 2012, suggesting that rubber expansion led to a substantial carbon loss of 42,229 Mg. On average, the net loss of carbon stocks of smallholder rubber plantations outside the natural reserves in XSBN in the past three decades was estimated to be about 21 Mg/ha/year. If assuming that smallholder rubber farmers would not have changed their land use patterns since 2012, then a carbon balance between smallholder rubber plantations and the replaced land was achieved for the first time by 2015. However, due to the rotation of rubber trees, the net gains of carbon stocks may only have lasted to 2020. Considering the relatively large carbon surpluses by reforesting from rubber plantations, it was recommended that carbon-trading schemes were used as a potential means to restore the natural environment threatened by rubber farming, while Payment for Ecosystem Service Schemes may also support the maintenance of ecosystem services (Klasen *et al.*, 2016).

The expansion of natural rubber has also influenced the labor market in local areas. The study by Yang et al. (2022) investigated the impacts of rubber expansion on the migration of rural labor in XSBN. The main findings are as follows: First, smallholders with a higher proportion of natural rubber planting on their land are less likely to migrate and have fewer migrant family members. Second, the price of natural rubber negatively affects migration, indicating that migration is a strategy used by smallholders to cope with income volatility when rubber prices decline. Finally, there is a positive interaction between the proportion of rubber planting area and rubber price on migration, meaning that the impact of rubber cultivation on migration decreases as rubber prices fall. These findings provide empirical evidence on the effects of rubber expansion on rural labor migration. They reveal that natural rubber cultivation generally hinders migration and can lead to natural resource degradation. However, periods of low rubber prices offer opportunities to promote labor migration, reduce resource curse, and foster local labor market development. While the study focuses on XSBN, its implications can be applied to other rubber planting regions in the Mekong area.

Outlook

China's agricultural sector has experienced profound transformations over the past decades, driven by continuous adjustments in land policies and rural development strategies. These changes have significantly improved agricultural productivity, rural economic structures, and the livelihoods of farmers. However, several key challenges remain, particularly in the context of sustainable development, ecological protection, and the need for further structural transformation.

Despite these achievements, China's agricultural sector faces several challenges. First, the "large country with small-scale farming" characteristic remains a fundamental constraint. While small-scale farming remains predominant among Chinese agricultural households, a trend of operational scale expansion has become evident. The vast majority of farmers in China have to rely on multiple income sources, with a declining share of income derived from agriculture, in order to make ends meet. The dualistic structure of agriculture, with small-scale farmers on one side and some large agribusinesses on the other, is slowly emerging. One example is New Hope Group, which is one of the largest private agribusinesses in China. New Hope Group is primarily engaged in livestock breeding, feed production, and meat processing. In 2024, its operating revenue exceeded US$20 billion. Second, the ecological impact of agricultural land use changes is a growing concern. For example, the rapid expansion of rubber plantations in Xishuangbanna (see Case Study) has led to significant ecological degradation and loss of carbon sequestration capacity. This highlights the need for sustainable land management practices and policies that balance economic benefits with environmental protection. Third, the development of off-farm labor markets, while providing new income opportunities for rural households, also poses challenges. On the one hand, the shift of labor from agriculture to non-farm employment can lead to labor shortages in the agricultural sector, severely affecting productivity. The reliance on off-farm income may reduce the incentive for farmers to invest in agricultural modernization and innovation. On the other hand, although the share of farmers engaged in off-farm employment has been increasing (see Figure 7.3), the contribution of wage income to total household income has remained stable at around 45% since 2013. This might be because more rural households engaged in off-farm self-employment under the policy support for farmers to start their own businesses in their hometowns.

To address these challenges and ensure sustainable agricultural development, several policy directions are suggested: First, promoting agricultural modernization and technological innovation. Encourage the adoption of advanced agricultural technologies, such as precision farming and smart agriculture, to improve productivity and resource efficiency. Support research and development in agricultural technologies tailored to small-scale farmers. Second, strengthening rural infrastructure and services. Invest in rural infrastructure, including transportation, irrigation, and digital connectivity, to enhance market access and reduce production costs. Provide training and education programs to improve farmers' skills and adaptability to new technologies. Third, enhancing land use policies and environmental sustainability. Implement policies that promote sustainable land use practices, such as agroforestry, conservation agriculture, and carbon sequestration projects. Encourage the development of ecosystem service markets to incentivize farmers to adopt environmentally friendly practices. Fourth, supporting diversified rural economies. Develop policies that support the diversification of rural economies, including the growth of non-farm industries and services in rural areas. Facilitate the integration of rural and urban economies to create more employment opportunities and reduce rural-urban income disparities. Fifth, implement comprehensive social protection programs for rural households, including health insurance, pension schemes, and unemployment benefits. Provide targeted support to vulnerable groups, such as elderly farmers and smallholders, to ensure their economic security and well-being.

References

Ahlheim, M., Börger, T. and Frör, O. (2015) Replacing rubber plantations by rain forest in Southwest China – who would gain and how much? *Environmental Monitoring and Assessment* 187(2), 1–20.

Ahrends, A., Hollingsworth, P.M., Ziegler, A.D., Fox, J.M., Chen, H. *et al.* (2015) Current trends of rubber plantation expansion may threaten biodiversity and livelihoods. *Global Environmental Change* 34, 48–58.

Cai, J.M. and Li, M.M. (2021) The history and logic of contemporary China's land institutional changes. *Economic Perspectives* (12), 40–51 (In Chinese).

Chen, C.W., Chen, J.P. and Tao, J.K. (2018) Industrial poverty alleviation: foreign experience and policy implications. *Economic Geography* 38(01), 127–134.

Chen, M.H., Zhang, B.X. and Li, Q. (2023) The evolution and formation mechanism of the urban-rural gap in the development of people's livelihood in China. *Urban Problems* 10, 85–93 + 103 (In Chinese).

Chen, S. (2011) A review and enlightenment of the Chinese famine from 1959 to 1961. *Journal of World Economy* 34(04), 80–103 (In Chinese).

de Blécourt, M., Brumme, R., Xu, J., Corre, M.D. and Veldkamp, E. (2013) Soil carbon stocks decrease following conversion of secondary forests to rubber (*Hevea brasiliensis*) plantations. *PLoS ONE* 8(7), e69357.

Du, H.M. (2012) Analysis on the coordinated development of agricultural structure and market economy. *Agricultural Economy* 12, 47–49 (In Chinese).

Du, R.S. (1995) China's agriculture in the period of economic transition. *Reform* 4, 15–23 (In Chinese).

Fox, J. and Castella, J. (2013) Expansion of rubber (*Hevea brasiliensis*) in mainland Southeast Asia: What are the prospects for smallholders? *Journal of Peasant Studies* 40, 155–170.

Fox, J., Castella, J., and Ziegler, A.D. (2014) Swidden, rubber and carbon: Can REDD+ work for people and the environment in Montane mainland Southeast Asia? *Global Environmental Change* 29, 318–326.

Fu Y., Chen, J., Guo, H., Chen, A., Cui, J., and Hu, H. (2009) The role of non-timber forest products during agroecosystem shift in Xishuangbanna, southwestern China. *Forest Policy and Economics* 11(1), 18–25.

Gnanavelrajah, N., Shrestha, R.P., Schmidt-Vogt, D. and Samarakoon, L. (2008) Carbon stock assessment and soil carbon management in agricultural land-uses in Thailand. *Land Degradation & Development* 19(3), 242–256.

Gu, H.X. (2016) On the deepening and innovation of the rural collective land system. *Journal of China Executive Leadership Academy Jinggangshan* 9(04), 130–137 (In Chinese).

Hu, C.P., Hu, X. and Lu J.Y. (2023) Agricultural economic growth, factor endowments and dual circulation of agricultural: International comparison based on input output tables of 52 economies. *Inquiry into Economic Issues* 6, 54–63 (In Chinese).

Ji, X. (2017) Analysis of the development of rural economic cooperation organizations based on agricultural modernization – Taking Guizhou Province as an example. *Agricultural Economy* 1, 53–55 (In Chinese).

Ji, X., Rozelle, S., Huang, J., Zhang, L. and Zhang, T. (2016) Are China's farms growing? *China & World Economy* 24(1), 41–62.

Jin, S. (2020) Microeconomic analysis of sustainable natural resources management innovations in rubber cultivation in China. PhD thesis, Leibniz University Hannover, Germany.

Jiang, X.J., Liu, W., Wu, J., Wang, P., Liu, C. *et al.* (2017) Land degradation controlled and mitigated by rubber-based agroforestry systems through optimizing soil physical conditions and water supply mechanisms: A case study in Xishuangbanna, China. *Land Degradation & Development* 28(7), 2277–2289.

Klasen, S., Meyer, K.M., Dislich, C., Euler, M., Faust, H. *et al.* (2016) Economic and ecological trade-offs of agricultural specialization at different spatial scales. *Ecological Economics* 122, 111–120.

Kong, X.Z. and Mu, N.N. (2018) Achieving an organic connection between small farmers and modern agricultural development. *Rural Economy* 2, 1–7 (In Chinese).

Li, J.T., Xiong, C. and Cai, J.M. (2020) The land reform is stepping into a new chapter: The equalization of urban-rural land poverty rights and the marketization of resource allocation. *Journal of Management World* 36(06), 93–105 + 247 (In Chinese).

Li, S., Dong, Y., Zhang, L. and Liu, C. (2021) Off-farm employment and poverty alleviation in rural China. *Journal of Integrative Agriculture* 20(4), 943–952.

Lian, W.M. and Shen, C.F. (2024) Awareness of potential dangers: An indispensable ideological variable in party history research. *Teaching and Research* 2, 58–69 (In Chinese).

Lin, Y.F. (1994) Major problems and prospects of China 's Rural Reform in the 1990s. *Journal of Management World* 3, 139–144 (In Chinese).
Liu, G.D., and Cheng J.M. (2007) The theory and practice of rural land system change in China since 1949. *China Rural Survey* 2, 70–80 (In Chinese).
Liu, M., Liu, X. and Yang, Z. (2016) An integrated indicator on regional ecological civilization construction in China. *International Journal of Sustainable Development & World Ecology* 23(1), 53–60.
Liu, W., Hu, H., Ma, Y. and Li, H. (2006) Environmental and socioeconomic impacts of increasing rubber plantations in Menglun township, southwest China. *Mountain Research and Development* 26(3), 245–253.
Maes, L. (2012) Conflicts between rubber plantations and nature reserves: A case study in Naban river nature reserve, China. Master's thesis, Ghent University, Belgium.
Min, S. (2016) Economic analysis of rubber land use systems in Southern China. PhD thesis, Leibniz University Hannover, Germany.
Min, S., Huang, J., Waibel, H., Yang, X. and Cadisch, G. (2019) Rubber boom, land use change and its implications for carbon balances in Xishuangbanna, Southwest China. *Ecological Economics* 156, 57–67.
Nizami, S.M., Yiping, Z., Liqing, S., Zhao, W. and Zhang, X. (2014) Managing carbon sinks in rubber (*Hevea brasiliensis*) plantation by changing rotation length in SW China. *PLoS ONE* 9(12), e115234.
Qiu, J. (2009) Where the rubber meets the garden. *Nature News* 457(7227), 246–247.
Qu, S., Lv X.Y., Hu, X.D. and Yuan L.J. (2021) The evolution law, hot issues and future perspectives of China's rural land system during the 100-year-history of CPC: Summary of the "High-level forum on centenary goal and rural land issues. *Issues in Agricultural Economy* 10, 139–144 (In Chinese).
Rural Department of the National Bureau of Statistics. (2019) *China Rural Statistical Yearbook 2019*. China Statistics Press (In Chinese), Beijing.
Sun, B. and Zhang R.J. (2024) Reform and opening up empower Chinese-style modernization: logic, dynamics and paths. *Economist* 8, 5–14 (In Chinese).
Statistics Bureau of China. (2020) *China Statistical Yearbook 2000*. China Statistics Press (In Chinese), Beijing.
Wauters, J.B., Coudert, S., Grallien, E., Jonard, M. and Ponette, Q. (2008) Carbon stock in rubber tree plantations in western Ghana and Matogrosso (Brazil). *Forest Ecology & Management* 255(7), 2347–2361.
Wei, G.C. and Kong X.Z. (2024) The logic, practice direction and policy value of the "Land consolidation" Reform: From the perspective of perfecting the basic rural business system. *Economist* 6, 107–116.
Xu, J., Fox, J., Vogler, J.B., Fu Yongshou, Z.P., Lixin, Y. *et al.* (2005) Land-use and land-cover change and farmer vulnerability in Xishuangbanna prefecture in southwestern China. *Environmental Management* 36(3), 404–413.
Yan, J.D. (1989) A discussion on the basis of national economy – a discussion with some colleagues. *Issues in Agricultural Economy* 6, 28–36 (In Chinese).
Yan, L., Zhao, J. and Wang, L. (2019) History, experience and focus of continuous promotion of China's poverty alleviation work. *Economic Daily, October 16:12* (In Chinese).
Yang, L. (2013) Since the founding of the People 's Republic of China, the Communist Party of China 's understanding of the problem of farmers and its enlightenment. *Journal of Socialist Theory Guide* 8, 73–75 (In Chinese).
Yang, N., Ding, Y., Min, S. and Bai, J. (2022) Does rubber expansion hinder the migration of rural labor? Evidence from southwest China? *Review of Development Economics* 26(2), 1108–1131.
Yang, X., Blagodatsky, S., Lippe, M., Liu, F., Hammond, J. *et al.* (2016) Land-use change impact on time-averaged carbon balances: Rubber expansion and reforestation in a biosphere reserve, South-West China. *Forest Ecology and Management* 372, 149–163. DOI: 10.1016/j.foreco.2016.04.009.
Zhang, M., Zhou, K. and Xue, D.Y. (2007) Rubber's influence on tropical rainforest in Xishuangbanna and how to reduce the impact. *Ecological Economy* 2, 377–378+439 (In Chinese).
Zhang, X.S. (2021) Promoting integration urban-rural development and rural vitalization on all fronts: Study the law on the promotion of rural vitalization. *Issues in Agricultural Economy* 11, 4–11 (In Chinese).
Zheng, Z.Q. and Zheng, G.C. (2021) Eradicating poverty: The Chinese miracle and the Chinese experience. *Journal of the Party School of the Central Committee of the C.P.C.(Chinese Academy of Governance)* 25(2), 39–48 (In Chinese).
Zhong X.P., Yu, X.H. and Tang, Z. (2020) The class attribute of rural land tenure and institutional evolution of the separation of "three rights" for agricultural land. *Issues in Agricultural Economy* 7, 47–57 (In Chinese).

Zhu, J., Zang, X.Y. and Li, T.X. (2021) China's food security risks and prevention strategy under the new development pattern. *Chinese Rural Economy* 9, 2–21 (In Chinese).

Zhuang H. (2022) Essays on the Transformation of the Rural Economy in Rubber producing areas in Southwest China. PhD thesis, Leibniz University Hannover, Germany.

Zuo, Y., Ma, L., Cai, H., Wu, T. and Xin, X. (2015) China's on-going debates over large-scale farming: what are the pros and cons? *China Agricultural Economic Review* 7(3), 338–343.

8 Indian Agricultural Policy Developments: A Review

Priyanka Parvathi

Abstract

This chapter traces Indian agricultural policy developments since post-independence. The effect of national development theories and policies on Indian agricultural policies is examined. Based on literature five phases of Indian agriculture from the 1950s–2020 are analysed. This ranges from the celebrated Indian Green Revolution encompassing both its pre- and post-phases and extends to the 1991 economic reforms and India`s growth story thereafter. Through these phases we identify patterns of structural changes in agriculture. Like the other chapters in this book, this approach provides insights into existing constraints and opportunities for a productive policy debate and target-orientated subsequent research.

Introduction

The Indian economy has grown hugely since independence in 1947. In 2024, India was the fifth largest economy in the world by GDP at US$3.41 trillion. With largely the service sector driving growth, the Indian agrarian sector employed 46% of the total labor force and contributed 18% to GDP in 2024. This shows that agriculture is still the primary source of livelihood for 661 million Indians. These indicate that although the Indian economy has grown enormously, agricultural transformation is still in its nascent stage. This structural gap also means that farm incomes are outpaced by income generated in other sectors (Thakur *et al.*, 2024).

The sections below give a sketch of the Indian agricultural journey from struggling to reach self-sufficiency post-independence to the booming staple food production during the Green Revolution to being the 8th largest exporter of agricultural products in the world amounting to US$51 billion in 2023. Based on literature five stages of Indian agriculture from 1950s to 2014 were identified. Table 8.1 encapsulates the five distinct stages of Indian agricultural growth, summarizing the landmark policies that have been instrumental in increasing productivity. These are: (1) the Pre-Green Revolution-Land Reform Period, (2) the Green Revolution Period, (3) the post-Green Revolution period, (4) the phase of economic reforms and (5) the post-Economics-Reforms-Period, until the present. A detailed description and discussion of the periods follow below.

1950–1965: Pre-Green Revolution

Indian agriculture was eclipsed by the Bengal famine of 1943 in the Bengal province of British India (includes the present-day Bangladesh and West Bengal state of India). Hence, the priority

DOI: 10.1079/9781800627338.0008

Table 8.1. Indian Agricultural Developments. (Source: Author's own table).

Decades/Stages (#)	General Development Policy	Agricultural Policies	Structure of Agriculture	Significant Events	Performance of Agriculture
1950–1965: Pre Green-Revolution Land Reforms (1)	Rapid industrialization over agricultural development	Land reform, tenancy reform, land ceilings act	75–80% employed in agriculture; share in GDP declined from 61.7% to 47.6%; 137–156 million ha total gross crop area; 22.6–30.9 million ha gross irrigated area	1961 Indo-China war & 1965 Indo-Pak war coupled with the 1965 & 1966 drought	Declining output; insufficient domestic supply to meet domestic demand
1965–1980: Green Revolution (2)	Interventionist industry strategy	Introduction of HYVs and chemical fertilizers	Approximately 67% employed in agriculture; share in GDP declined from 41.7% to 35.7%; 159–171 million ha total gross crop area; 38.2–49.8 million ha gross irrigated area	Nationalization of banks, Establishment of NABARD	Self-sufficiency in food-grain achieved
1980–1991: Post Green Revolution (3)	New Computer Policy 1984	Oilseed Mission	Labor share 63%; GDP share 29.5%, 175–178 million ha total gross crop area; 54.3–63.2 million ha gross irrigated area	Balance of Payment crisis	Focus expanded to fruits, vegetables and livestock
1991–2010 Economic Reforms (4)	Liberalization, Privatization, and Globalization	First National Agricultural Policy with detailed components for agricultural growth	51% employed in agriculture in 2010; share in GDP 14.4% in 2010; 185–192 million ha total gross crop area; 71.4–88.9 million ha gross irrigated area	Rapid economic growth	Sustained agricultural growth at an average annual rate of 2.5%
2011–2024: Post Economic Reforms (5)	Liberalized Foreign Direct Investment	Various government schemes encompassing crop insurance, historic rise in minimum support price (MSP)	44% employed in agriculture in 2023; share in GDP was 16% in 2024, 159–180 million ha total gross crop area; 95.8–120.3 million ha gross irrigated area	Demonetization, Covid pandemic	Rural farming stagnation

in post-independence in 1947, was to improve food security and to pave the way for several land reforms. It included first and foremost the abolition of the "zamindari system," i.e. intermediaries who were rent collectors in the pre-independence land revenue administration. This brought the cultivators in direct contact with the government. This was followed by the tenancy reform to regulate rent, improve the contractual term, and provide security of tenure. The third major reform was the redistribution of surplus land to the landless based on Land Ceiling Acts to prevent the ownerships of agrarian land in the hands of a few. But as a new nation, the Government's first Five-Year plan (1951–1956) centered on industrial development over agriculture; but stressed on keeping food prices down.

The government believed that the need of the hour was to produce sufficient food to support and feed its industrial labor. In keeping with its industrialization targets, it focused on jute rather than wheat or rice exports and on social equity rather than increased production. The focus was more on food distribution and in keeping the cost of living down and affordable. Although there was an increase in domestic grain production it was still insufficient to meet the needs of the growing population. The downside of prioritizing industry over agriculture resulted in a food crisis compelling urgent food imports. The decision to import around 3 million tons of grain from the USA under the PL-480 scheme ("Food for Peace") in 1956 was critiqued. Nonetheless, the volume of food imports kept increasing and was 10 million tons in 1966.

In 1951, the agricultural sector employed 82% of the labor force and contributed 54% to the GDP. But its share was reduced to 47.6% by 1960–61 and the total labor force employed in agriculture reduced to 75%. The total food grain production increased from 51 million tons in 1950–51 to 82 million tonnes in 1960–61. Between 1953–54, 95% of the agricultural landholdings were individually operated with an average size of 2.11 ha (5.21 acres) and 5% were jointly operated (managed jointly by two or more households) with landholdings averaging 4.01 ha (9.92 acres) in rural India. The equipment owned were mostly wooden ploughs and harrows. Tractors and other power-driven engines were a rarity. The total net irrigated area in 1950–51 was 20 million hectares; with 8% from canals, 4% from tanks, 6% from wells, and 3% from other sources. It increased to 26.3 million hectares by 1965 (Modak, 2018, Page 33, table 3).

The significant events of the 1961 war against China and the 1965 war against Pakistan, coupled with the successive droughts in 1965 and 1966 led to a 17% fall in domestic crop production. The pre-Green Revolution period was defined by a sharp fall in agricultural GDP, with decadal growth rates plunging from 2.78% to 1.06% between the period 1950–51 and 1967–68 (Ministry of Statistics and Programme Implementation, Government of India).

Overall, the pre-Green Revolution period was marked with significant regional differences; on average, modest agricultural growth, driven mainly by area expansion facilitated through irrigation investment (Rao, 1971). During the early to mid-60s, performance of the Indian economy was poor, while agriculture remained the major contributor to growth. During this period poverty and inequality especially in rural areas increased noticeably (Shetty, 1978).

1965–1980: Green Revolution

With falling grain output (rice and wheat), Indian government decided to import and distribute high-yielding varieties of wheat and rice which defined the Indian Green Revolution. This was initially introduced in the irrigated areas, mostly belonging to large-scale farmers. Farmers were also provided with subsidized chemical fertilizers and other agrochemicals to enhance yields. A number of extension services in support of knowledge transfer and the adoption and usage of HYV were also established. The strategy was to achieve self-sufficiency and despite the rising input costs, it did provide a breakthrough in Indian grain production. During this period the average yield of rice escalated by 60% and wheat by 97% (Somvanshi *et al.*, 2020). The Green Revolution is largely credited in making India today an important player in the wheat export market and the largest exporter of rice in the world.

There was a massive surge in private tube well construction in 1965 boosted by government loans, subsidies, and rural electrification. By 1969, as indicated in Subramanian (2015) and Stone (2019), there were a quarter million operational tube wells as farmer adoption of HYVs was dependent on their accessibility to tube wells, especially in the rice belt, especially as it removed the dependency on monsoons. The gross irrigated area increased by 27.2 million ha, from 22.6 million ha in 1950–51 to 49.8 million ha by 1980.

A major policy that strengthened the Green Revolution was the nationalization of 14 commercial banks in 1969 to enhance credit flow to the agrarian sector and the establishment of National Bank for Agriculture and Rural Development (NABARD).

The general development policy towards industry during this period was of interventionist. As noted in Adhia (2015) the policy was wary of large firms. Consumer goods were reserved by law to be manufactured only by small firms at the cost of economies of scale. Furthermore, delays in license issuance created additional bottlenecks. Foreign companies were asked to reduce their stake in Indian subsidiaries with many multinationals like Coca-Cola and IBM exiting the Indian market.

In agriculture, the Green Revolution also started the advent of underutilized agrarian workers shifting to off-farm activities. The share of labor force employed in agriculture during this period declined to 67%. In 1970–71 the share of agriculture in GDP was 41.7% and it fell by −6% points to 35.7% by 1980–81. Grain production increased from 108.42 million tons in 1970–71 to 129.59 million tons in 1980–81. The average farm size decreased from 2.28 ha in 1970–71 to 1.84 ha in 1980–81.

1980–1991: Post Green Revolution

Growth in agricultural output reached its peak during the Green Revolution but showed a declining trend during the 1980s. As observed by Rath (1989) this can be attributed to a declining trend in the share of public investment in agriculture.

As noted by Arora (2013), the post Green Revolution phase was also the introduction of high-yielding varieties to other crops and regions. There was a policy shift beyond grains towards high value crops like fruits, vegetables, dairy products, fish, poultry and others (Chand, 2003) which accelerated output and growth in agricultural GDP. This period also witnessed further increase in subsidies, extension services and credit support, while public sector spending for rural infrastructure started to decline. At the same time, private investments by farmers had increased (Tripati and Prasad, 2010).

India was importing $1 billion of cooking oil each year despite domestic land well suited for growing oil crops. Hence, to achieve self-reliance in edible oil production an "Oilseed Mission" program was launched in 1987 to make planting oilseeds profitable for farmers. The marketing logistics were re-structured to ensure that small producers are not defrauded by middlemen. The intervention was successful, and India started to export oil by the early 1990s. This was also popularly refereed as the "Yellow Revolution."

In 1990–91 the share of agriculture in GDP fell to 29.5% and the share of labor in agriculture was 63%. The grain production was at 176.39 million tons and average farm size was approximately 1.57 ha. The gross irrigated area jumped 21% from 1980–81 to reach 63.2 million ha in 1990–91 indicating the continued investment by the government in public irrigation facilities to facilitate small-scale farmers.

The New Computer Policy 1984 sowed the seeds of the Indian IT boom that emerged in the 2000s. It encouraged private investments in the information technology (IT) sector, provided tax benefits, eased import restrictions and incentivized software exports. Also, there was trade liberalization in high value goods like automobiles, electronics, etc. This resulted in significant imports. It enabled GDP to grow by 5.6% on average annually between 1980–1990 (Panagariya, 2005). But as the exchange rate was fixed, the government had to maintain parity, leading to the depletion of its foreign reserves. This coupled with the rise in oil prices due to the first Gulf War between Iran and Iraq (1980–1988) led to the Balance of Payment (BoP) crisis. Despite measures promoting exports it was hit by the 1980–1983 global recession. All this resulted in a BoP deficit of INR 11,384 crores in 1985 forcing the Indian government to seek external assistance and borrowing from the International Monetary Fund (IMF) (Panagariya, 2005).

By the end of the post-Green Revolution period, the rural population totaled 683 million, 20% higher than in 1981 and equivalent to almost 75% of the Indian population (Agricultural Statistics of India, cited in Tripati and Prasad (2010). Likewise, the share of agricultural labor increased to 40.3%, while GDP share declined. So, Indian agriculture experienced what Binswanger-Mkhize (2013) called "stunted transformation of Indian economy."

1991–2010: Economic Reforms

The growing BoP deficit worsened with the Gulf War between Kuwait and Iraq in 1990 increasing oil prices. By 1991 the Indian foreign reserve had drastically reduced with the country barely able to finance imports. This crisis forced India to give birth to a New Industry Policy which centered on economic liberalization, privatization, and globalization. Although initially it did not benefit agriculture, the trade policy reforms that followed including the Uruguay Round Agreement on Agriculture in 1994 eased agricultural trade. But falling agro prices in 1997 led to tariff rollbacks; nevertheless, the 1990s transitioned Indian agriculture from a state-led to a market-driven growth (Gulati and Mullen, 2003).

In 2000, the government for the first time published a comprehensive agricultural policy. The National Agricultural Policy (NAP) aimed for more than 4% annual growth through efficient natural resource utilization that included farmers across regions in the growth process. It encouraged private investments through contract farming to ensure a demand-driven growth to exploit the benefits of the export market while still offering price protection to domestic farmers. The policy also involved transforming wastelands for agriculture and afforestation purposes. A major focus was horticulture, floriculture, roots and tubers, plantation crops, black pepper and medicinal plants and beekeeping. Higher emphasis was also placed on raising the production of animal and fish products (Arora, 2013).

With regards to the agricultural structure during this period, average farm holdings remained small. They were around 1.57 ha in 1991 which declined to 1.33 ha in 2001, and further declined to 1.23 in 2005–06 and to 1.16 ha in 2010–11. As pointed out by Acharya (2002), agricultural productivity has been affected by land fragmentation due to the slow shift of labor from farm to non-farm income sources. A contributing factor to this is the sluggish growth of the manufacturing sector inhibited by strict labor laws in the industry sector and small-scale industry reservations. But the total operational landholdings increased across the country from 129 million ha in 2005–06 to 138 million ha in 2010–11. The share of labor force employed in the agrarian sector was 62% in 1995, fell to 55% in 2005 and further to 51% in 2010. The share of agriculture in the nation's GDP also decreased from 22.3% in 2000–01 to 14.4% in 2010–11. The food grain production during this period rose 39% and reached 244.49 million tons by 2010–11. The gross irrigated area increased from 71.4 million ha in 1995–96 to 88.9 million ha by 2010–11.

The most significant event during this period in Indian history was the economy's dream run between 2003 and 2008 growing at 9% annually. This was a private corporate led growth that tapped into the exceptional rise in world trade. It was led by investment financed by private savings enhanced by unprecedented influx of foreign private capital (Nagaraj, 2013).

As noted by Binswanger-Mkhize and D'souza (2015), improvement in comparative advantage of agriculture and better trade opportunities were important drivers of change in the composition of agricultural output.

2011–2014: Post Economic Reforms

We largely use verbatim from Ramakumar (2024) on the agrarian developments wherein four conclusions on India's agricultural economy between 2014 and 2024 are drawn. He notes that during this period (a) there was an overall economic slowdown and poor growth of employment, particularly in rural India. Incentives to invest deteriorated. Agricultural prices grew slower than the costs of cultivation, leading to a fall in profitability rates and the real incomes of farmers. Also, the Government failed to fulfill a series of promises it had made to the rural electorate during its election campaigns in 2014 and 2019, particularly the promises to double farm incomes and raise minimum support prices. Furthermore, the Government's efforts to amend or reform laws in the domain of agricultural marketing were met with united resistance by farmers' organizations, and, finally, a series of ill-thought measures of the Government – especially the restrictions on cattle trade demonetization, reform of the GST system, and the inadequate response to the Covid-19 pandemic – led to enormous problems among farmers and adversely affected the health of the agrarian economy.

The share of agriculture in GDP was 13.9% in 2015 and increased to 20.2% in 2020–21 but declined to 16% in 2024. Agriculture still employed 44% of the total labor force in 2023. The food grain production increased six-fold from 50.82 million tons in 1950–51 to 314.51 million tons in 2021–22. The gross irrigated area reached 120.3 million ha in 2021–22 largely covered by tube wells. The average land-holdings decreased from 1.08 ha in 2016–17 to 0.74 ha in 2021–22 but the monthly household income increased from INR 8k in 2016–17 to INR 12k in 2021–22. The institutional lending to rural families has increased to 75.5% in 2021–22 indicating that rural farmers too have access to formal financial avenues.

The GDP continued to grow at an average of 7% annually during this period. India went from ninth largest to the fifth largest economy in the world by nominal GDP in 2019. As per the European Commission report 2024, India emerged as the fastest growing G20 economy post the pandemic crisis. Although the economy shrank by 6.6% during the first wave of Covid-19, it rebounded with a 7.9% average growth rate in 2021. India´s financial sector has strengthened to meet the demands of investments in an economy growing at a rapid pace. Despite India having a current account deficit since 2000, it is lower compared with its peers and was 1% of GDP in 2023 (European Commission, 2024).

Summary

Summarizing the five phases of agricultural development in India highlights the major transition of farming and rural areas. In phase I, the pre-Green Revolution period was marked by poor performance of the agricultural sector despite agrarian reforms, institutional changes, investment in irrigation, and strengthening of credit institutions. However, domestic supply fell short of demand for food in India. Phase II, from 1965–1980, witnessed growth in agricultural productivity due to widespread adoption of Green Revolution technology with high-yielding rice and wheat varieties. The biggest achievement of this strategy has been attainment of self-sufficiency in food grains. Phase III was marked with the process of diversification which resulted in fast growth in non-food grains commodities which accelerated growth in agricultural GDP. The increase in subsidies and support to the agricultural sector caused agriculture to "move" further, despite the reduced public sector spending for infrastructure. In phase IV, for the first time in the history of India, a comprehensive agricultural policy, the New Agricultural Policy launched in 2000 along with the economic reform process, including deregulation, liberalization, and opening of the economy with WTO membership as a major step. Finally, in the post-reform phase, to date, agriculture fell behind the overall positive growth picture of the economy as a result of stunted rural transformation and an increasing rural-urban divide. A major event was Covid-19 which as almost everywhere in the world made the economy slump. But the Indian economy rebounded quickly afterwards.

Case Study: Organic Pepper. Source: Own Compilation Based on Parvathi (2014)

This is a brief case study of a non-grain commodity as an example of the diversification of Indian agriculture away from cereal food crops.

During the early 2000s India was confronted with a pepper crisis. From being a top world exporter, India suddenly had to import pepper in 2003–04. The dropping pepper yields and supply shortage particularly affected smallholder farmers. One of the coping strategies was the promotion of organic pepper production to improve soil fertility in combination with a fair-trade certification to minimize price risk and safeguard smallholders from global volatility in pepper prices. In the thesis case study, panel data from 300 smallholder pepper farmers in Idukki district, Kerala in 2011 and 2012 were collected. The objective of the study was to analyze the adoption and impact of organic agriculture and fair-trade certification in combination on pepper farmer welfare and poverty reduction. A critical lesson of the study was that organic farming can be used as a strategy by vulnerable households with less capacity and skills to close the productivity gap with more efficient households. This could help bridge domestic production gaps and reduce dependency on imports. The study also ascertained that there are potential benefits if organic and fair-trade certifications are adopted in combination. Nevertheless, it needs to be noted that additional benefits for organic farmers on adoption of fair-trade certifications are not immediate. Fair-trade has the potential to extend additional benefits to organic farmers with time leading to long-term welfare. Hence, the study recommended that policymakers should create more awareness and accessibility to these systems especially in remote and rural areas.

The Way Forward

Indian agriculture has been successful to overcome widespread poverty and food insecurity of the past. With the help of the Green Revolution, India has achieved self-sufficiency in food grain production and is now a major exporter of rice, spices, tea, cashew nuts and also cotton. Following the general path of economic development GDP shares for agriculture have declined significantly from 60% in the 1950s to 16% in 2024. Labor shares have declined as well but much less than income shares, i.e. from around 70% during the 1950s to 44% in 2024. In this regard, India is well in line with the agricultural transformation pattern of other Asian countries. But the share of agrarian income in national income has witnessed a rapid fall. The stunted shift of labor from farm to non-farm has widened the productivity and income gap between agriculture and non-agricultural sectors. Also, among rural households, income from farming has gone down, i.e. the ratio of farm to non-farm income was 51% in 1950–51 and 29% in 2020–21 (Chand and Singh, 2023). Another indicator of the "stunted agricultural transformation" is the development of the structure of farm sizes. Roughly 85% of the landholdings are below 2 ha. Using data of the 70th round of the National Statistical Survey in 2013, Das and Ganesh-Kumar (2018) report that 67% of Indian farms were marginal, with an average farm size of 0.38 ha, about 18% were small, i.e. average farm size 1.37 ha. A further 10% were labeled as semi-medium with an average farm size of 2.53 ha, some 4.5% are called medium size farms, i.e. 5.4 ha and less than 1% are considered large farms with a farm size of 15.1 ha on average. The fragmentation of farm size has several reasons. One of them is inheritance practice of dividing land among descendants, the second reason is that rural urban migration rates in India are low. Households with migrants stick to their land as a safety net for economic crises. Thirdly, the government is reluctant to actively promote land rental and land sales markets for ideological reasons. Lastly, population growth in rural areas outpaces urban areas due to more effective population control measures in the latter. Consequently, mechanization is hampered, and manual labor is widespread. The lack of land consolidation has inhibited economies of scale in the Indian farming system resulting in low negotiating power in the hands of farmers relative to traders and middlemen.

On the plus side, as submitted by Binswanger-Mkhize (2013), urban-rural spillovers are now important drivers of a growing rural non-farm sector, which now generates the largest number of jobs in India. Rural non-farm has become the major source of income growth. Hence if the labor force remains in rural areas farm sizes will continue to decline. Also, there is likely to be a trend towards female-headed households and part-time farming will be the dominant farm model for a long time to come.

The government of India has a vision called "Amrit Kaal," to conclude with 100 years of Indian independence in 2047. This vision includes agriculture, and its roadmap focuses on a sustainable and resilient agrarian sector to drive growth, rural prosperity and achieve global leadership in agricultural exports. It centers on the involvement of technological advancements including Artificial Intelligence, Internet of Things (IoT) and precision farming for enhanced productivity. It also encompasses drone technology for pesticide spraying and crop monitoring. It emphasizes climate-resilient farming and inclusion of agroforestry and bio-fertilizers as well as drought and flood-resistant crop varieties. It aims to provide better market access for farmer producer organization and promote agri-start-ups. It includes plans to adopt drip and sprinkler irrigation systems and strives for an efficient water management system (Chand and Singh, 2023).

The United Nations 2030 agenda is impossible to achieve without sustainable agricultural solutions. Although steps have been taken to build a future vision for Indian agriculture, it remains to be seen on the extent of implementation of this vision on the farm and its effectiveness in lifting rural agrarian per capita incomes. The agrarian sector is fundamental for inclusive growth, employment generation, sourcing renewable energy, and meeting the goals of sustainable development.

References

Acharya, S. (2002) India's Medium-Term Growth Prospects. *Economic and Political Weekly* 37(28), 2897–2906.

Adhia, N. (2015) The history of economic development in India since independence. *Education About Asia* 20(3), 18–22.

Arora, V.P.S. (2013) Agricultural policies in India: retrospect and prospect. *Agricultural Economics Research Review* 26(2), 135–157.

Binswanger-Mkhize, H.P. (2013) The stunted structural transformation of the Indian economy: Agriculture, manufacturing and the rural non-farm sector. *Economic and Political Weekly* 48, 5–13.

Binswanger-Mkhize, H.P. and D'souza, A. (2015) Structural change and agricultural performance at the state level in India: 1980–2010. *Agricultural Economics Research Review* 28(1), 27–38.

Chand, R. and Singh, J. (2023) From green revolution to amrit kaal. *Natl Inst Transf India* 3, 67–89.

Chand, R. (2003) *Government Intervention in Foodgrain Markets in the New Context*, Policy Papers 344977, ICAR National Institute of Agricultural Economics and Policy Research (NIAP). DOI: 10.22004/ag.econ.344977.

Das, V.K. and Ganesh-Kumar, A. (2018) Farm size, livelihood diversification and farmer's income in India. *Decision* 45, 185–201.

European Commission (2024) India's Economic Surge: From Regional to Global Economic Player. European Commission. Available at: https://economy-finance.ec.europa.eu/economic-forecast-and-surveys/economic-forecasts/autumn-2024-economic-forecast-gradual-rebound-adverse-environment/indias-economic-surge-regional-global-economic-player_en.

Gulati, A. and Mullen, K. (2003) Responding to policy reform: Indian agriculture during the 1990s and after. In: *Fourth Annual Conference on Indian Economic Policy Reform, Stanford University, California*, 5–7 June 2003. Available at: https://kingcenter.stanford.edu/sites/g/files/sbiybj16611/files/media/file/189wp_0.pdf.

Modak, T.S. (2018) From public to private irrigation: Implications for equity in access to water. *Review of Agrarian Studies* 8(1). DOI: 10.25003/RAS.08.01.0003.

Nagaraj, R. (2013) India's dream run, 2003–08: Understanding the boom and its aftermath. *Economic and Political Weekly* 48(20), 39–51.

Panagariya, A. (2005) India in the 1980s and the 1990s: A Triumph of Reforms. In *India's and China's Recent Experience with Reform and Growth*. Palgrave Macmillan, London, 170–200.

Parvathi, P. (2014) Assessing the adoption and impact of organic and fair trade certification of pepper in India. PhD thesis, Leibniz University Hannover, Germany.

Ramakumar, R. (2024) India's agricultural economy, 2014 to 2024: Policies and outcomes. *Review of Agrarian Studies* 14(1), 11–37.

Rao, S.K. (1971) Inter-regional variations in agricultural growth, 1952–53 to 1964–65: A tentative analysis in relation to irrigation. *Economic and Political Weekly* 6(27), 1333–1346.

Rath, N. (1989) Agricultural growth and investment in India. *Journal of Indian School of Political Economy* 1(1), 64–83.

Shetty, S.L. (1978) Structural retrogression in the Indian economy since the mid-sixties. *Economic and Political Weekly* 13(6/7), 185–244.

Somvanshi, P.S., Pandiaraj, T. and Singh, R. P. (2020). An unexplored story of successful green revolution of India and steps towards ever green revolution. *Journal of Pharmacognosy and Phytochemistry* 9(1), 1270–1273.

Stone, G.D. (2019) Commentary: New histories of the Indian green revolution. *The Geographical Journal* 185(2), 243–250.

Subramanian, K. (2015) Revisiting the green revolution: Irrigation and food production in 20th century in India. PhD thesis, Kings College, London. https://pib.gov.in/newsite/PrintRelease.aspx?relid=78369

Thakur, N., Sharma, A. and Sharma, R. (2024) Structurally transforming Indian economy with special reference to the agriculture sector. *Economic Affairs* 69(3), 1227–1235.

Tripathi, A. and Prasad, A.R. (2010) Agricultural development in India since independence: A study on progress, performance, and determinants. *Journal of Emerging Knowledge on Emerging Markets* 1(1), 8.

9 Review of Agricultural Development and Policy with a Case Study on the Role of Homestead Fish Ponds for Food Security in Bangladesh

Badrun Nessa Ahmed

Abstract

This chapter outlines Bangladesh's structural transformations and policy reforms in the agriculture sector from the country's independence in 1971 to the present. Over the past 35 years, agriculture in Bangladesh has evolved considerably, achieving significant progress. This chapter presents a detailed analysis of the evolution of the agricultural sector in Bangladesh. Patterns of structural changes using historical data are presented which help to highlight challenges and opportunities and offer a foundation for impactful policy analysis. A case study on homestead pond aquaculture in Bangladesh is included as an example of recent developments. The last section shows a detailed summary, including some statistics on agricultural growth and employment.

Introduction

Agriculture has been the backbone of Bangladesh's economy for over half a century, with the sector playing a significant role in shaping the country's socio-economic development. Since its independence in 1971, Bangladesh has undergone significant agricultural transformations, reflecting both the challenges and successes of policy interventions. Significant changes in agricultural policies and development have been observed in transitioning from a largely subsistence agricultural economy to one that emphasizes commercialization, diversification, and modernization and at the same time resilience and sustainability. Agricultural policies have evolved to meet the challenges posed by population growth, food security, climate change, and economic transformation. This review explores Bangladesh's agricultural development and policy implementation over the last decades, highlighting key phases of transformation, challenges faced, the strategic initiatives undertaken, and the impact on food security, productivity, and sustainability.

Using five categories, namely: (i) general development policy, (ii) agricultural policies, (iii) structure of agriculture, (iv) significant events, and (v) the performance of the agricultural sector, the history of sectoral development is presented in Table 9.1. However, the evolution of agricultural policy has two main phases: pre-independence and post-independence. Both phases have been discussed briefly, followed by a detailed analysis of the development decade after the independence of Bangladesh, summarized by keywords in Table 9.1.

Agricultural Policy and Development in Bangladesh

There are two distinct periods in the evolution of agricultural policy: pre-independence and post-independence.

DOI: 10.1079/9781800627338.0009

Table 9.1. Development Overview by Decades. (Author's own table).

Decades	General Development Policy	Agricultural Policies	Structure of Agriculture	Significant Events	Performance of Agriculture
First Decade (1971–1980)	Extremely restrictive trade and exchange rate policies	Inward-oriented and import-substituting trade policy	Dominated by rice production and cash crop jute	Effect of Great Cyclone of 1970; devastating famine of 1974; food shortages	Stagnant with low labor productivity and lack of mechanization
Second Decade (1981–1990)	Structural Adjustment, trade liberalization, agricultural input market deregulation, and public enterprise reforms	Government support for agriculture; economic reforms	Dominated by rice and jute; diversification into wheat and other crops; Green Revolution technologies	Devastating floods in 1987 and 1988	High vulnerability to natural disasters; gradual technological adoption, including mechanization
Third Decade (1991–2000)	Trade liberalization; export promotion	Agricultural diversification; government support; input management; trade policy reforms	Diversified farming systems such as fruits, vegetables, and pulses; expansion of livestock and fisheries	Political unrest; 1991 cyclone; 1998 flood	Rice production increased; limited access to markets; inadequate infrastructure; vulnerability to climate change
Fourth Decade (2001–2010)	Investment in infrastructure; promotion of export-oriented industries; liberalization of agricultural input markets	Mechanization and technological adoption; government investment in farm productivity, irrigation systems, and agricultural extension services	Increased diversification and modernization; Smallholder farming predominated; greater use of technology, improved irrigation, and crop varieties; Government programs and international aid for growth	Devastating effect of Cyclone Sidr; political unrest; exports of ready-made garments surged	Mixed outcomes; self-sufficient in rice production; boosting GDP growth; disrupted progress for natural disaster
Fifth Decade (2011–2020)	Economic growth, poverty reduction, and sustainable development	Increase productivity; improve food security and foster sustainability	Diversification and modernization; agro-processing and value addition; fostering economic resilience; changing dietary patterns; urbanization	2017 floods; damaged crops across farmland impacting food security; land degradation and urban expansion	Declining in GDP share of agriculture, advancements in crop production, and growth in agricultural sub-sectors

Before 1971: The Pre-independence Period

Until 1971, the macroeconomic indicators and overall trade of unified Pakistan significantly impacted the agricultural trade policy and price incentives mechanism in East Pakistan (now Bangladesh). Pakistan employed an import-substitution trade strategy between the 1950s and 1960s, which included imposing import tariffs to safeguard domestic industries and taxing agricultural exports. Instead of opting for currency devaluations, the policy primarily focused on implementing quantitative import controls to manage effective demand for foreign exchange at the official exchange rate (Lewis and Guisinger, 1971). Pakistan launched quantitative import limits in the early 1950s via an import licensing system, which prioritized foreign exchange for capital and intermediate products while restricting imports of consumer items.

Simultaneously, cotton and jute exports were subject to export tariffs. Between 1952 and 1958, the Pakistan rupee was overvalued with these explicit export taxes, resulting in a 70% decline in the real value of total exports (Dorosh and Valdes, 1990).

After 1971: The Post-independence Period

Following independence, Bangladesh adopted a very restrictive trade and exchange rate policy that mirrored the policies of the 1960s. This policy involved export taxes, strict import regulations, high import tariffs, ongoing quantitative restrictions, and an overvalued exchange rate. The policy regime implicitly taxed the agricultural sector with the restrictions of export of raw jute, a major agricultural product. At the same time, the government kept monopoly on the import of most agricultural commodities, resulting in a significant decrease in agricultural price incentives (Rahman, 1994).

From the early 1980s to the mid-1990s, agricultural reforms in Bangladesh were undertaken in two main waves. In the first wave, domestic input trading was liberalized, and input subsidies were reduced (the early to mid-1980s). In the late 1980s and early 1990s, the second wave of reforms was undertaken through the liberalization of the import of inputs, private trading in local and international grain markets, and a significant reduction in public distribution of grains for long-standing programs. In short, comprehensive and widespread agricultural policy reforms were implemented over 15 years. In contrast to earlier periods of significant and widespread state interventions and regulations, the market for inputs and outputs was almost fully liberalized. At that time Bangladesh stood alone amongst its South Asian neighbors regarding liberalization, considering its breadth and depth of agricultural reforms (Ahmed, 1996).

Due to the reduction of public transfers to farmers and food consumers and the reasonably satisfactory performance of private markets for input and outputs, the reform can be considered as successful. The consistent increase in rice production and decrease in the wholesale and retail price of rice indicated the sign of reform making progress.

Development of Bangladesh's Agriculture by Decade

The First Decade: Foundation of Agricultural Development

After independence, Bangladesh faced a serious balance of payments crisis and thus implemented extremely restrictive trade and exchange rate policies. High tariff rates, a fixed and overvalued exchange rate policy, and quantitative import restrictions were all applied as general development policies to protect domestic industries. Even after the balance of payment had improved during the middle of the 1970s, following a 70% devaluation of the Bangladeshi currency against the US dollar, the political government's protectionist stance remained intact for the trade regime. Even a rebound in export revenue, a rise in foreign aid, and a plunge in the global price of grains and other commodities were all part of it.

The government focused on reconstructing the agricultural sector by introducing the First Five-Year Plan (1973–1978). This plan emphasized the importance of food self-sufficiency, focusing particularly on the production of rice through increased irrigation and high-yielding variety (HYV) seeds (GoB, 1973). The Bangladesh

Agricultural Development Corporation (BADC) was established in 1972 to provide farmers with agricultural inputs such as seeds, fertilizers, and irrigation systems. A major achievement of this era was the introduction of HYV rice varieties, such as IR8, developed by the International Rice Research Institute (IRRI), which significantly increased rice productivity and laid the groundwork for Bangladesh's progress in food security.

Bangladesh followed an inward-oriented and import-substituting agricultural trade policy during this decade. Most agricultural products were either prohibited or restricted from import. It was intended to shield manufacturers from outside competition and guarantee fair pricing. At the same time, exports of agricultural products were also subject to different restrictions. Export duties on some agricultural products were put in place. The aim was to ensure an adequate supply of agricultural products in the domestic market. To encourage the export of non-traditional goods, certain financial incentives were provided in 1976. Non-traditional export revenue was doubled from 15–20% to 30–40% under the 1976 Export Policy Order. Duty drawback schemes were implemented to encourage textile product exports using imported raw materials.

However, compared with the rapid growth of East Asian economies that adopted a more outward-oriented development strategy, the results of autarkic trade and exchange rate policy were disappointing regarding export growth, the balance of payments, and overall economic growth. The export promotion processes gradually introduced in the late 1970s, were narrow in scope and implemented slowly. Finally, it can be said that the economy continued to be over-regulated, and trade policies were primarily inward-looking during that decade.

The structure of agriculture was dominated by the production of staple crops, with rice being the primary product and main food grain for the population. Other significant crops included jute, often called "golden fibre," a major export commodity. With government support, wheat cultivation began gaining importance during this period. Pulses, oilseeds (such as mustard), and sugarcane were expanding. Tea was introduced in the north-eastern regions. During this time, the government was the main supplier of agricultural machinery for pump irrigation and tillage was provided through state-owned manufacturers. The focus was mainly on rice self-sufficiency.

Regarding significant events, this decade was a critical period for agriculture in Bangladesh, highlighted by efforts to recover from the devastation of the 1971 Liberation War and the effect of the 1970 Bhola cyclone known as the Great Cyclone of 1970 (Longshore, 2007). Moreover, the subsequent devastating 1974 famine led to food shortages and highlighted the inefficiencies in agricultural production (Crow, 1984).

Agricultural performance was marked by stagnation in many respects. Although rice and jute dominated production, growth was constrained by old technology. During the 1970s, labor productivity in Bangladesh's agriculture was low, hindered by traditional farming methods, lack of mechanization, and frequent natural disasters. Although Green Revolution technologies began to improve yields later in the decade, per-worker output remained insufficient to meet population growth, reflecting the sector's slow modernization (Alauddin and Tisdell, 1991).

The Second Decade: Initial Agricultural Policy Reform with Limited Diversification and Low Level of Commercialization

The general development policies in the 1980s focused on addressing economic and social challenges through reforms and strategic interventions. Structural adjustment programs were implemented, emphasizing trade liberalization, agricultural input market deregulation, and public enterprise reforms including privatization of state companies, particularly in the jute and textile sectors. These efforts aimed to improve productivity and enhance export competitiveness, albeit with moderate success. Ultimately, trade liberalization made very slow progress in the 1980s, especially regarding the reduction in import tariffs (Bakht, 2001).

Agriculture during the second and third Five-Year Plans, i.e. from 1980–1985 and from 1985–1990, underwent significant adjustments as part of agricultural trade policy reform (GoB, 1980, 1985). For example, pesticide subsidies were abolished and importation liberalized. In fertilizer, the state monopoly of the Bangladesh

Agriculture Development Corporation (BADC) stopped selling retail goods and wholesaling at the sub-district level. Except for the 8 km border zone, limitations on movement were relaxed, licensing was cancelled, prices were deregulated, and subsidy was reduced from 50% to 21% from 1979 to 1982. Moreover, in 1987, private traders were permitted to purchase fertilizers at ports and industrial gates. Furthermore, in irrigation, from 1980–1985 the diffusion of low-lift pumps and tube wells to private entities, supported with special credit arrangements took place. In 1987, private dealers were also permitted to import pumps and engines for irrigation. Liberalizations for agricultural exports, included, for example, the removal of export taxes on raw jute, tea and other products in 1981. On the other hand, export restrictions persisted on wheat, pulses, non-frozen shrimp and some other commodities, even after 1995. During this decade, Bangladesh's agricultural structure remained dominated by rice and jute cultivation, gradually diversifying into wheat and other crops. Smallholder farms prevailed, while Green Revolution technologies like high-yielding varieties (HYVs) and irrigation expanded unevenly. Limited mechanization and vulnerability to natural disasters shaped agricultural productivity (Sultana and Thompson, 2017; Mahmood, 2021).

Major events in agriculture between 1980 and 1990, were the devastating floods in 1987 and 1988. Especially the latter was catastrophic, inundating up to 57% of the country and causing extensive damage to infrastructure, crops, and livelihoods. Additionally, coastal and flash floods, driven by heavy monsoon rains and tidal surges, regularly impacted rural and urban areas. These events underscored Bangladesh's vulnerability to extreme weather events, prompting initiatives to enhance disaster management and infrastructure resilience (Sultana and Thompson, 2017; Baten *et al.*, 2018).

Agricultural performance gradually improved during the Second Decade, driven by policy reforms and technological adoption. The liberalization of agricultural input markets, including irrigation expansion and higher use of fertilizers, and particularly the spread of high-yielding varieties (HYVs) in rice contributed to modest increases in crop yields. However, frequent natural disasters, inadequate mechanization, and challenges accessing technology and credit limited progress (Gotur, 1991; Mahmood, 2021). Investment in rural infrastructure, including roads and irrigation, partially mitigated these barriers, laying a foundation for future improvements (Sultana and Thompson, 2017).

The Third Decade: Major Agricultural Trade Liberalization Policy with Diversification and Commercialization

Between 1991 and 2000 Bangladesh shifted from military rule towards a multi-party system and parliamentary democracy, first led by Khaleda Zia as Prime Minister and followed by Sheikh Hasina, in 1996. The democratic governments were more serious about trade liberalization. During this time, the general development policies included significant reduction of tariffs, as well as implementing a simpler import tax structure, gradually removing non-tariff import restrictions, and encouraging exports through bonded warehouses, income tax exemptions, and flexible exchange rate management. On the other hand, even though customs duties decreased para-tariffs such as value-added tax, license fees, regulatory duties, surcharges, and supplemental duties increased. As a result, total protective import duty rates practically did not change much and even increased for some commodities (World Bank, 2004).

The 1990s also marked a shift in Bangladesh's agricultural policy towards integrating agriculture into global markets, emphasizing export promotion and agricultural diversification. This period saw the growth of the private sector in agriculture, particularly in the supply of inputs, credit, and marketing. The National Agricultural Policy 1999 was introduced to promote sustainable agricultural practices and encourage diversification. This policy aimed to increase the production of non-rice crops, enhance livestock and poultry farming, and improve rural infrastructure to support agricultural growth. The policy also addressed the need for irrigation systems, rural credit, and technological innovation. Moreover, establishing the Bangladesh Fisheries Development Corporation (BFDC) and promoting aquaculture led to significant growth in the fisheries sector, making

Bangladesh one of the world's leading producers of freshwater fish. Simultaneously, the private sector began playing an increasing role in agricultural development, particularly in the marketing and distribution of agricultural inputs. Particularly, the 1990s saw the rise of microfinance institutions, such as the famous Grameen Bank, which provided much-needed credit to small farmers (IFAD, 1999).

As part of the fourth Five-Year Plan between 1990–1995, domestic agricultural support was adjusted in the context of agricultural trade policy reform (GoB, 1990). For example, private companies were allowed to import fertilizer from the global market since 1992. However, controls were re-imposed in 1994 after prices of phosphate and potash fertilizers rose sharply as a result of India lifting price controls. On the other hand, duty-free imports of power tillers and tractors were allowed in 1995, and customers were given credit support. In marketing, rationing was eliminated, and public procurement also declined from 1991 to 1993. The changes had positive effects on the agricultural sector. Ahmed (1995) submits that rice production increased between 20% and 32%. At the same time, agriculture diversified into high-value crops, such as horticultural crops as well as livestock and aquaculture (MOA, 2017).

Among the significant events, aside from political conflicts, ending autocratic rule, and moving towards more democratic structures, the 1991 cyclone is most remarkable. With over 135,000 deaths and millions of people displaced, this was a catastrophic event with severe impact on agriculture and infrastructure (Oxford Research Encyclopedia, 2018; History.com, 2019).

The performance of the agricultural sector during the 1990s had improved significantly over previous periods, as growth in food production outpaced population growth and national self-sufficiency was achieved in several food commodities.

The Fourth Decade: Modernization and Mechanization

During the 2000s, the general development policies of Bangladesh focused on reducing poverty and fostering economic growth through targeted development policies. Key initiatives included investments in rural infrastructure, promoting export-oriented industries, and the further liberalization of agricultural input markets. Programs such as microfinance expanded access to credit, empowering rural populations and boosting incomes. The government also emphasized social development, achieving significant progress in health, education, and gender equality. Despite challenges like weak governance and environmental vulnerabilities, these policies contributed to a steady reduction in poverty rates and GDP growth averaging 6% annually (World Bank, 2003; Mahmood, 2021). Moreover, ready-made garment exports surged, boosting Bangladesh's GDP growth (World Bank, 2009).

The key agricultural policy includes agricultural credit and subsidies for agricultural machinery, such as tractors, harvesters, and irrigation pumps, helping reduce labor costs and improve efficiency. National Agricultural Policy (2006) focused on expanding agricultural research, enhancing rural credit systems and rural infrastructure, especially rural electrification to facilitate the use of electric pumps for irrigation. Additionally, Vision 2021 plan (GOB, 2021), developed in 2000, aimed to ensure that by 2021, the entire population's basic needs will be met sustainably.

Bangladesh's agricultural structure significantly transformed during the Fourth Decade, with increased diversification and modernization. Smallholder farming predominated, but the sector shifted towards greater use of technology, improved irrigation, and crop varieties. Growth was supported by government programs and international aid (Rahman, 2011).

Among the significant events during this time, Cyclone Sidr in 2007 caused immense devastation, affecting millions of people with significant loss of life and economic damage (Karim and Mimura, 2008). The event once again underlined the climate vulnerability (Islam *et al.*, 2022). Politically, the 2007–2008 emergency rule under a caretaker government marked efforts to curb corruption and reform electoral processes, creating a period of political uncertainty (Riaz, 2008).

As regards agricultural performance, rice production increased due to improved irrigation and high-yield varieties (Hossain, 2009a, b). Diversification took place, as seen in the increased production of high-value crops, non-crop agricultural products, and the export of various high-value crops. Gradually, agriculture shifted from subsistence farming to more commercial production due to improvements in marketing facilities. However, floods and cyclones disrupted progress (Islam and Sarker, 2010) and smallholder farmers continued to face challenges, including limited credit access and rising input costs (Rahman, 2008).

The Fifth Decade: Climate Resilience and Sustainable Agriculture

During the Fifth development decade (see Table 9.1), general development policies continued to emphasize economic growth, poverty reduction, and sustainable development. The 6th Five-Year Plan (2011–2015) focused on equitable growth and achieving the Millennium Development Goals (MDGs), while the 7th Five-Year Plan (2016–2020) aimed to align with the Sustainable Development Goals (SDGs), emphasizing inclusivity, employment generation, and resilience to climate change (GoB, 2011, 2016). Key initiatives included infrastructure, education, healthcare investments, and efforts to empower women and enhance social protection. These policies contributed to GDP growth averaging over 7% annually and significant reductions in poverty and inequality (Financial Express, 2019; GED, 2023).

Key agricultural policies aimed at increasing productivity, ensuring food security and promoting sustainability. Crop diversification was the central policy during this period with the promotion of high-value horticultural crops underscored in a seven-year plan. The Bangladesh Wheat and Maize Research Institute (BWMRI) was established in 2017 to boost wheat and maize production. Use of agricultural machinery continued to increase and it is now widely applied in land preparation, seeding/transplanting, fertilizer application, irrigation, harvesting, and post-harvest processes. The National Agriculture Policy 2018 prioritized crop diversification, efficient resource use, and increasing farm incomes (MoA, 2018). Additionally, the National Agricultural Mechanization Policy 2020 supported the adoption of modern machinery to facilitate labor transfers out of agriculture. The influence of NGOs and private-sector microfinance institutions continued to grow (Bhuiya *et al.*, 2016). Hereby, the Central Bank, has played a key role in enhancing access to rural finance through various policies. Group lending to Common Interest Groups (CIGs) and Farmer Producer Groups (FPGs) has been particularly effective.

The structure of agriculture transitioned toward diversification and modernization while retaining rice as the dominant crop. High-yielding rice varieties continued to enhance food security, contributing to near self-sufficiency. Crop diversification into high-value products, such as fruits, vegetables, and shrimp, expanded due to rising domestic demand and export opportunities. Mechanization increased significantly, particularly for irrigation and harvesting, reducing labor reliance. The agricultural sector now also emphasized agro-processing and value addition, responding to changing dietary patterns and urbanization (World Bank, 2020; Bangladesh Ministry of Agriculture, 2021).

Severe significant natural and economic events continued to affect agriculture in Bangladesh. The increasing frequency of extreme climatic events, such as floods and cyclones, caused substantial agricultural losses, particularly in coastal regions. The 2017 floods damaged crops across large swathes of farmland, impacting food security. On the economic front, government initiatives to modernize agriculture, including promoting climate-resilient seeds and expanding irrigation, helped mitigate some risks. Furthermore, investments in cold storage for perishable produce reduced post-harvest losses and increased export potential (World Bank, 2016; ADB, 2023). However, land degradation and urban expansion threatened sustainability, necessitating strategic policy interventions (World Bank, 2016).

Notable achievements marked the performance of the agricultural sector during 2010–2020. Despite a declining share in GDP, agriculture remained a cornerstone of the economy, employing around 40% of the workforce (Ahmed, 2024). Crop yields in rice, pulses, and vegetables continued to increase due to the adoption of high-yielding varieties and better farming practices (Tabassum and Rezwana, 2021). Growth in aquaculture, poultry, and dairy also contributed to agricultural diversification (Seraj, 2022). The continued development of agro-processing industries also played a crucial role, turning agriculture into a key contributor to rural employment and economic growth (WB, 2016; Ahmed *et al.*, 2020). However, land degradation, climate change impacts, and urbanization are reducing arable land (Tabassum and Rezwana, 2021).

After 2020: Post-COVID Resilience and Future Directions

After 2020, Bangladesh's development policies have focused on responding to the after-effects of Covid-19 with economic transformation, resilience to climate change, and social inclusion. The eighth Five-Year Plan (2020–2025) prioritized industrialization, infrastructure development, and fostering a green growth agenda (GoB, 2020), aligning with the Perspective Plan 2041, which envisions Bangladesh as a high-income country by 2041. The government has also emphasized improving governance, reducing poverty, and achieving sustainability through initiatives like the Delta Plan 2100, which integrates environmental concerns with development (USAID, 2020). However, challenges remain in integrating these policies effectively due to financing constraints and policy coordination (Ahmed, 2021).

The performance of the agricultural sector was hampered by the Covid-19 pandemic. Lockdowns and movement restrictions disrupted agricultural supply chains, leading to labor shortages, reduced agricultural production, and limited market access (World Bank, 2021). For example, Gatto and Islam (2021) using household panel data, showed that rice production and the share of rice sold declined. Furthermore, diversity of diet and education expenditure was reduced. Recently, by means of increased investment in agricultural modernization and digital systems in rural areas, recovery has progressed (FAO, 2021). However, there are large differences among sub-sectors with vegetables and poultry recovered wholly, the dairy sub-sector stabilized, while other sub-sectors are still struggling (Islam *et al.*, 2023).

Following this historical review of agricultural development in Bangladesh, the following is a case study on homestead aquaculture. This serves as an example of how household food and nutrition security can be improved in a country that has achieved considerable progress in spite of its vulnerability to natural disasters.

Case Study: Homestead Pond Aquaculture. Source: Author's Own Compilation

This case study on homestead pond aquaculture was carried out in mid-2016 as a PhD research work (Ahmed, 2019). Primary data from homestead aquaculture producers in Bangladesh was collected by means of a socioeconomic survey. In Bangladesh, pond culture represents the mainstay of aquaculture, accounting for more than 80% of the total recorded production and over 55% of the area under culture in 2014–15 (Shamsuzzaman et al., 2017). Pond aquaculture includes commercial aquaculture and homestead ponds. The latter captures over half of the recorded aquaculture production (DoF, 2016). In rural Bangladesh, when an owner constructs a house, the basement is higher than the surrounding area to protect the house from flooding during the rainy season. Therefore, many households have homestead areas with small or medium-sized ponds and ditches, which allow for growing fish (Belton et al., 2011; WF, 2015). Homestead ponds usually have multiple uses besides fish farming, including bathing, washing clothes, and growing crops such as vegetables on the dikes. Hence, homestead fish production is a low external input system. Production mainly depends on the natural productivity of the water body for growing fish. Most inputs are farm by-products, e.g., rice bran, rice products, and mustard oil cake as supplementary feed. The production system is polyculture with different large fish species (e.g. carp), together with a range of small indigenous species (SIS), which are especially rich in micronutrients (Castine et al., 2017). Most of the fish from homestead production is for home consumption, and any surplus is supplied to the local markets. For home consumption, fish (e.g. small fish species) are considered an important food item for low-income households when other sources of micronutrients are expensive (Islam, 2007). The study found that homestead aquaculture increases household food consumption and improves dietary diversity by generating additional cash income and stimulating higher fish consumption from home production. Moreover, aquaculture income helps poor farmers improve the quality of households' diets by purchasing more calories from the market associated with protein-rich and energy-dense food items. Results of this study suggest that homestead pond aquaculture remains an important source of protein and micronutrients despite the emergence and spread of large-scale commercial aquaculture in Bangladesh. Farmers integrate different micronutrient-rich fish species in homestead fish production for easy household access. Therefore, this study recommends that the Department of Fisheries should reconsider its view on the role of homestead pond production and give homestead ponds more recognition in its extension activities. Moreover, better market access and improved infrastructure are needed if more homestead fish producers gradually develop into a semi- or fully commercial small-scale aquaculture production system. This study has several important implications for nutrition, health, and agricultural policy in Bangladesh and other developing countries with low dietary diversity. First, aquaculture income from homestead ponds contributes to increased food consumption expenditures and calorie intake and improves dietary diversity. Second, the gender of the household head is important for household food and nutrition security because households with female heads have lower food consumption and dietary diversity. Third, farm income from crops and livestock also play an important role in fish farming households. This implies that income from homestead aquaculture is complementary to other sources of income, albeit its contribution is significant. Fourth, households with better market access derive more income from selling fish. Fifth, access to credit is important to enable fish farming households to move gradually from subsistence homestead production to more commercial aquaculture. Sixth, homestead fish production increases home consumption of fish and thereby its share in total calorie intake. Seventh, additional aquaculture income helps improve the quality of a household's diet by consuming more calories from pulses, meat, eggs, and fish. Therefore, the food consumption structure of poor farmers is diversified with high-quality protein and energy-dense food items. In conclusion, despite an increasing trend of rural households engaging in non-farm income-generating activities in Bangladesh and other developing countries, this study submits that homestead fish ponds will remain important for small and marginal farmers for many years.

Summary and Outlook

To summarize the development of the agricultural sector in Bangladesh, some basic statistics are presented. In Table 9.2, sector shares and sectoral growth rates for seven financial years (FY) between 1985 and 2023 are displayed. As per development theory principles, the share of agriculture in GDP in 2023/24 was only one third of the value in 1985/86. The shares of the service sector were always high due to Bangladesh's trade activities and services related to the export-oriented jute industry. Furthermore, growth in agriculture sector was always below those of services and especially industry.

Following the pattern of other Asian economies, the share of labor in agriculture has declined much less than the GDP share. For example, while in 2012 over 45% were engaged in agriculture, 10 years later, still close to 37% were engaged in agriculture, indicating low labor productivity (Figure 9.1). Remarkably, industry shares did not rise much since 2012 while shares of the service sector did, being almost double those of industry.

Rice remains the major crop in Bangladesh, with an almost constant area since 2015/16 of some 29 million acres or 11.6 million ha. Average yields are in the order of 3.4 t/ha, slightly over 10% higher compared with 2016 (see Table 9.3). Significant changes can be observed in maize production with an almost doubling of output, while wheat production has declined by almost 15%.

Diversification has taken place with an expansion of higher-value crops such as vegetables, fruits, and especially aquaculture products (Rahim and Moniruzzaman, 2022). By 2023, with almost 3 million metric tons, Bangladesh was the second largest producer of freshwater fish in the world, surpassing China, and just below India (Table 9.4). In terms of vegetable production, Bangladesh ranked third globally and also fruit production has grown considerably since 2016.

Table 9.2. Sectoral Shares in GDP and Growth Rates from 1985 to 2023.

Period	1985–86	1995–96	2005–06	2010–11	2016–17	2020–21	2023–24*
Sector				**Sector Shares**			
Agriculture	31.15	25.68	19.01	18.01	13.62	12.07	11.02
Industry	19.13	24.87	25.4	27.38	32.98	36.01	37.95
Service	49.73	49.45	55.59	54.61	53.4	51.92	51.04
				Growth Rates			
Agriculture	3.31	3.1	5.5	4.46	3.2	3.17	3.21
Industry	6.72	6.98	9.8	9.02	8.27	10.29	6.66
Service	4.1	3.96	6.6	6.22	6.37	5.73	5.8
GDP**	3.34	4.47	7.18	6.64	6.54	7.00	5.83

Source: Bangladesh Bureau of Statistics (BBS). *Provisional; **at producer prices.
Note: Data up to financial year (FY) 2000–01 are based on FY 1995–96, data from FY 2005–06 to FY 2014–15 are based on 2005–06, and data since FY 2016–17 are based on FY 2015–16.

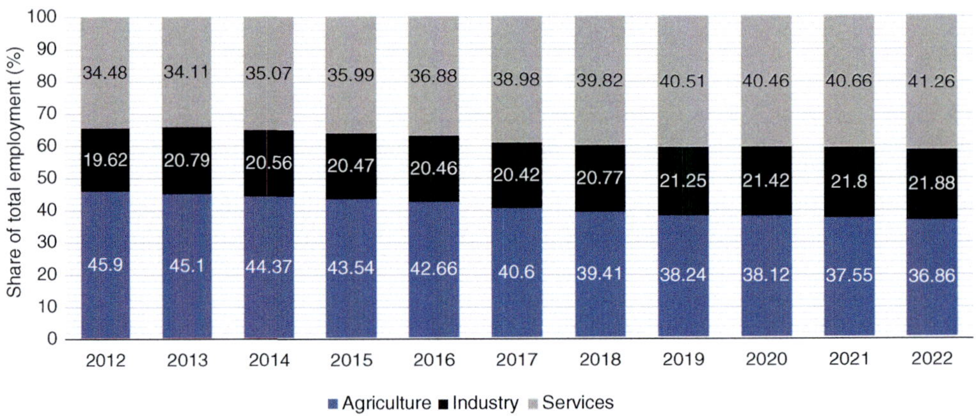

Figure 9.1. Distribution of Employment by Economic Sector. Source: World Development Indicators, World Bank.

Table 9.3. Food Grains Production During 2015–23 in Bangladesh.

	Rice			Wheat	Maize
Crop Year	Area (Thousands of acres)	Production (in lakh MT)	Yield (kg/acre)	Production (in lakh MT)	Production (in lakh MT)
2015–16	28124	347.10	1234	13.48	27.59
2016–17	27184	338.04	1244	13.12	35.78
2017–18	28699	362.79	1264	10.99	38.93
2018–19	28455	374.64	1220	11.48	46.99
2019–20	28800	386.95	1280	12.46	54.02
2020–21	28912	376.08	1301	10.85	56.63
2021–22	28891	391.80	1320	10.86	56.30
2022–23	28754	391.02	1362	11.60	57.68

Source: Bangladesh Bureau of Statistics (BBS), Ministry of Agriculture.

Table 9.4. Production of Vegetables, Fruits, and Aquaculture since 2015/16.

Crop Year	Vegetables (Thousands of MT)	Fruits (Thousands of MT)	Aquaculture (Thousands of MT)
2015–16	3892	4807	2204
2016–17	4076	5068	2333
2017–18	4121	4948	2405
2018–19	4336	4948	2488
2019–20	4575	4973	2583
2020–21	4814	5143	2639
2021–22	5010	5334	2731
2022–23	5365	5659	2852
2023–24*	–	–	2988

Source: Department of Fisheries, Ministry of Fisheries and Livestock. *Projected.

An important feature of structural transformation is the change in average farm size and its distribution over different size categories. Despite rural urban migration, farm sizes in Bangladesh have remained small since 2005 as a result of population growth and land fragmentation (GED, 2014; World Bank, 2016). Table 9.5 presents the distribution of cultivated land in rural areas. The data reveal that the percentage of landless households has been increasing. Also, the vast majority of households, i.e. almost 90% operate on less than 2.49 acres of land, i.e. about 1 ha. Less than 1% of households operate more than 3 ha. These figures basically did not change between 2005 and 2022 (see Table 9.5). This shows that dominantly rural households in Bangladesh cannot live on farming alone but must increasingly complement their household income by non-farm sources. In effect, farmers in Bangladesh today, are part-time farming households. Increasingly, agriculture plays the role of a safety net for a large number of rural households.

Table 9.5. Land operated by rural households, 2005–2022, in %, by size category (acres).

Farm Size (in acres)	2005	2010	2016	2022
Landless	4.1	3.6	6.4	5.0
0.01–0.49	52.9	55.4	64.5	62.3
0.50–0.99	14.2	14.2	13.1	13.8
1.00–2.49	19.1	18.3	12.4	13.9
2.50–7.49	8.6	7.8	3.1	4.1
>7.50	0.9	0.7	0.6	0.8

Source: Household Income and Expenditure Survey in Bangladesh, Bangladesh Bureau of Statistics (BBS), different years.

Clearly, the emergence of large-scale or corporate farming, as often aimed for by policymakers and scholars, so far, has not taken place in Bangladesh. The agricultural sector is predominantly composed of smallholders and subsistence farmers. Structural challenges, including fragmented landholdings due to institutional constraints, imperfect land rental markets, cultural attachment to land and lack of sufficiently secure non-farm alternatives, limit the emergence of commercial, professional and larger-scale farming. Notable exceptions are contract farming and private-sector investments in aquaculture and poultry farming (World Bank, 2016). However, to date, these are a small share of the overall agricultural sector in Bangladesh. On the other hand, small-scale and subsistence farming does not mean backwardness and traditional farming. Mechanization has advanced significantly, with widespread adoption of power tillers, irrigation pumps, and threshers, reducing reliance on manual labor. Investments in high-yielding varieties and irrigation have contributed to productivity gains in some regions, although disparities are large (GED, 2014; World Bank, 2016).

Generally, rural households in Bangladesh struggle to sustain livelihoods through farming alone. Commonly, they supplement incomes through off-farm employment and remittances, particularly from domestic urban migrants and overseas workers. Thus, remittances have become a critical part of rural household income, resembling Southeast Asian trends (GED, 2014; IOM, 2018).

While agricultural transformation continues to take place, challenges like climate vulnerability, land fragmentation, and reliance on low-productivity farming practices are still widespread. Off-farm income and remittances are playing increasing roles in rural livelihoods. To facilitate structural transformation, a comprehensive rural development strategy is needed in order to foster diversification and modernization of the rural and peri-urban space. Such a strategy must do both, modernize agriculture to facilitate the gradual development of full-time, commercial farming and, at the same time, the generation of rural non-farm employment through rural industrialization, for example, agro-processing. In this regard, education will be crucial so that those who engage in part-time farming or exit farming can find decent jobs that can sustain their livelihoods. It is clear that the transition towards an agricultural system where professional and commercial farmers form the majority of farmers in Bangladesh still has a long way to go. Smallholders, with multiple income sources, simultaneously engaged in agriculture and non-agriculture will be the dominant system for another generation at least. Good policies that create an enabling institutional setting together with well-targeted investment in climate-resilient and digital technologies that at the same time ensure the inclusivity of smallholder farmers and open up opportunities for larger farms, can effectively guide this transition.

References

Ahmed, R. (1996) Agricultural market reforms in South Asia. *American Journal of Agricultural Economics* 78(3), 815–819.

Ahmed, A.U. (2024) *Agricultural Growth Key to Accelerated Poverty Reduction in Bangladesh*. The Financial Express, Dhaka.

Ahmed, B. (2019) *The Economics of Aquaculture Production and Consumption Emphasizing the Nutritional Benefit of the Poor in Bangladesh*. Gottfried Wilhelm Leibniz Universität Hannover, Germany.

Ahmed, M., Albert, E.J. and Wanjiru, G. (2020) *Bangladesh – Promoting Agri-Food Sector Transformation in Bangladesh: Policy and Investment Priorities (English)*. World Bank Group, Washington, D.C. Available at: http://documents.worldbank.org/curated/en/475301590995321523/Bangladesh-Promoting-Agri-Food-Sector-Transformation-in-Bangladesh-Policy-and-Investment-Priorities (accessed 30 March 2025).

Ahmed, R. (1995) Liberalization of agricultural input markets in Bangladesh: Process, impact, and lessons. *Agricultural Economics* 12(2), 115–128.

Ahmed, S. (2021) Ensuring sustainability: Bangladesh's need for a green growth strategy. *Policy Insights*. Available at: https://policyinsightsonline.com/2021/09/ensuring-sustainability-bangladeshs-need-for-a-green-growth-strategy/ (accessed 30 March 2025).

Alauddin, M. and Tisdell, C. (1991) The 'Green Revolution' and labour absorption in Bangladesh agriculture: The relevance of the East Asian experience. *The Pakistan Development Review* 30(2), 173–188.

Bakht, Z. (2001) Trade liberalization, exports and growth of manufacturing industries in Bangladesh. In: Huq, M.M. and Love J. (eds.) *Strategies for Industrialization: The Case of Bangladesh*. University Press Ltd, Dhaka.

Bangladesh Ministry of Agriculture (2021) *Bangladesh Economic Review 2021: Agriculture Chapter*. Bangladesh Ministry of Agriculture.

Baten, A., González, P.A. and Delgado, R.C. (2018) Natural disasters and management systems of Bangladesh from 1972 to 2017: Special focus on flood. *OmniScience: A Multi-disciplinary Journal* 8(3), 35–47.

Belton, B., Karim, M., Thilsted, S., Collis, W. and Phillips, M. (2011) *Review of aquaculture and fish consumption in Bangladesh. Studies and Reviews 2011–2053*. The WorldFish Center, Penang, Malaysia.

Bhuiya, M.M., Khanam, R. and Rahman, M.M. (2016) Microfinance operations in Bangladesh: An overview. *Journal of Applied Business and Economics* 18(3), 52–26.

Castine, S.A., Bogard, J.R., Barman, B.K., Karim, M. *et al.* (2017) Homestead pond polyculture can improve access to nutritious small fish. *Food Security* 9(4), 785–801.

Crow, B. (1984) Warnings of famine in Bangladesh. *Economic and Political Weekly* 19(40), 1754–1758.

Department of Fisheries (DoF) (2016) Fisheries statistics in Bangladesh: Issues, challenges and plan. In: *Paper Presented at the Meeting of Asia and Pacific Commission on Agricultural Statistics, Thimphu, Bhutan, February*. Department of Fisheries (DoF), Dhaka.

Dorosh, P. and Valdès, A. (1990) *Effects of Exchange rate and Trade Policies on Agriculture in Pakistan, Research Report 84*. International Food Policy Research Institute, Washington, DC.

FAO (2021) *Agricultural Transformation in Asia – Policy and Institutional Experiences*. FAO, Bangkok. DOI: 10.4060/CB4946EN

Food and Agriculture Organization (FAO) (2021) *COVID-19 in Bangladesh*. Available at: FAO (accessed 30 March 2025).

Gatto, M. and Islam, A.H.M.S. (2021) Impacts of COVID-19 on rural livelihoods in Bangladesh: Evidence using panel data. *PLoS ONE*, 16(11), e0259264.

GED (General Economics Division), Bangladesh Planning Commission (2014) Achievements and challenges in agriculture under the Sixth Five-Year Plan. Available at: GED KPGEDKP (accessed 22 November 2024).

GED Knowledge Portal (2023) *End Evaluation of the 7th Five-Year Plan*. General Economics Division, Government of Bangladesh, Dhaka, Bangladesh.

GoB (1973) *The First Five Year Plan: 1973–78*. Dhaka: Planning Commission.

GoB (1980) *The Second Five Year Plan: 1980–85*. Dhaka: Planning Commission.

GoB (1985) *The Third Five Year Plan: 1985–90*. Dhaka: Planning Commission.

GoB (1990) *The Fourth Five Year Plan: 1990–95*. Dhaka: Planning Commission.

GoB (2011) *Sixth Five Year Plan: FY2011–FY2015*. Dhaka: Planning Commission.

GoB (2016) *Seventh FIVE YEAR PLAN: FY2016–FY2020*. Dhaka: Planning Commission.

GoB (2020) *Eighth Five Year Plan (July 2020–June 2025)*. Dhaka: Planning Commission.

GoB (2021) Available at: https://bhtpa.portal.gov.bd/sites/default/files/files/bhtp (accessed 1 June 2025).

Gotur, M.P. (1991) Bangladesh: Economic Reform Measures and the Poor. *IMF Working Papers 1991, 039*.

History.com Editors (2019) 'Bangladesh cyclone of 1991', History, 29 April. Available at: https://www.history.com/topics/natural-disasters-and-environment/bangladesh-cyclone-of-1991 (accessed 22 November 2024).

Hossain, M. (2009a) *Agricultural Development in Bangladesh: Issues and challenges*. University Press Limited, Dhaka.

Hossain, M. (2009b) *Rice Research and Development in Bangladesh: Progress and Challenges*. International Rice Research Institute, Los Baños, Philippines.

IFAD (1999) *The Role of Microfinance in Rural Development in Bangladesh*. IFAD, Rome, Italy.

IOM (International Organization for Migration) (2018) *The Role of Migration in Bangladesh's Rural Economy*. IOM Publications, Geneva, Switzerland.

Islam, F. (2007) Self-recruiting species (SRS) in aquaculture: Their role in rural livelihoods in two areas of Bangladesh. Unpublished doctoral dissertation, University of Stirling, Scotland.

Islam, M.S. and Sarker, M. (2010) Natural disasters and their impact on agricultural development in Bangladesh. *Journal of Environmental Studies* 25(3), 45–58.

Islam, M.S., Samreth, S., Islam, A.H.M.S. and Sato, M. (2022) Climate change, climatic extremes, and households' food consumption in Bangladesh: A longitudinal data analysis. *Environmental Challenges* 7, 100495.

Islam, M.S., Hossain, M.E., Bekun, F.V. and Sujan, M.H.K. (2023) Impact of COVID-19 on Bangladesh's agriculture sector and the ways forward to recovery: An overview. *Journal of Public Affairs* 23(2), e2862.

Karim, M.F. and Mimura, N. (2008) Impacts of climate change and sea-level rise on cyclonic storm surge floods in Bangladesh. *Global Environmental Change* 18(3), 490–500. DOI: 10.1016/j.gloenvcha.2008.05.002.

Lewis, S. and Guisinger, S.E. (1971) The structure of protection in Pakistan in the structure of protection in developing countries. In Balassa, B. (ed.) *The Structure of Protection in Developing Countries*. Johns Hopkins Press, Baltimore.

Longshore, D. (2007) *Encyclopedia of Hurricanes, Typhoons, and Cyclones* (2nd edn). Facts on File Inc. p. 215. ISBN 978-0-8160-6295-9. New York, NY.

Mahmood, A. (2021) Bangladesh's remarkable development journey: Government had an important role too. Available at: https://www.brookings.edu/articles/bangladeshs-remarkable-development-journey-government-had-an-important-role-too/ (accessed 22 November 2024).

Ministry of Agriculture, Bangladesh (2017) *Bangladesh Economic Review 2017: Agriculture Chapter*. Available at: Ministry of Agriculture Portal Ministry of Finance Bangladesh, Dhaka (accessed 22 November 2024).

Ministry of Agriculture, Government of Bangladesh (2018). National Agriculture Policy. Available at: https://www.bluegoldwiki.com/images/1/14/GoB_National_Agricultural_Policy_2018.pdf (accessed 22 November 2024).

Oxford Research Encyclopedia of Natural Hazard Science (2018) *Livelihoods in Bangladesh Floodplains*. Oxford University Press, Oxford, UK. Available at: https://oxfordre.com/naturalhazardscience/article/10.1093/acrefore/9780199389407.013.462 (accessed 22 November 2024).

Rahim, M.A. and Moniruzzaman, M. (2022) Fruit production in Bangladesh – a unique contribution to nutritional food security, household income, poverty reduction, women participation, biodiversity, sustainable development, environmental protection and immunity enhancement to manage COVID-19. In: *Proceedings of the International Symposium on Tropical Fruits*. International Tropical Fruits Network (TFNet), Selangor, Malaysia.

Rahman, A. (2011) *The Structure and Dynamics of Agriculture in Bangladesh*. Agribusiness Institute Press, Dhaka.

Rahman, S. (2008) Determinants of agricultural productivity and smallholder development in Bangladesh. *Agricultural Economics Research Review* 21, 55–68.

Rahman, S.H. (1994) *The Impact of Trade and Exchange Rate Policies on Economic Incentives in Bangladesh Agriculture*. Working Paper No. 8, International Food Policy Research Institute, Washington DC.

Riaz, A. (2008) *Inconvenient Truths about Bangladesh's Democracy*. Palgrave Macmillan, New York.

Seraj, S. (2022) 'Role of agriculture in Bangladesh's economic growth', The Daily Star. Available at: https://www.thedailystar.net/recovering-covid-reinventing-our-future/blueprint-brighter-tomorrow/news/role-agriculture-bangladeshs-economic-growth-2960736 (accessed 30 March 2025).

Shamsuzzaman, M.M., Islam, M.M., Tania, N.J., Al-Mamun, M.A., Barman, P.P. *et al.* (2017) Fisheries resources of Bangladesh: Present status and future direction. *Aquaculture and Fisheries* 2(4), 145–156.

Sultana, P. and Thompson, P. (2017) Livelihoods in Bangladesh floodplains. In: *Oxford Research Encyclopedia of Natural Hazard Science*. Oxford University Press, Oxford, UK.

Tabassum, N. and Rezwana, F. (2021) Bangladesh agriculture: A review of modern practices and proposal of a sustainable method. *Engineering Proceedings* 11(1), 12. DOI:10.3390/ASEC2021-11190.

The Financial Express (2019) Bangladesh Development Perspectives: Planning Strategies. Available at: https://thefinancialexpress.com.bd/views/analysis/bangladesh-development-perspectives-planning-strategies-1575038474 (accessed 30 March 2025).

USAID (2020) Country Development Cooperation Strategy (CDCS) December 2020 – December 2025. Available at: https://www.usaid.gov/bangladesh/cdcs/pdf (accessed 30 March 2025).

World Bank (2003) Bangladesh – Development Policy Review: Impressive Achievements but Continuing Challenges. World Bank, Washington DC.

World Bank (2004) *Trade Policies in South Asia: An Overview (Volume II: An Overview)*. World Bank, Washington DC.

World Bank (2009) *Bangladesh: Country Economic Memorandum – Unlocking the Potential of the Garment Sector*. World Bank, Washington DC.

World Bank (2016) Bangladesh: Growing the economy through advances in agriculture, D.C. World Bank Group. Available at: https://www.worldbank.org/en/results/2016/10/07/bangladesh-growing-economy-through-advances-in-agriculture (accessed 30 March 2025).

World Bank (2020) Promoting agrifood sector transformation in Bangladesh: Policy and investment priorities. Available at: World Bank Document (accessed 30 March 2025).

World Bank (2021) 'Bangladesh economy shows early signs of recovery amid uncertainties'. Available at: World Bank (accessed 30 March 2025).

WorldFish (WF) (2015) Aquaculture for improved income, food security, and livelihood: Initiative and success. CSISA Performance Report, Dhaka, Bangladesh.

10 Agricultural Development and Policy in Indonesia since Independence with a Case Study of the Oil Palm Sector

Eko Ruddy Cahyadi and Ernah

Abstract

This chapter reviews the evolution of agricultural development, corresponding policy and policy changes in Indonesia. The time frame ranges from independence in 1945 to the present. It traces key policy phases, including post-independence food security efforts, the Green Revolution under the New Order, and the decentralization and liberalization of the Reform period. Recent policies have emphasized sustainable practices, digital transformation, and global market competitiveness, reflecting a shift from food security to food sovereignty. The chapter also addresses emerging challenges and their implications for future directions, including climate change, the digitalization of agriculture, and possible disruptions in global supply chains due to geopolitical tensions. A case study on the oil palm industry highlights the sector's significance to Indonesia's economy, while examining its environmental impacts and social complexities. Overall, the chapter offers insights into Indonesia's agricultural trajectory, highlighting policy directions that prioritize innovation, resilience, value added and for a sustainable future of Indonesia's agricultural sector in a complex global landscape.

Introduction

Indonesia is the largest economy in Southeast Asia and is one of the emerging market economies. On the global level Indonesia is the world's 17th largest economy by nominal GDP and the 7th largest in terms of Purchasing Power Parity (PPP) GDP. Agriculture has long been a cornerstone of Indonesia's economy, contributing significantly to employment, food security, and rural livelihoods. As of 2022, agriculture accounts for approximately 13.5% of Indonesia's Gross Domestic Product (GDP) and employs about 30% of the nation's workforce (Badan Pusat Statistik [BPS], 2023). Agriculture is not only essential for ensuring national food security but also plays a crucial role in alleviating poverty and supporting rural economies. However, Indonesia's agricultural landscape has undergone significant transformations, driven by shifts in policy, environmental pressures, and global economic changes (Saragih, 2001). Understanding the dynamics and evolution of agricultural development and policy over time is essential to learn from past successes and challenges, shaping informed strategies for the future of agriculture in Indonesia.

In Table 10.1, a historical framework is presented that can facilitate the analysis of agricultural transformation in Indonesia. By connecting key elements of general development policy, with agricultural policy, together with a description of the structure of the sector and concluding its performance by taking into account significant events, a quick overview of the development decades is achieved.

Starting out with a short overview upfront, it can be stated that agriculture has undergone

 Asian Agriculture in a Climate Change-driven and Risky World (H. Waibel, S. Liebenehm and Ernah eds)
DOI: 10.1079/9781800627338.0010

drastic changes since 1945. In the early years after Indonesia's independence, productivity was low and food shortages were common. By the 1970s–1980s, Green Revolution technologies were promoted and officially Indonesia achieved the politically proclaimed goal of self-sufficiency in rice which came at high social and environmental costs. Furthermore, during the 1990s–2000s, palm oil plantations, supported by government incentives and transmigration programs, became a landmark of Indonesian agriculture, making palm oil Indonesia's top commodity. However, this growth came with high costs to society, e.g. deforestation and land-use conflicts. In response to agriculture's social and environmental problems, during the second decade of the 21st century, the government gave higher priority to sustainability considerations with stricter regulations and economic incentives such as carbon credit programs in line with the SDGs of the United Nations.

For the future it can be expected that agriculture will continue to modernize, with more information technology and even AI powered autonomous farming, biotech crops, and block chain-integrated traceability systems in the commercial farming sector. On the other hand, economic and environmental risks together with constraints to rapid structural transformation, mean that the vast majority of small-scale farmers are actually part-time farmers with a growing share of non-farm income and agricultural income as a basis, backup and safety net.

Indonesia's agricultural policy over time has experienced shifts in national priorities, economic goals, and external influences. Each era brought a distinct focus, i.e. from food security and rice self-sufficiency due to productivity gains with Green Revolution technology adoption, from market liberalization and regional autonomy to the current emphasis on sustainability, value addition, and global competitiveness. This evolution has been shaped by both domestic needs and global trends.

A closer examination of these eras is given in the text following the summary presented in Table 10.1.

Post-Independence Era (1945–1960s): Old Order: Food Security, Self-Sufficiency, and Controlled Trade

In historical periods, productivity of agriculture in Indonesia was relatively high providing a good basis for the necessary growth in food production after independence (Maat, 2014). Yet, Geertz (1963) in his popular book *Agricultural Involution: The Process of Ecological Change in Indonesia*, was quite pessimistic about the ability of the country to feed its rapidly growing population in the post-colonial period. Although self-reliance in staple foods as a means of economic stability and national sovereignty was a policy priority during the Sukarno government, success was limited (Booth, 1989). From 1960–1965, rice output per capita fell as population increased by more than 10% while production was static. Land reform, implemented to reduce disparities in land ownership and provide smallholders with secure land tenure, faced bureaucratic inefficiencies and strong resistance from the politically influential landowners. Price controls and input subsidies in rice production were introduced but were insufficient to significantly improve productivity (Van der Eng, 2000; Nawiyanto, 2017). Rice yields remained low, averaging 1.5 tons per hectare during this period (BPS, 2023). Similarly, trade policies were aligned with the goal of food self-sufficiency. The government exercised strict control of imports and exports, even for non-food crops, such as rubber. Another constraint was the conflict of central versus local governments in implementing agricultural development programs. Furthermore, agricultural productivity growth immediately after independence was hampered by the war against the Dutch who tried to re-establish colonial power after World War II. Furthermore, during the conflict with Malaysia from 1963 to 1966, the government encouraged consumers to substitute domestically produced maize for imported rice to manage the country's foreign exchange constraints.

The Green Revolution and New Order Era (1960s and 1970s): Productivity, Self-Sufficiency, and Managed Trade

When Sukarno, Indonesia's first president after independence, had been replaced by Suharto, the "New Order Regime" was introduced. Hence, economic policy focused on economic growth and was shaped by a group of American-educated Indonesian economists ("*Berkeley Mafia*"). Subsidies were cut and hyperinflation was brought under control. Food aid from the USA

Table 10.1. Development Overview by Decades. (Authors' own table).

Decades	General Development Policy	Agricultural Policies	Structure of Agriculture	Significant Events	Performance of Agriculture
1945–1960	Nation-building, economic recovery; post-independence, nationalization of foreign assets	Land reform with food security focus, mainly in rice	Traditional, smallholder subsistence- agriculture	Independence; Dutch-Indonesian War 1945–1949, First Agrarian Law, 1950	Low productivity, food shortages, dependence on imports, risk of famine
1960s and 1970s	New Order Regime; economic stabilization, transmigration program intensified	First Five-Year Plan, food price controls; Green Revolution programs, BIMAS and INMAS; infrastructure investment; input subsidies;	Dual structure of agriculture; small-scale farming in rice and other food crops; large-scale farming in plantation crops	US Food Aid; peak in oil prices generate high government revenues; growth of palm oil plantations	Starting out with shortages of rice, near self-sufficiency achieved by end of 1970s
1980s	Industrialization, Export-oriented growth of manufactured goods	Further intensification of rice production, transmigration program, export-oriented agriculture	Expansion of agricultural land use at expense of forest land; commercialization of agriculture	Pest outbreaks in rice; World Bank and ADB funding for palm oil plantations	Strong productivity increase; environmental concerns emerge
1990s	Economic liberalization, infrastructure expansion	Reduced public investment in agriculture; reduced fertilizer subsidy, abolished pesticide subsidy; shift towards agribusiness and export markets	Dual structure maintained; small scale in rice and large-scale in plantation crops	Oil palm boom; El Niño causing drought; Asian Financial Crisis in 1997	Productivity growth slowing down, environmental externalities occur
2000s	Adjusting Fiscal and Monetary Policy; Stabilization; Decentralization of governance, export-oriented policies	Certification schemes in palm oil; promotion of biofuel, protection of domestic rice market, high food prices	Large-scale plantation expansion, smallholder integration into supply chains	Tsunami in Aceh, 2004	Strong palm oil production growth, deforestation
2010s	Continue sustainable development agenda with policy reforms and trade agreements; foster digital economy growth	Moratorium on new palm oil plantations; more enforcement of sustainability standards, introduction of digital technologies	More corporate-smallholder partnerships; dualistic structure remains	EU trade restrictions for palm oil; litigation process under WTO; Covid-19	Moderate productivity growth, constrained by sustainability standards
2020s	International Economic Openness including investments by China; continue digital transformation, and green economy policy	Agro-industry development and smart agriculture in Medium-Term Development Plan 2020–2024	Information technology in large plantations, digital platforms for smallholders	Emergence of AI-driven agriculture, expansion of traceability in supply chains	Gradual shift to precision farming, rise in certified sustainable palm oil

was received in large quantities, helping to prevent famine. In agriculture, a *Green Revolution* program, especially for rice, with massive investment in irrigation which allowed to introduce double cropping and planting IR8, a high-yielding semi-dwarf rice variety of the International Rice Research Institute (IRRI) was introduced, aimed to achieve self-sufficiency in rice (Hansen, 1972). The area served by technical and semi-technical irrigation increased by nearly 40% from 1966 to 1989. Under the so-called BIMAS (mass guidance) program, farmers received subsidized seeds, fertilizers, pesticides, and access to credit, which contributed to significant productivity gains (Timmer, 2004). The first version of the program, BIMAS Gotong Royong (mutual self-help), relied on multinational and local agrochemical companies to deliver modern inputs to farmer groups financed by government credit. The input packages were not specific to local conditions with the result that impact on productivity was negligible. An "improved" BIMAS program established village-level entities to provide fertilizers, credit, technical help, and inputs to farm households. The government set floor prices linked to the cost of farm inputs to incentivize intensification, and managed food imports through the state buffer stock agency BULOG. Rice yields doubled from less than 2 tons per ha to 4 tons to the mid-1980s, and self-sufficiency in rice was briefly attained (Timmer, 2004).

Export-oriented Growth, Industrialization, Monetary Expansion and Financial Crisis (1980s and 1990s)

President Suharto was awarded the *Ceres Medal* by the FAO, in recognition of the country's journey from import dependence to rice self-sufficiency. However, productivity growth slowed, partly as a result of pest infestations caused by overuse of agrochemicals. In response, the government eliminated pesticide subsidies, banned some broad-spectrum insecticides and promoted integrated pest management in the main rice producing areas (Theirs, 1997; Praseyoto *et al.*, 2012). Nevertheless, rising demand associated with population growth and slower growth of cropping area and intensity resulted in a steady rise in imports, culminating in record imports of nearly 5 million tons in the wake of the 1999 El Niño drought.

The decade also saw the emergence of palm oil as the dominant export crop, largely at the expense of forest destruction. Transmigration programs helped to expand oil palm farming in Sumatra. However, in general, these policies benefited larger landholders who had easier access to credit and inputs than smallholder farmers. Trade policies focused on import substitution and controlled exports. Reliance on rice imports was reduced while the export of cash crops such as palm oil, rubber, coffee, and spices was promoted, yet regulated, to avoid domestic shortages.

Post Asian Financial Crisis (until 2010): Decentralization, Market Liberalization, Regional Autonomy, and Open Trade

The Asian financial crisis had a dramatic impact on the Indonesian economy because it was accompanied with a political crisis. The 30 years of Suharto rule came to an end and transition to a broader based democratically legitimized government system took place. In economics, the key factor was the return to macroeconomic discipline and to maintain an open economy policy with more careful financial management (Hill and Shiraishi, 2007). The centralized model of government gradually transited into decentralization and more regional autonomy with local governments given increased authority over agricultural planning and tailoring policies to regional needs (Firman, 2009).

As part of broader structural adjustment programs, Indonesia continued to reduce state intervention, liberalized agricultural markets, and encouraged foreign investment. Subsidies for inputs like fertilizers were scaled back, exposing smallholders to market volatility and reducing their access to affordable inputs. Economic policy facilitated foreign investment in cash crop sectors such as palm oil and cocoa. These policies catalysed growth in agribusiness, particularly in palm oil production, which saw a 10% annual growth rate in the area planted by 2010 (BPS, 2023).

The liberalization policy shift was not without consequences. While export-oriented crops like palm oil and cocoa saw rapid growth, the reduction in subsidies left smallholders vulnerable, and income disparities widened across regions, as shown by an increase in the Gini coefficient from 0.309 in 1998 to 0.389 in 2013 (World Bank, 2024). Regional autonomy allowed some areas to flourish, but inconsistent policy implementation and varying resource levels led to uneven development (Suryahadi *et al.*, 2009). For example, resource-rich provinces like East Kalimantan and Jakarta showed per capita incomes significantly higher than the national average while the eastern Indonesian provinces like East Nusa Tenggara were falling behind (Vidyattama, 2010).

International trade policies during this period were oriented towards open trade and export expansion. The government reduced trade barriers, aligning with IMF recommendations, and prioritized cash crop exports to support economic recovery. This openness positioned Indonesia as the world's largest palm oil exporter, contributing significantly to rural economic growth. However, the expansion of palm oil brought environmental and social challenges, such as deforestation and conflicts over land rights, causing increasing societal debates (Obizinsky *et al.*, 2012; Qaim *et al.*, 2020, Phami *et al.*, 2023). Stakeholders in this era included regional governments, empowered by decentralization, and foreign investors, who became key players in agribusiness. Also, NGOs gained prominence, advocating for smallholders affected by liberalization and highlighting environmental and social impacts (e.g. Laraswati *et al.*, 2022).

A major event during this period was the tsunami in the province of Aceh in 2004, with the loss of some 165,708 lives. This catastrophic event had long-term economic effects for this politically sensitive region (Vidiyattama *et al.*, 2021). Furthermore, the 2006 Yogyakarta-Central Java earthquake claimed 5,716 lives, with material losses reaching US$4.4 million and US$3 million, respectively. Also, the 2018 earthquake in Palu, Central Sulawesi, and the Lombok earthquake in West Nusa Tenggara caused losses of up to 18.4 trillion rupiah and 7.45 trillion rupiah, respectively (Musyarofah and Indira, 2022). Nearly half of Indonesia's deforestation from 2000 to 2010 took place within concessions for oil palm, timber, logging, and mining, while around 2.3 million hectares of peatlands in Sumatra and Kalimantan were transformed into industrial plantations – most of it after 2000 – often using fire as a low-cost land clearing method (Marlier *et al.*, 2015).

The Sixth Development Decade (2010–2020): Digitalization, Global Competitiveness, and Covid-19

Until the global pandemic had fully hit Indonesia in late 2019, the rather stable annual GDP growth of around 5% had continued from the decade before. After 2010, the country further improved its positive investment climate for both domestic and foreign investors. Policy reforms focused on simplifying regulatory procedures and further consolidating the banking system. Also, trade agreements were entered. Prominent examples are the ASEAN-Australia-New Zealand Free Trade Area (AANZFTA), the Indonesia-EU Comprehensive Economic Partnership Agreement (IEU-CEPA), and the Regional Comprehensive Economic Partnership (RCEP), which includes all ASEAN countries plus Australia, China, Japan, South Korea and New Zealand. Furthermore, the government implemented a digital technology push by launching Agriculture 4.0 with better internet coverage and promoting low-cost devices, which has led to an expansion of online activities. As noted by Jurriens and Tapsell (2017), digital technologies in Indonesia are spreading fast in all strands of life, with an increasing share of the population using social media and internet-based transport systems like Gojak and Grab, for example. Agriculture is no exception, with considerable opportunities, but also challenges, as shown in a review article by Deichmann *et al.* (2016). The digital transformation of agriculture in Indonesia has started, especially in the supply chains and among large-scale producers while opportunities among smallholders are yet to be exploited and several constraints must be overcome (e.g. Johan *et al.*, 2024).

In agricultural policy, a moratorium on new palm oil plantations was established which restricts new concessions in forests and peatlands (Presidential Instruction No. 6/2013). Hereby, the Indonesian Government demonstrated that it wants to give a higher weight on environmental protection. In the still dominant smallholder sector, policies on sustainable practices such as the sustainable rice platform (SRP) with the promotion of climate-change-adapted new varieties, water-saving technologies as well as partnerships between smallholders and marketing corporations, for example in vegetables (Widadi *et al.*, 2021) were endorsed. The period was also marked by trade issues with the EU, in relation to the EU's revised renewable energy policy of 2016. While Indonesia brought the case for litigation at the WHO, the government, with Presidential Regulation No. 44 incentivized investments in domestic capacities of palm oil processing.

Overall, since the 2010s, Indonesia's agricultural policy has focused on sustainability, value addition, and global competitiveness without neglecting food security (Limenta and Chandra, 2017) but with less flexibility under WTO's Agreement on Agriculture (AoA) as pointed out by Thow *et al.* (2019). Agricultural growth during this period was steady but modest, compared with the previous phase, except for the Covid-19 event. By 2020, GDP share of agriculture was 13.7%, almost the same as 10 years earlier. However, labor shares declined from 39% in 2010 to 29.7% in 2020.

After 2020

Indonesia's economy recovered rapidly from the Covid-19 shock and soon reached the GDP growth rates of around 5% of the pre-Covid-19 period. In 2024, the "Jokowi government" was replaced in democratic elections with President Prabowo, who basically continued the economic policy of his predecessor. To realize Indonesia's economic potential on its pathway to become an industrialized high-income country by 2040, foreign capital inflow is maintained. Joint ventures with Chinese capital, as evidenced, for example, by the US$7 billion "Whoosh" fast train system, continue to expand. Domestically, social programs such as a costly nationwide free school feeding program are going to increase public debt beyond 3% of GDP. Advancements in the digitalization, and, at the same time, greening, of the economy, are being continued. A prominent example is the development of Indonesia's new capital, Nusantara in East Kalimantan. It is a model for sustainable urban development, with advanced digital technologies, renewable energy, and climate-resilient arrangements, with the aim to become carbon-free by 2045.

As regards agricultural policies, food security, agricultural competitiveness, farmer empowerment, land use protection, and sustainable practices are still priority areas. For example, the widely popular Farmer Field School model, first introduced in Indonesia in the late 1980s (van der Fliert, 1993) is still relevant today, as shown in a study of Mariyono *et al.* (2021) who found that successfully empowered farmer communities augment livelihoods. Also, "Agriculture 4.0," launched in early 2018, with the aim to integrate information technology, was intensified. Examples are, the use of IoT sensors for monitoring soil conditions and plant health, drone applications for crop mapping and input application, blockchain technologies for transparency in supply chains and AI-powered tools for data analysis and forecasting. While in Indonesia's still vast small-scale farming sector, digitalization of agriculture is still in its infancy, it has advanced rapidly in large oil palm plantations (Akhatar *et al.*, 2023). The National Medium-Term Development Plan (RPJMN) 2020–2024 identifies agro-industrial development as a driver for rural economic growth, for downstream processing of commodities like palm oil, rubber, cocoa, and coffee (Ministry of National Development Planning, 2020). Recently, Indonesia has moved from exporting primarily crude palm oil to producing and exporting higher-value products such as oleochemicals, biofuels, and other derivatives (World Bank, 2021; Ernah *et al.*, 2024) with considerable success. For example, in 2021, the export value of processed palm oil products reached $20 billion, compared with $8 billion in 2015 (Badan Pusat Statistik [BPS], 2022).

On the downside, extreme weather-related natural disasters, including El Niño, continue to rise with negative effects on agriculture. For example, erratic weather conditions, with extremely high temperatures affected cocoa production, causing significant production slumps which resulted in noticeable price hikes. Overall, in the early 2020s, growth in agriculture was modest and structural change was small, as shown by the GDP and labor shares, which are still 12.5% and 28.2%, respectively.

Case Study: Oil Palm Production

Before drawing conclusions and giving an outlook on the transition of agriculture in Indonesia, the oil palm issue, being the most prominent example of agricultural development in Indonesia, is highlighted by means of this case study. Both chapter authors undertook their PhD thesis on this topic in 2013 and 2015, respectively. Therefore, the case study draws heavily on the results of the two PhD theses and is complemented by other relevant literatures.

Historically, oil palm plantations started in the early 20th century, when the Dutch colonial administration was looking for a replacement of natural rubber plantations, after prices collapsed during the world economic crisis of 1929. After independence, colonial land use rules were adopted by the Government of Indonesia which had favored the establishment of large-scale oil palm plantations, some 30 years later (Berenschot and Dhiaulhaq, 2023). The government strongly encouraged the expansion of oil palm plantations in the 1970s and 1980s, backed by incentives such as subsidized land leases, low-interest loans, and foreign investment. This period saw a rapid increase in plantation development, particularly in Sumatra and Kalimantan which made Indonesia become the major producer of palm oil in the world. Until the late 1970s, oil palm production was dominated by large-scale plantations of private corporations and state farms. In response to equity concerns, the government started to promote oil palm plantations by smallholders which rapidly took off. To date, corporate plantation and smallholders are almost at par with state farms being constant with just over 10% of the area (MOA, 2023).

Oil palm has been a key driver of economic growth in Indonesia, contributing approximately 3.5% to Indonesia's GDP and directly or indirectly employs around 16 million people (BPS, 2023). Economic studies (e.g. Qaim *et al.*, 2020) suggest that oil palm has contributed considerably to rural income growth and has reduced poverty among farmers and plantation workers. However, a rigorous impact study in Kalimantan (Santika et al., 2019) also found that the social impacts of oil palm plantations varied with locations and baseline socioeconomic conditions, i.e. plantations in remote villages, with high forest cover, experienced little improvement compared with non-oil palm villages. Furthermore, the downsides of the oil palm boom soon became apparent, as documented in many studies (e.g. Nathan and Tisdell, 2009; Obizinski et al., 2012; Ayompe et al., 2021; Phami et al. 2023). Deforestation and loss of peatland, leading to reduced biodiversity, including wildlife, as portrayed by the endangerment of orangutan populations. Improper management practices lead to soil erosion, water pollution, pesticide poisoning, and other occupational health hazards. Land clearing by burning vegetation causes forest and peatland fires which in 2015 alone, caused over $16 billion in economic losses in Indonesia and neighboring countries (Wijaya *et al.*, 2017). Deforestation linked to oil palm expansion is estimated to be responsible for 25% of total deforestation in Indonesia between 2001 and 2019 (Austin *et al.*, 2019).

In response to public pressure, primarily in Europe, the palm oil industry established a voluntary label called the Round Table of Sustainable Palm Oil (RSPO) in 2004. Producers under the RSPO label have to undergo certification and in return, theoretically, can sell their products at a higher price. Indonesia, however, had established its own label, i.e. the Indonesian Sustainable Palm Oil (ISPO), currently mandatory for large-scale producers and, by the end of 2025, for smallholders as well.

While mandatory certification of oil palm production may be one way to make palm oil production more sustainable, the question remains, what actually are the possibilities for farmers, especially small-scale farmers, to produce palm oil with significantly less negative environmental and human health effects? This is addressed below. Following this, another equally relevant question is addressed, namely to what extent have smallholders benefited from contract schemes with oil palm companies? Both topics have been studied empirically by the authors in Jambi, Sumatra and hence are briefly summarized here.

As regards the first topic, an empirical study investigated oil palm management practices and their compatibility with ISPO standards, among 233 smallholder oil palm farmers in Jambi province, Sumatra (Ernah *et al.*, 2016). Drawing upon a three-year panel data set, factors that explain the use of these practices were identified by means of a bivariate probit model, for use and non-use of practices, and a count model for the intensity of practices. Results showed that the use of ISPO-type crop management was only modest. Frequency and intensity of use declined with the number of practices. For example, only 40% of the farmers used six or more practices. Household characteristics such as age, household size, and gender are positively correlated with their use, as well as shocks experienced in the past and the perceived risk of diminishing oil palm productivity. The use of sustainable practices was positively correlated with yields. The study showed the need for investment in extension to foster the adoption of ISPO practices. This issue was further investigated by means of a cost-benefit study comparing the potential impact of two extension strategies, namely, (1) a conventional extension campaign and (2) a farmer field school approach (Ernah *et al.*, 2020). Results showed that investment in extension are highly profitable, whereby the farmer field school approach was dominant. A major conclusion was that a purely regulatory approach, making ISPO practices mandatory for small-scale farmers, is unlikely to be successful. Instead, government should seriously invest in training and extension. Results were confirmed in a recent experimental study, using the discrete choice method with 251 Indonesian oil palm smallholders (Reich and Musshoff, 2025). The authors showed that smallholders generally have a positive attitude towards certification schemes and show willingness to adopt these. However, willingness to adoption significantly increases with the provision of more farm management training and financial subsidies.

The second study assessed the impact of contract farming on smallholders' well-being and poverty reduction, using data from 245 smallholders in the province of Jambi, Sumatra (Cahyadi and Waibel, 2013). A two-stage treatment effects model was applied in order to correct for endogeneity for assessing the impact of contract participation on household income. Results show that participation in contract farming is significantly associated with household characteristics, the size of oil palm plot, and the time of plantation establishment. Farmers provided with a contract with oil palm companies, tend to be the better-off farmers and contractual conditions discriminate against the poorer smallholders. A follow-up study (Cahyadi and Waibel, 2016) confirmed and re-emphasized these findings. The study used an asset-based vulnerability to poverty approach (Chiwaula *et al.*, 2011), i.e. considering a household's assets as a determinant of future income. Propensity score matching, comparing contract and non-contract smallholders, was carried out and hereby the risk of falling into poverty was for contract and non-contract farmers was assessed. The data base were the same 245 smallholders out of which 126 were contract farmers. The study found that approximately 40% of oil palm smallholders can be classified as stochastic poor, i.e. their poverty status depends on stochastic events. Contract smallholders remain vulnerable to poverty as a result of uninsured production shocks. Consistent with these results, Euler *et al.* (2016) found that independent, non-contract farmers are expanding faster than smallholders with contracts.

Reaching a final conclusion of the oil palm issue in Indonesia is impossible despite the hundreds of studies that have been conducted. Certification, initially an industry response to criticisms by NGOs and later embraced by governments, did not solve the problem. Neither voluntary nor mandatory certification have been a great success. For example, to date, the market share of RSPO-certified and other voluntary sustainability standards (e.g. Rainforest Alliance, organic production, etc.) products are just 17% of the total market in 2016 (Voora *et al.*, 2019) with price premiums not exceeding 2% of non-certified products (Tey *et al.*, 2020). The RSPO impact report 2024 (RSPO, 2024) itself, stated the area of certified palm oil with 5.2 million ha in 2023, which, again, is barely 17% of the total. In Indonesia, where ISPO certification is mandatory, it remains open what percentage of the oil palm area actually follows sustainability standards. Besides, the EU still only recognizes the RSPO but not the ISPO standard. Furthermore, consumers outside Europe and the USA, show little concern for the way palm oil is produced, what they care about is a low price. Even within the EU, for consumers, as shown in a choice experiment conducted in Germany (Richartz and Abdullai, 2022) the attribute 'organic' for palm oil products, is clearly preferred over other sustainability standards.

To close, and, as nicely summarized by Lusiana *et al.* (2023), the oil palm debate is torn between business interests on the one hand and worries about nature on the other hand. Unfortunately, this has resulted in demonizing the crop itself while paying less attention to healthier and more sustainable ways of producing it. It is hoped that this case study within the context of the chapter, has shed some light on the "darkness" of the "oil palm forest."

The Way Ahead

The review of agricultural policies and rural transformation in Indonesia since independence serves as a basis for looking ahead. The historical shifts in agricultural policy, as summarized in Table 10.1, illustrate that there was a gradual transition from state-controlled self-sufficiency to a market-oriented growth paradigm, and more recently, to a more diversified, competitive and sustainable agriculture with a focus on value addition without neglecting food security.

Looking ahead, agriculture in Indonesia must be seen in the context of the country's vision of becoming a high-income economy by 2045. To achieve these goals, sustainably high growth rates must be maintained over the next 20 years. Indonesia will need to focus on aggregate supply by improving productivity, increasing investment, enhancing its global competitiveness, and establishing strong institutions. The agricultural sector must be aligned with these ambitious goals. The agricultural sector will remain of utmost importance even though the share of primary agricultural production in GDP will decline further. However, the food industry as a whole, including processing, transportation, distribution, and storage of processed, more food and other agricultural products will grow.

With the agro-industry complex, oil palm is likely to maintain a dominant role. Hereby, key developments are to move beyond crude palm oil and advance the development of processed products, including bioplastics and renewable chemicals, thus creating higher-value products. Recently, Indonesia has already moved from exporting primarily crude palm oil to higher-value products such as oleochemicals, biofuels, and other derivatives for domestic consumption and export (World Bank, 2021; Ernah *et al.*, 2024). For example, in 2021, the export value of processed palm oil products reached $20 billion, compared with $8 billion in 2015 (BPS, 2022). Also, the palm oil industry is a major employer with approximately 16 million people directly and indirectly engaged (MoA, 2023). The agroindustry strategy is extended to other crops as well. The National Medium-Term Development Plan (RPJMN) 2020–2024 identifies agro-industrial development as a critical driver for shifting away from reliance on raw exports focusing on downstream processing for commodities like palm oil, rubber, cocoa, and coffee and hereby augmenting rural economic growth (Ministry of National Development Planning, 2020). Aside from this, the development, promotion, and actual diffusion of sustainable practices must be maintained as a priority. for the sake of reducing societal costs as negative environmental and human health externalities. Likewise, working towards social equity and reducing land conflicts rights and through strengthening smallholder rights remain a challenge but are essential to reducing social tensions that could inhibit a stable business development.

In the foreseeable future, i.e. the next 20–30 years, Indonesian agriculture is likely to retain its dualistic structure, i.e. large-scale agroindustry corporations mostly engaged in permanent crops on the one hand and the vast majority of small-scale farmers on the other hand, mostly growing rice, maize, cassava, vegetables, and fruits as well as raising livestock. Increasingly, however, in rural households, income from farming will decline relative to income from non-farm sources. Table 10.2 shows that while in 2007 income from farming was over 70%, it is estimated to be on par with non-farm income in 2024, following a steady decline. This means that, more and more, rural household members will have to engage in non-farm occupations either as migrants in urban areas or through wage work and self- employment in their villages. To a greater extent, small-scale

Table 10.2. Rural Household Income Shares (%).

Year	Farm	Non-Farm
2007	73	27
2010	61	39
2016	55	45
2024*	50	50

Source: Based on Badan Pusat Statistik (Central Bureau of Statistics) Indonesia, several years. *Estimate.

farming will become a part-time job. Since Indonesia has not yet entered the phase of aging populations, but still has a relatively high share of young people, this farming system could be around for a long time to come.

As in most other Asian countries, farms have remained small in Indonesia and most likely will be so in the near future. The FAO's small family farms country factsheet (FAO, 2018) reveals that the average farm size is 0.6 ha for 93% of all farms, and 5 ha for all other farms which results in a weighted average farm size of 0,91 ha. While average farm size does not tell too much about structural change in agriculture, the structure of farm sizes is more revealing. Table 10.3 shows that the share of farms with less than 1 ha, has practically not changed during the past 30 years, i.e. 75% in 1993 and 76% in 2023. Similarly, the share of farms with more than 2 ha has only increased marginally from 10% to 12% in the same period. Note, these data refer to all farms in Indonesia, not just small family farms. While the micro- and small farm pattern does not hold for perennial crops like oil palm and cocoa for example, the vast majority of farmers in Indonesia are far below a level at which their families could make a living from farming alone. Hence, it can be expected that the part-time type of farming system will exist for a long time to come. A rapid transition into an agricultural sector where large-scale, full-time farms will become the dominant farming model is hardly foreseeable at the moment. Recognizing these facts requires adjusted rural development policies that facilitate the existence of multi-enterprise farm household systems. New technologies, whether in biology or in the field of information, can help to increase productivity and improve environmental conservation and human health in either farming system, given the correct institutional setting and incentives are put in place. Small-scale farming supported by EU-style rural development programs (e.g. Pelucha and Kveton, 2017), including much improved physical, digital, and social infrastructure could make rural areas a more attractive work and living space. Diversified income of rural households from farming and from non-farm sources can increase the resilience of rural areas and at the same time help to reduce the pressure on the already overcrowded urban agglomerates such as Jakarta, Yogyakarta, Bandung, and Surabaya. At the same time, the government should also provide incentives for the gradual transition towards a higher share of full-time, professional farms in the future. In the end, policies must recognize the dynamics of systems development and facilitate their change in the course of demographic and technological transitions.

Table 10.3. Distribution of Farm Size in Indonesia (%).

Year	<0.5 ha	0.5–0.99 ha	1–1.99 ha	≥2 ha
1993	50	25	15	10
2003	55	20	15	10
2013	55	18	14	13
2018	58	17	13	12
2024*	60	16	12	12

Source: Based on Badan Pusat Statistik (Central Bureau of Statistics) Indonesia, several years. *Estimate.

As regards food crops, rice will remain as the main crop with about half of arable, non-plantation crop land. The importance of rice is demonstrated by the Government's plan to add one million ha of new rice land in Papua. Drawing on experiences of the past, however, it remains doubtful whether such a project will be successful. Nevertheless, fruits, vegetables, and animal-based protein products will gain in importance with rising per capita incomes. Ultimately, this will increase the share of processing and distribution in the food system value chain (Reardon *et al.*, 2015). Related to the modernization of the food system, there are benefits to be gained from reducing post-harvest losses and food waste, some of which can be converted into bioenergy such as bio-oil, bio-diesel, bio-hydrogen, bio-char, and other solid energy products (Suhartini *et al.*, 2022). This adds new roles to the agricultural sector, making it an integral part of the bio-economy. Already Indonesia has made progress in the development and implementation of climate-smart farming practices and technologies (Savelli *et al.*, 2021).

This also brings organic farming into the picture of Indonesia's agriculture of the future, although still very small to date (David and Ardinasyah, 2017). The world statistics of agriculture (FiBL, 2019) reported just 6000 producers by 2016. On the other hand Indonesia is exporting organic products such as coffee and tea On a producer site, a pilot project with 1000 organic villages initiated under the previous (Jokowi)

Table 10.4. Major Challenges, Opportunities and Constraints of Agricultural Development in Indonesia. (Authors' compilation).

Challenges	Indications	Opportunities	Constraints
Climate change and extreme weather events	1.5–2°C temperature increase, higher frequency and severity of droughts, floods, and storms	Climate-smart agriculture; carbon pricing; stop deforestation	Institutions, incentives, farmer participation
Digitalization and Information Technologies	Internet penetration in rural areas 72% in 2022, online marketing platforms	Smartphone Apps; IoT, precision farming, blockchain technology	IT education, user-friendliness, costs
Bio-industry development	Share of primary sector decreasing. Crop diversification	Trade-agreements, alignment of standards, certification; new products, waste recycling	Geopolitical tensions, Divergence in standards and consumer preferences
Farming systems transition	Farm size structure, age structure, rural population	Land markets, secure non-farm jobs, technological change	Attractiveness of farming and living in rural space, heritage laws

government showed only limited success due to poor implementation by the Ministry of Agriculture (Keilbart, 2021). A more optimistic view is portrayed by Fritz *et al.* (2021), provided better coordination among relevant actors and a more conducive institutional environment with supportive incentive schemes, can be put in place. Positive signals, in this regard, come from the consumer side. Najib *et al.* (2021) found that middle class urban consumers show a positive attitude towards purchasing organic products. Another opportunity to improve the efficiency of small-scale farming is the enhancement of their management capacities. For example, in a study with 200 potato farmers in West Java, Wulandari *et al.* (2023) show that farmers who regularly record their financial transactions using commercial software, perform better in terms of income.

To sum up the challenges, opportunities, and constraints for agricultural development in Indonesia in the near future, Table 10.4 may serve as a final overview and take-away message. In short, there are four major challenges as listed, namely Climate Change, Digitalization, Bio-industry development, and Farming systems transition, with the first challenge mentioned perhaps being the biggest.

Moving agriculture forward, Indonesia must integrate innovation and new technologies with past lessons to build resilience, expand value chains, and ensure equitable land access, securing a sustainable and competitive agricultural in the future.

References

Akhtar, M.N., Ansari, E., Alhady, S.S.N. and Abu Bakar, E. (2023) Leveraging on advanced remote sensing-and artificial intelligence-based technologies to manage palm oil plantation for current global scenario: A review. *Agriculture* 13(2), 504.

Austin, K., Mosnier, A., Pirker, J., McCallum, I., Fritz, S. *et al.* (2019) Shifting patterns of oil palm driven deforestation in Indonesia and implications for zero-deforestation commitments. *Environmental Research Letters* 14(4), 1–9.

Ayompe, L.M., Schaafsma, M. and Egoh, B.N. (2021) Towards sustainable palm oil production: The positive and negative impacts on ecosystem services and human wellbeing. *Journal of Cleaner Production* 278, 123914.

Badan Pusat Statistik (BPS). (2022) *Ekspor Produk Pertanian 2021*. BPS Indonesia, Jakarta.

Badan Pusat Statistik (BPS). (2023) *Statistik Indonesia 2023*. BPS Indonesia, Jakarta.

Booth, A. (1989) Indonesian agricultural development in comparative perspective. *World Development* 17(8), 1235–1254.

Cahyadi, E.R. and Waibel, H. (2013) Is contract farming in the Indonesian oil palm industry pro-poor? *Journal of Southeast Asian Economies* 30 (1), 62–76.
Cahyadi, E.R and Waibel, H. (2016) Contract farming and vulnerability to poverty among oil palm smallholders in Indonesia. *Journal of Development Studies* 52(5), 681–695.
Chiwaula, L.S., Witt, R. and Waibel, H. (2011) An asset-based approach to vulnerability: The case of small-scale fishing areas in Cameroon and Nigeria. *Journal of Development Studies* 47(2), 338–353.
David, W. and Ardiansyah. (2017) Organic agriculture in Indonesia: challenges and opportunities. *Organic Agriculture* 7, 329–338.
Deichmann, U., Goyal, A. and Mishra, D. (2016) Will digital technologies transform agriculture in developing countries?. *Agricultural economics* 47(S1), 21–33.
Ernah, E., Priyanka, P. and Waibel, H. (2020) Will teaching sustainability standards to oil palm smallholders in Indonesia pay off? *International Journal of Agricultural Sustainability* 18, 1–17.
Ernah, E., Priyanka, P. and Waibel, H. (2016) Adoption of sustainable palm oil practices by Indonesian smallholder farmers. *Journal of Southeast Asian Economies* 33(3), 291–316.
Ernah, E., Elvina, G., Eliana, W. and Eti, S. (2024) The factors influencing on demand of biodiesel in Indonesia. *ISSAAS Journal* 30(2), 185–194.
Euler, M., Schwarze, S., Siregar, H. and Qaim, M. (2016) Oil palm expansion among smallholder farmers in Sumatra, Indonesia. *Journal of Agricultural Economics* 67(3), 658–676.
FAO. (2018) Smallholders data portrait. Available at www.fao.org/family-farming/data-sources/dataportrait/farm-size/en (accessed 13 September 2025).
FiBL. (2019) *The World of Organic Agriculture: Statistic and Emerging Trends 2019*. Routledge, London.
Firman, T. (2009) Decentralization reform and local government proliferation in Indonesia: Towards a fragmentation of regional development. *Review of Urban & Regional Development Studies* 21(2–3), 143–157.
Fritz, M., Grimm, M., Keilbart, P., Laksmana, D. D., Luck, N., Padmanabhan, M., Subandi, N. and Tamtomo, K. (2021) Turning Indonesia organic: Insights from transdisciplinary research on the challenges of a societal transformation. *Sustainability* 13(23), 13011.
Geertz, C. (1963) *Agricultural involution: the processes of ecological change in Indonesia* (No. 11). Univ of California Press, Oakland, USA.
Hansen, G.E. (1972) Indonesia's green revolution: The abandonment of a non-market strategy toward change. *Asian Survey* 12(11), 932–946.
Hill, H. and Shiraishi, T. (2007) Indonesia after the Asian crisis. *Asian Economic Policy Review* 2(1), 123–141.
Johan, D., Syamsul Maarif, M., Zulbainarni, N. and Yulianto, B. (2024) Agricultural digitalization in Indonesia: challenges and opportunities for sustainable development. *Educational Administration: Theory and Practice* 30(7), 640–648. DOI: 10.53555/kuey.v30i7.6599
Jurriens, E. and Tapsell, R. (2017) Challenges and opportunities of the digital 'revolution' in Indonesia. *Digital Indonesia: Connectivity and Divergence* 2020, 275–288.
Keilbart, P. (2021) Integrating organic farming into the Indonesian bioeconomy? Sustainable agriculture between productivism and deep ecology. *ASIEN: The German Journal on Contemporary Asia* (160/161), 87–117.
Laraswati, D., Krott, M., Soraya, E., Rahayu, S., Fisher, M.R. *et al.* (2022) Nongovernmental organizations as interest groups and their roles in policy processes: Insights from Indonesian forest and environmental governance. *Forest and Society* 6(2), 570–589.
Limenta, M.E. and Chandra, S. (2017) Indonesian Food Security Policy. *Indonesia Law Review* 7(2), 245–265.
Lusiana, B., Slingerland, M., Miccolis, A., Khasanah, N.M., Leimona, B., *et al.* (2023) Oil palm production, instrumental and relational values: the public relations battle for hearts, heads, and hands along the value chain. *Current Opinion in Environmental Sustainability* 64, 101321.
Maat, H. (2014) Agriculture in Indonesia. In: Selin, H. (eds) Encyclopaedia of the History of Science, Technology, and Medicine in Non-Western Cultures. Springer, Dordrecht. Available at: https://doi.org/10.1007/978-94-007-3934-5_10229-1 (accessed 13 September 2025).
Mariyono, J., Waskito, J., Suwandi, Tabrani, K.A., Latifah, E. and Suswati, E. (2021) Farmer field school: Non-formal education to enhance livelihoods of Indonesian farmer communities. *Community Development* 52(2), 153–168.
Ministry of Agriculture. (MoA). (2023) *Laporan Tahunan Kementerian Pertanian 2023*. MoA Indonesia, Jakarta.
Ministry of National Development Planning. (2020) *National Medium-Term Development Plan (RPJMN) 2020–2024*. Ministry of National Development Planning, Jakarta.

Marlier, M.E., Ruth DeFries, S., Patrick Kim, S., Shannon Koplitz, N., Daniel Jacob, J. *et al.* (2015) Fire emissions and regional air quality impacts from fires in oil palm, timber, and logging concessions in Indonesia. *Environmental Research Letters* 10, 1–9.

Musyarof, Z. and Indira, N.Q. (2022) Perekonomian Provinsi-Provinsi di Indonesia Pasca Terdampak Gempa Bumi. *Jurnal Ekonomi-Qu* 12(1), 86–105.

Nawiyanto, N. (2017) The politics of food and food security during Indonesia's old order (1945–1965). *Lembaran Sejarah* 10(1), 65–73.

Najib, M., Sumarwan, U., Septiani, S., Waibel, H., Suhartanto, D. *et al.* (2022) Individual and socio-cultural factors as driving forces of the purchase intention for organic food by middle class consumers in Indonesia. *Journal of International Food & Agribusiness Marketing* 34(3), 320–341.

Nanthan, H.S. and Tisdell, C. (2009) The orangutan-oil palm conflict: Economic constraints and opportunities for conservation. *Biodiversity and Conservation* 18(2), 4.

Obidzinski, K.R. Andriani, H., Komanidin, A. and Andrianto. (2012) Environmental and social impact of oil palm plantations and their implications for biofuel production in Indonesia. *Ecology and Society* 17, 1–20.

Pelucha, M. and Kveton, V. (2017) The role of EU rural development policy in the neo-productivist agricultural paradigm. *Regional Studies* 51(12), 1860–1870.

Phami, P., Gunawan, B., Iskandar, J. and Soemarwoto, R. (2023) Reverse social impact of oil palm plantation expansion: A study of three communities in Jambi, Indonesia. *Forest and Society* 7(1), 61–75.

Prasetyo, S.Y.J., Winarko, E. and Daryono, B.S. (2012) Endemic outbreaks of brown planthopper (*Nilaparvata lugens* Stal.) in Indonesia using exploratory spatial data analysis. *International Journal of Computer Science Issues (IJCSI)* 9(5), 162.

Presidential Instruction No. 6/2013. Jakarta, Indonesia. Available at: https://lpr.adb.org/sites/default/files/resource/809/indonesia-extension-to-the-moratorium-on-issuing-new-peatland-and-forest-utilization-license.pdf (accessed 04 March 2026).

Presidential Regulation No. 44 of 2016. (2016) On Downstream Processing Incentives for the Palm Oil. President of the Republic of Indonesia. Jakarta. https://pnblawfirm.com/wp-content/uploads/2020/06/Presidential_Regulation_44_of_2016_2016_Negative_List_English_151.pdf (accessed 04 March 2026).

Qaim, M. Sibhatu, K.T. Siregar, H, and Grass, I (2020) Environmental, economic, and social consequences of the oil palm boom. *Annual Review on Resource Economics* 12, 321–344.

Reardon, T., Stringer, R., Timmer, C. P., Minot, N. and Daryanto, A. (2015) Transformation of the Indonesian agrifood system and the future beyond rice: A special issue. *Bulletin of Indonesian Economic Studies* 51(3), 369–373.

Reich, C. and Musshoff, O. (2025) Oil palm smallholders and the road to certification: Insights from Indonesia. *Journal of Environmental Management* 375, 124303.

Richartz, C. and Abdulai, A. (2022) The role of information in consumer preferences for sustainable certified palm oil products in Germany. *PLoS ONE* 17(7), e0271198.

RSPO. (2024) Impact Report. 2024. Available at: https://rspo.org/wp-content/uploads/RSPO_ImpactReport_2024.pdf (accessed 13 September 2025).

Saragih, B. (2001) Agricultural policies and performance in Indonesia. *ASEAN Economic Bulletin* 18(1), 36–50.

Savelli, A., Atieno, M., Giles, J., Santos, J., Leyte, J., Nguyen, N.V.B., Koostanto, H., Sulaeman, Y., Douxchamps, S. and Grosjean, G. (2021) Climate-Smart Agriculture in Indonesia. CSA Country Profiles for Asia Series. Hanoi (Vietnam): The Alliance of Bioversity and CIAT; The World Bank Group, Washington, DC, 88.

Suhartini, S., Rohma, N.A., Elviliana, S.I., Paul, R., Listiningrum, P. *et al.* (2022) Food waste to bioenergy: current status and role in future circular economies in Indonesia. *Energy, Ecology and Environment* 7(4), 297–339.

Suryahadi, A., Suryadarma, D. and Sumarto, S. (2009) The effects of location and sectoral components of economic growth on poverty in Indonesia. *Journal of Development Studies* 45(9), 1344–1362.

Tey, Y.S., Brindal, M., Darham, S., Sidique, S.F.A. and Djama, M. (2020) Early mover advantage in Round on Sustainable Palm Oil certification: A panel evidence of plantation companies. *Journal of Cleaner Production* 252, 119775.

Thiers, P. (1997) Successful pesticide reduction policies: Learning from Indonesia, *Society & Natural Resources* 10(3), 319–328, DOI: 10.1080/08941929709381030.

Thow, A.M., Sharma, S.K. and Rachmi, C.N. (2019) An analysis of Indonesia's shrinking food security policy space under the WTO. *Food Security* 11(6), 1275–1287.

Timmer, C.P. (2004) The road to pro-poor growth: The Indonesian experience in regional perspective. *Bulletin of Indonesian Economic Studies* 40(2), 177–207.

Van de Fliert, E. (1993) Integrated pest management: Farmer field schools generate sustainable practices: a case study in Central Java evaluating IPM training. *Wageningen University Papers* 93, 304.

Van der Eng, P. (2000) Food for growth: Trends in Indonesia's food supply, 1880–1995. *Journal of Interdisciplinary History* 30(4), 591–616.

Vidyattama, Y. (2010) A search for Indonesia's regional growth determinants. *ASEAN Economic Bulletin* 27(3), 281–294.

Vidyattama, Y., Merdikawati, N. and Tadjoeddin, M.Z. (2021) Aceh tsunami: Long-term economic recovery after the disaster. *International Journal of Disaster Risk Reduction* 66, 102606.

Voora, V., Larrea, C., Bermudez, S. and Baliño, S. (2019) *Global Market Report: Palm Oil*. International Institute for Sustainable Development, Manitoba, Canada, p. 16.

Widadie, F., Bijman, J. and Trienekens, J. (2021) Value chain upgrading through producer organisations: linking smallholder vegetable farmers with modern retail markets in Indonesia. *International Journal on Food System Dynamics* 12(1), 68–82.

Wijaya, A., Chrysolite, H., Ge, M., Wibowo, C., Pradana, A. *et al.* (2017) *How can Indonesia Achieve its Climate Change Mitigation Goal? An Analysis of Potential Emissions Reductions from Energy and Land-Use Policies*. World Resources Institute.

World Bank (2024) Available at: https://data.worldbank.org/indicator/SI.POV.GINI?locations=ID&utm_source=chatgpt.com (accessed 30 July 2025)

World Bank (2021) *Indonesia Economic Update: Maintaining Resilience Amidst Uncertainty*. World Bank, Washington, DC.

World Bank (2021) *World Bank East Asia and Pacific Economic Update*. World Bank, Washington, DC.

Wulandari, E., Karyani, T., Ernah, Trizaldi, R. and Alamsyah, P. (2023) What makes farmers record farm financial Transactions? Empirical evidence from potato farmers in Indonesia. *International Journal of Financial Studies* 11, 19. DOI: 10.3390/ijfs11010019.

11 Agricultural Development and Policy in Laos

Phouvong Phami, Priyanka Parvathi and Hermann Waibel

Abstract

This chapter analyses the development of economy and agriculture in Laos since its independence from France, to date. The profound transition from a command and control economic system towards a communist governed market economy, is presented, using a historical framework. During the course of economic reforms and better connectivity with other Asian economies, especially China and Thailand, the country started to industrialize. Mining products and hydropower electricity have become major export items. However, by and large, Laos is still an agricultural country, where the vast majority of households are engaged in farming. Due to low productivity in agriculture, farm income is supplemented by natural resource extraction. A brief case study on income from natural resources is included in the chapter. While structural transformation of agriculture is gradually taking place, the sector is heading towards a similar kind of dualistic structure to that observed in other Asian countries. The chapter closes with an outlook considering regional and global risks ahead.

Introduction

Laos is a mountainous, landlocked country, with low population density and a high rate of forest cover which is still at some 70% (FAO, 2023). Laos is an agricultural country, with over 70% of the population living in rural areas. Consequently, agriculture, including forestry, plays an important role in the economy, both in terms of GDP and especially labor share which is still close to 70%. Per capita income is among the lowest in Southeast Asia, with US$2066.9 in 2023 (World Bank, 2025). On the other hand, between 2000 and 2015, the Laos economy enjoyed quite impressive rates of economic growth of some 8% p.a. However, since then the growth rate has been falling constantly, even before the Covid-19 pandemic, and reached just 3.7% in 2023 (see Table 11.1). Currently, the country suffers from high financial debt and high inflation rates of some 20%. Public debt has almost reached 120% of GDP, far further than its ASEAN partners. Besides, many Lao workers migrate abroad, in search of better paid jobs and as a response to the domestic economic difficulties (World Bank, 2024).

In this chapter, we provide a historical review of agricultural development by applying a framework (see Table 11.2) that briefly outlines general development, agricultural policies, the structure of the agricultural sector and its performance along with significant events in as much as they have played a role. The time periods covered include the period after Lao's independence from France as stipulated in the Geneva convention of 1954 until the end of monarchy and the communist takeover in 1975. The third period is defined thereafter until the

 Asian Agriculture in a Climate Change-driven and Risky World (H. Waibel, S. Liebenehm and Ernah eds)
DOI: 10.1079/9781800627338.0011

Table 11.1. GDP growth, share of agriculture and labor share over time.

Year/Parameter	1990	1995	2000	2005	2010	2015	2020	2023
GDP growth (%)	6.7	7	5.8	7.1	8.5	7.3	0.5	3.7
Share Agriculture value added (%)	46.5	40.5	33.6	28.3	22.6	17.6	16.3	16.1
Labor share Agriculture (%)	86[1)]	85	82	77	71	64	66	70

Source: https://data.worldbank.org/indicator/NV.AGR.TOTL.ZS?locations=LA.

country started to open-up economically, in particular, to its neighboring countries China and Thailand. Hereby, periods are further divided into the development decades, as defined by the UN.

The Three Princes Period (1954–1975)

The first development period considered in this review was popularly labeled as the "Three Princes Period" (Time Magazine, 1961). Immediately after independence, Lao politics was marked by severe instability. Three groups fighting for power, namely the neutralists under *Prince Souvanna Phouma*, the right-wing party under *Prince Boun Oum of Champassak* and the left-wing, Pathet Lao under *Prince Souphanouvong*. The latter group, operating in Northern Laos was supported by North Vietnam and the Soviet Union, with the aim to establish a communist state. The non-communist groups received massive financial assistance from the USA, mostly military aid. By 1957, the US had spent about US$150 per person, twice the average annual per capita income at that time. Basically, Laos was a split country, between the area under the control of the Royal Lao government and a larger but less populated part, under the control of the communist Pathet Lao. Hence, refugee movements from the communist controlled areas to Central and Southern Laos became a major burden to the economy. The civil war had withdrawn farmers from productive activities. Investment in economic and social development was impaired by heavy military expenditures. The few development projects focused on the exploitation of the country's vast forest resources, some initial steps to exploit the hydroelectric power potential and a few projects on fisheries, aquaculture and navigation for border trade with Thailand along the Mekong River. However, within-country trade was severely constrained due to lack of infrastructure and security risks. Smuggling and illegal trade was fueling a shadow economy. Planned agricultural development basically did not take place. Instead, agricultural production was frequently troubled by wet season floods of the Mekong and its tributaries, thereby augmenting food insecurity (e.g. Jain, 1972). To illustrate the situation in the public health sector, for example, in 1960, on average, there was just one physician per 54,000 people, and in rural areas there were over 9000 persons per available hospital bed, according to a World Bank mission report (World Bank, 1983).

The Birth of Lao People's Democratic Republic (1975–1986)

When in 1975, South Vietnam was defeated by the North, and after the USA had removed its presence and support, the monarchy in Laos was overthrown as well. The shift in political power to the Vietnam-backed communist Pathet Lao resulted in fundamental changes in economic policy. Under the strong influence of Vietnam, their central command and control economic system was adopted. The private sector was largely dismantled and replaced by state enterprises and cooperatives. Investment, production, trade, pricing, payments and wages were put under central control. Barriers to internal and foreign trade were introduced and official trade was confined to the Soviet-controlled COMECON (Council for Mutual Economic Assistances) countries, especially Vietnam. Consequently, support from western countries ended abruptly. The switch from a laissez-faire to a state-controlled economy, with, at the same time, loss of, especially American foreign assistance, the performance of the economy went from bad to worse. Initially, however, there was a positive growth effect, simply because the war had ended and soldiers went back to the labor force. However, overall, socioeconomic development of the

Table 11.2. Overview of development in Laos by time periods and development decades. (Authors' own table).

#	Period	General Development Policy	Agricultural Policies	Structure of Agriculture	Significant Events	Performance of Agriculture
1	1953–1975; Independence, the Three Princes Period	Political instability; frequent change in government; some infrastructure projects	Laissez-faire, basically none, due to civil war	Village-based subsistence agriculture	Wet season floods at Mekong and Tributaries; Civil War	Low yields, heavy reliance on natural resource extraction
2	1975–1985; Communist take-over: The Lao People's Democratic Republic	Centrally based command and control economy; First 5-year development plan; dependence on Vietnam	Redistribution of land from large land owners to small-scale peasants; village-based collectivized farming	Collectivized production	Refugees fleeing to Thailand	Food deficit; small exports of coffee
3	1986–1994 NEM; 1st Friendship Bridge	New Economic Mechanism, maintain state-owned enterprises (SOEs)	Land reform, rubber plantations; resettling of mountain farmers to the lowlands	Shift from collectivized to individual-based family farming;	n.a.	Gradual increase in agricultural productivity; cash crop production (coffee, maize)
4	1995–2010 Land-linked	Continued NEM; focus on poverty reduction and sustainable development; Join ASEAN in 1997	Promotion of diversification; Agricultural Development Strategy (ADS)	Increased reliance on smallholder farming	The Asian financial crisis	Growth in agricultural value-added; rise in rice production; coffee and rubber as key cash crops
5	2011–2020	Rising regional player as a hydroelectric power supplier to its neighbors China, Thailand and Vietnam. Foreign direct investment to attract capital from overseas. Reforms and education	Implementation of agricultural master and investment plans aligned with the NSEDP. Emphasis on public and private investment in agriculture and food sector by permitting land concessions to plant industrial trees, for livestock production and agricultural cash crops	Smallholder farming systems dominant; lowland rainfed/ irrigator, upland, swidden agriculture	Covid-19	Agricultural sector growth proves effective for poverty reduction; agriculture contributes 42% of GDP and employs 67% of the workforce. Annual growth rates of around 4.16% noted
6	After 2020	Rigid Covid-19 measures; labor shortage economy	Agriculture included in 2030 sector vision; to modernize agriculture and make it more competitive	Rice still major crop. Cassava, maize, coffee and rubber, livestock, products for export	Covid-19, Laos–China express train project open, Russia–Ukraine war and inflation	Input prices (fertilizers and fuel) increases, sluggish sector growth; labor share in agriculture increasing again

country was low. As indicated in a World Bank mission report (World Bank, 1983), the per capita income for 1979 was estimated to be only $80 per person p.a. At the same time, population growth had gone up to about 3% p.a. with rates of over 5% in rural areas; 86% of the population were living in rural areas and only 6% of the labor force were in industry. On average, food supply was below minimum caloric requirement (World Bank, 1983). Natural calamities such as drought in 1976 and 1977, followed by floods in 1978, hampered agricultural recovery and caused serious food shortages.

On the plus side, Laos had maintained its membership with the IMF and continued to collaborate with the World Bank. Furthermore, the government managed to implement a multiple donor strategy despite the mastery of the Soviet Union and Vietnam. Considering the collapse of the USSR, some 15 years later, Lao's strategy of serving two masters, i.e. the Soviet Union and the multilateral institutions, was quite successful. The exposure to expertise from international organizations may also have facilitated a policy turn away from the purely ideologically derived Vietnamese economic model. The latter had led to mismanagement and corruption and turned out to be rather unsuitable for Lao conditions. Perhaps, a "game changer" was the Lao–Thai conflict in 1979, with shootings on the Mekong River and the closing of the border. This caused severe shortages of vital goods in Laos as the largely illegal but widely tolerated trade across the river came to a sudden end. In December 1979, a series of reforms were implemented which reversed the policies of total state control and collectivization. Furthermore, currency and trade reforms were implemented. While the role of private agriculture was recognized, officially, the socialist-type of economy was maintained. The policy changes had an immediate effect and the economy grew by 10% in 1980 (Arndt, 1992). However, it was also due to re-opening the border with Thailand. Between 1982–1985, average GDP growth was 6.1% and even 7% in 1986 but dropped to 2.2% a year later (IBRD, 1988). Much of the growth was fueled by official development assistance which had reached almost US$170 million by 1986. Yet, Laos was running into financial difficulties, due to a worsening of its terms of trade, with drastically falling export and increasing import prices. The Government therefore was challenged to take further steps for economic reforms.

From New Economic Mechanism to Lao–Thai Friendship Bridge (1986–1994)

Initiated in 1986, the "new economic mechanism" (NEM) was initiated, following developments in Vietnam, where it became known as "Doi Moi" (see Chapter 4). As described in the papers of Worner (1989), Rosser (2006) and Phimphanthavong (2012), the NEM reforms were designed to create more favorable conditions for the private sector to develop. Prices set by market forces replaced government-determined prices and farmers were permitted to own land and sell crops in the open market. State firms were granted increased decision-making authority and subsidies and pricing advantages were removed. In macroeconomic policies, the exchange rates were associated with market rates and import barriers were replaced with tariffs. Private firms were given direct access to imports and credit without the need for state permission.

The implementation of the reforms was in three phases. In the first phase, between 1986 and 1989, it involved measures to liberalize prices and trade, including putting more pressure on state-owned enterprises to operate more efficiently, promote the role of private enterprises, reform the taxation system and liberalize the foreign investment regime. In the second phase – between 1990 and 1991 – the reform focused on monetary and fiscal policies, including stricter financial controls of state-owned enterprises, reducing government spending and absorbing excess liquidity in the economy as well as encouraging higher savings through the issuance of government bonds. In the third phase, from 1991 onwards, when communism in Eastern Europe and the Soviet Union had collapsed, NEM focused on the introduction of structural economic reforms, primarily the privatization campaign of all state-owned, strategically less important enterprises, and the establishment of a central bank with a system of commercial banks.

An important milestone was the inauguration of the Thai–Lao Friendship Bridge, connecting Vientiane Prefecture with Nong Khai Province in Thailand in April 1994 (Pholsena, 2005; Nonthapot and Wattankul, 2014). The first bridge across the Mekong River, it was a strong symbol of Laos becoming linked to the world economy.

The NEM had also significant effects on the agricultural sector. A major policy change was land reform. The core element of the reform was contracts between villages and the state with regard to land use. The study of Ducourtieux *et al.* (2005) found that while the reform gave more responsibility to the villages, power relations were asymmetric, to the disadvantage of the villages. Formal rights and traditional practices clashed with land laws, officially aimed to serve interests of the community. A more recent study (Broegaard *et al.*, 2017) found that, in the end, land reform gave more control to the state, rather than the communities, thereby facilitating foreign investors for plantation agriculture. Especially, rubber plantations started to spread from neighboring Southwest China to Northern Laos.

Another element of the reform with the support of international donor agencies was the resettlement of ethnic minorities groups from remote highlands to lowland areas. Officially the resettlement programs were justified with security concerns, access and service delivery, cultural integration and nation building, the reduction of swidden agriculture as well as eradication of opium cultivation. The actual aim was to boost agricultural output, mostly rice in lowland and river valley areas (Baird and Shoemaker, 2007).

The short-term effect of NEM on agricultural output and productivity was mixed. Area harvested, total output of rice declined between 1985 and 1988. Rice yields in 1988 were around 0.5 t/ha but went down to just over 200 kg per ha in 1993, while the area had gone up to 5.5 million ha. Rice area fluctuated considerably since the launching of the NEM, perhaps as a result of the transmigration program, moving farmers from uplands to lowlands. For example, the area harvested was almost 8 million ha in 1981, 6.2 in 1986 and down to 5.9 million in 1994, some 10 years after the reform (FAO, 2024).

From Land-locked to Land-linked: 1995–2010

Ten years after the inauguration of the first bridge, the second one, connecting Mukdahan in Thailand with Savannakhet in Laos, had begun. With soon five bridges across the river to be completed (the opening of the fifth bridge is scheduled for December 2025), Lao is changing from a land-locked to a land-linked country. The road network is connecting Thailand with the Mohan border in Southwest China and several border points with Vietnam. GDP growth between 1996 and 2005 showed annual growth rates of slightly over 6% and around 8% until the end of the first decade after the millennium. This was translated into increasing per capita income as population growth fell from 2.7% in the early 1990s to around 1.5% until 2010. This is also the result of Lao becoming a member in regional and multilateral trade organizations (Kyophilavong and Lamphayphan, 2014).

As with other countries in Asia, the 1997 financial crisis was a major shock for the Lao economy with high rates of inflation and depreciation of the currency. In response, the government imposed stricter monetary policies to bring inflation under control. The downside, however, was a significant rise in foreign debt, in order to finance public programs. In agriculture, the government tried to modernize the rice sector with the aim to achieve food security. This included market liberalization, irrigation investments and the introduction of high-yielding varieties (HYVs) through cooperation with the International Rice Research Institute (IRRI). As a result, by 2010, rice area had almost reached 9 million ha, yet rice yields remained low, just a little bit above 1 t/ha on average. Other structural changes took place, such as the conversion of swidden agriculture and fallow forests to rubber plantations. Companies from China, Thailand and Vietnam established monoculture rubber estates in Laos. By 2010, over 180,000 hectares of land were planted to rubber in northern Laos alone (Baird, 2010). Overall, the performance of Lao agriculture was modest despite the attempt to promote contract farming for higher productivity (Goto and Douangngeune, 2017).

2011–2020 (Sixth Development Decade)

After 2010, until the time that Covid-19 brought the global economy down, the economy of Laos performed exceptionally well. With an annual growth of around 8% it was among the top in Asia. The government focused on economic growth and poverty reduction with the aim to overcome the Least Developed Country (LDC) status by 2020. One of the downsides, however, was a growing regional split into three economic zones. The western part, mostly along the Mekong River, was strongly influenced by the economy of Thailand. Here, economic development, urbanization, poverty reduction and an increase in wealth were highest. In the eastern part, with economic connections to Vietnam, development was more modest, while the central part of the country was lagging behind the most. On the other hand, Northern Laos has been increasingly influenced by China, albeit with few positive spillover effects for the socioeconomic development of Laos.

GDP growth has been driven by continuing foreign direct investment (FDI) primarily by China, Thailand and Vietnam. Major areas of investment were hydropower electricity, mining, including bauxite and potash, plantation agriculture and contract farming. Between 2008 and 2013, annual FDI has more than doubled, i.e. from about US$1 billion to over US$2.5 billion (Creak, 2014). Some projects, however, were of questionable economic viability and significantly increased the debt burden of Laos. Examples are the Xayaburi Hydropower Dam, costing some US$4 billion and the Kunming–Vietianne express train with an investment outlay of over US$7 billion.

In terms of agriculture and natural resources, deforestation continued to take place. Yet, Laos still has the highest forest cover percentage of countries in the region, with a calculated capital value of timber and non-timber forest products of some US$60 billion. To maintain this precious resource, the Laos Government in the context of its forest sector reforms, aims to expand participatory sustainable forest management (SFM) and forest restoration, to create green jobs and back rural livelihoods, support local industry and thereby meet Laos' contribution to climate action. Following the practice in other Asian countries, private commercial plantations are part of the plan, thus raising some questions as regards the outcome (Danyo, 2019).

Agricultural Development in the Post-Covid-19 Period

As summarized by Wilcox (2022) Laos, while initially coping well with Covid-19, suffered the full negative effects only in 2021 and recovery from the pandemic was sluggish. In fact, in 2022, the economic and debt situation, with high inflation escalated into a massive inflation and debt crisis. Nevertheless, infrastructural developments, including the launching of the prestigious Laos–China Railway in December 2021, have progressed. In fiscal policy, while the Lao government undertook measures to curb debt and inflation as well as to stabilize its currency, the economy remained in troubled waters. In fact, real GDP per capita has fallen to below US$2000 in 2023 and GDP growth was just 3.7% in 2023 with little prospect for improvement in 2024 and 2025 (World Bank, 2024). The large debt repayments are consuming most of the government revenues from exporting hydropower electricity, natural resources and income from tourism.

Growth in agriculture has been behind industry and service at less than 1% per year. Rice output and productivity have grown only modestly, as before, while self-sufficiency in rice was maintained. Also, there was a growing demand for non-rice crops such as maize, sugarcane and cassava (see Table 11.3).

The focus of modernizing agriculture and forestry has been on woody perennials and plantation crops, such as teak wood, eucalyptus/acacia and rubber. As reported by Smith *et al.* (2021), while teakwood is dominantly smallholder production, eucalyptus/acacia plantations are mostly companies. Almost half of rubber is companies, while the remainder is smallholder and contract farming with nearly equal shares. In agricultural statistics, this is shown in Table 11.3; while natural forest declined, plantation forest has increased more than 16-fold, i.e. from 18,000 ha in 2000 to 296,000 ha in 2022 (see Table 11.4).

Increasingly, agriculture in Laos is developing similar dualistic structures as those observed in Thailand. On the one hand, a partially subsistence sector with small-scale, low productivity, part-time farming and decreasing incomes. On the other hand, a commercial, high-tech sector

Table 11.3. Area by major crops over time in ha.

Year/Crop	1990	1995	2000	2004[1)]	2010	2015	2021	2022
Total paddy[2)]	663,746	559,889	700,000	770,320	927,114	968,638	939,602	952,084
Maize	48,050	28.,312	49,010	67,500	212,745	254,025	136,277	119,618
Root crops (cassava)	19,764	14,590	8000	23,908	40,840	101,885	121,578	210,702
Vegetables and fruits	8.599	13,715	40,000	107,150	126,270	179,690	n.a.	n.a.
Sugarcane	3,395	2,688	4,730	7,030	15,355	36,130	n. a.	n.a.
Coffee	17,277	20,158	42,250	37,576	50,595	77,535	83,320	89,400

Source: Ministry of Planning and Investment, Statical Yearbook 2015. Access online: https://laosis.lsb.gov.la/; Committee for Planning and Investment, 1975 – 2005 BASIC STATISTICS THE SOCIO-ECONOMIC DEVELOPMENT IN THE LAO P.D.R. Access online: https://laosis.lsb.gov.la/; Ministry of Planning and Investment, Statical Yearbook 2010. Access online: https://laosis.lsb.gov.la/. Notes: 1) 2005 data not available; 2) Including upland, lowland and dry season paddy; n.a. = not available.

Table 11.4. Area of natural and plantation forests since 2000, in thousand ha.

Year	2000	2005	2010	2015	2019	2022
Natural forest	14,032	13,852	13,551	13,231	13,034	12,846
Plantation forest [1)]	18	24	110	138	214	296
Total	20,281	20,034	19,593	19,531	19,420	19,387

Source: Department of Forestry, Ministry of Agriculture and Forestry. Available online: https://laosis.lsb.gov.la/; Note: 1) includes natural rubber.

with contract or company farming for mostly coffee and rubber. It remains to be seen if in the future, the sector will further split or eventually converge.

As shown in this chapter, agricultural productivity in Laos is low. On the other hand, the country is rich in natural resources, especially forests. Forests offer a range of income sources which are accessible to rural households in spite of government regulations that are often poorly enforced.

Case Study: How Much Household Income from the Forests? Source: based on Parvathi and Nguyen (2018)

In the following the summary of a case study by Parvathi and Nguyen (2018) is presented that estimated how much income rural households may actually take from the natural environment in which they live.

Since natural resource extraction is under strict regulation from the state, rural households tend to underreport income from natural resources, primarily forest products. Sometimes, in rural Laos, one can observe farm households with a lot of wood stored under the house. When asked, people say that this is for repairing the house.

In this study, the authors used panel data of some 600 households from a household survey in Savannakhet province, in 2013 and 2014. Income from natural resources was estimated following the principles of demand theory. Hereby, reported levels of consumption expenditures were used as a reference to predict the true income of rural households. The study found that when more than 25% of household income originated from natural resources, underreporting was more likely. The extend of underreporting was estimated at around 60%. The major reason for high natural resource extraction was found to be lack of other income opportunities, such as micro- and small-scale businesses and the high costs of implementing these, relative to just going to the forest. This especially applies to households with a high dependency ratio as well as poor households who do not have much choice other than engaging in illegal, although rarely sanctioned, natural resource extractions, in order to make ends meet. A consequence of distorted income data is that poverty measures are misguided. The study underlines the importance of consumption data for a correct assessment of welfare outcomes and poverty assessments. The study suggested that socioeconomic research in countries such as Lao PDR should examine the conditions under which poor households are inclined to provide incorrect information in household surveys. This should include more spatial and temporal coverage. Important policy implications of the study are that environmental conservation and rural development must be treated simultaneously. For example, establishing environment protection rules under a weak enforcement system will fail. Instead, rural households must be offered alternative income opportunities in the context of a comprehensive rural development strategy. The alternative, which is to let villages die and move people to the towns, as has happened in some cases, contradicts the concept of sustainable development with equity, fairness and environmental conservation.

Summary and Outlook

This review of economic development of Laos has shown that agriculture is facing many challenges but at the same time has opportunities. The transition from a command and control economy, under the heavy influence of Vietnam, towards a socialist-based market economy, has solved some old problems but has also brought along new ones. Foremost, there is the debt problem, as a result of heavy borrowing for mega projects that may not always have yielded the necessary high returns. Furthermore, unlike in Vietnam, economic reforms did not unlock agricultural growth with massive increases in productivity and export of rice and other agricultural commodities. There just has not been enough investment in rural infrastructure like roads or irrigation and agricultural research, and there has not been enough incentive and pressure to intensify agriculture. One reason for this, as shown in the case study in Box 11.1, is that there are still a lot of opportunities to extract natural resources and thereby supplement rural household income. By and large, Laos has remained an agricultural country where most households are still engaged in farming. While commercial farming has been developing, most farm households are subsistence oriented. In Table 11.5, the share of households who are engaged in farming and the percentage of villages where all households are in farming, is shown. The data are from agricultural censuses, unfortunately available only for 2005 and 2011.

Another indicator of structural change, or the lack thereof, is the development of farm size. In Table 11.5 average and standard deviation in 2005 and 2011 are shown. The data suggest that, similar to other Asian countries, average farms are getting smaller. However, some changes occurred in farm size distribution. The share of

Table 11.5. Farm size and farm size distribution in 2005 and 2011.

	Farm Size (ha)		Farm Size Distribution (%)	
Year/Parameter	2005	2011	2005	2011
Average (ha) **[<1 ha]**	1.85	1.84	33.61	17.79
STD **[1–2 ha]**	21.08	2.31	37.29	48.69
Min (ha)[1)] **[2–5 ha]**	0	0	26.05	32.04
Max (ha) **[>5 ha]**	17.85	8.50	3.02	1.47

Source: Ministry of Agriculture, Laos; Agricultural Census 2005 and 2011; Note: brackets in first column refers to farm size distribution presented in the third and fourth columns. 1) Households with no farming are included in the list and may lower the farm size numbers.

farms with less than 1 ha in 2011 was only one half of that in 2005. Yet the share of farms larger than 5 ha was half in 2011 as compared with 2005.

In its *Agricultural Development Strategy 2025*, the Ministry of Agriculture and Forestry (MAF) coined the vision that the agricultural sector can ensure food security, be competitive in the production of agricultural commodities, develop a clean, safe and sustainable agriculture which is modern, resilient and well-connected with overall rural development, and thereby contribute to the national economy (Bounthong and Manikam, 2021). Indeed, since agriculture in Laos did not follow the high intensification path of its neighbors, this also gives opportunities. As pointed out by Burana *et al.* (2023), crop production in rural Laos is "clean agriculture" by default. The use of chemical fertilizers is just 12 kg per hectare on average and in most remote areas no chemical inputs are used at all. This offers potential for organic and natural farming systems with rice, fruits and vegetables for export to higher income Asian countries, America and the EU. Environment- and health-conscious consumers may also be a growing market segment in Lao PDR. This requires government action to install the necessary institutional framework in order to facilitate these transitions.

References

Arndt, D. (1992) Foreign assistance and economic policies in Laos, 1976–86. *Contemporary Southeast Asia* 14(2), 188–210.

Baird, I.G. (2010) Land, rubber and people: Rapid agrarian changes and responses in Southern Laos. *Journal of Lao Studies* 1(1), 1–47.

Baird, I.G. and Shoemaker, B. (2007) Unsettling experiences: Internal resettlement and international aid agencies in Laos. *Development and Change* 38(5), 865–888.

Broegaard, R.B., Vongvisouk, T. and Mertz, O. (2017) Contradictory land use plans and policies in Laos: Tenure security and the threat of exclusion. *World Development* 89, 170–183.

Bounthong, B. and Manikam, D. (2021) Agriculture policy and institutional reforms in Lao-PDR: Experiences, impacts, and lessons. In: *Southeast Asian Agriculture and Development Primer, 2nd Edn*. SEARCA. College, Los Baños, Laguna, Philippines.

Burana, C., Sinavong, P. and Siripholvat, V. (2023) Clean agriculture development in Lao PDR: opportunities and challenges for food safety and market access. *Journal of ASEAN PLUS Studies* 4(1), 35–45.

Creak, S. (2014) Laos in 2013: International controversies, economic concerns and the post-socialist rhetoric of rule. *Southeast Asian Affairs* 1, 149–171.

Danyo, S. (2019) *Partnerships and Opportunities for a New Green Forest Economy in Lao PDR: Sustaining Forest Landscapes and Livelihoods (Vol. 2): Executive Summary*. World Bank Group. Available at: https://coilink.org/20.500.12592/kwxsfk (accessed 21 February 2025).

Department of Forestry, Ministry of Agriculture and Forestry, Laos (2022) Available online: https://laosis.lsb.gov.la/.

Ducoutoix, O., Laffort, J.-R. and Sacklokham, S. (2005) Policy and farming practices in Laos. *Development and Change* 36(3), 499–526.

FAO (2023) Lao PDR Investment Plans and Opportunities HiH Investment Forum 2023. https://www.fao.org/hand-in-hand/previous-editions/hih-IF-2023/laos-pdr/en? (accessed 05 March 2026).

FAO (2024). Statistical Yearbook. https://openknowledge.fao.org/server/api/core/bitstreams/d784864f-7f28-49d2-903e-6680d09a9d97/content/cd2971en.html (accessed 05 March 2026).

Goto, K. and Douangngeune, B. (2017) Agricultural modernisation and rural livelihood strategies: The case of rice farming in Laos. *Canadian Journal of Development Studies/Revue canadienne d'études du développement* 38(4), 467–486.

IBRD (1988). World Development Report 1988. International Bank for Reconstruction and Development; The World Bank. Washington, DC, USA.

Jain, O.P. (1972) Economy of Laos. *India Quarterly* 28(3), 236–254.

Kyophilavong, P. and Lamphayphan, T. (2014) Lao PDR country Report. In Zen, F. and M. Regan (eds.) *Financing ASEAN Connectivity, ERIA Research Project Report FY2013, No.15*. ERIA, Jakarta, pp. 131–165.

Ministry of Agriculture and Forestry, Laos; Agricultural Census 2005 and 2011. Available at: https://laosis.lsb.gov.la/.

Ministry of Agriculture and Forestry, Laos (2025) Agricultural Development Strategy 2025. Available at: https://www.maf.gov.la/wp-content/uploads/2016/01/MDS-2025-and-Vision-to-2030-Eng.pdf.

Nonthapot, S. and Wattanakul, T. (2014) Analysis of passenger transportation demand: Case study of the Customs House, the Thai-Laos Friendship Bridge, Nong Khai Province. *Journal of Social and Development Sciences* 5(1), 1–6.

Parvathi, P. and Nguyen, T.T. (2018). Is environmental income reporting evasive in household surveys? Evidence from rural poor in Laos. *Ecological Economics* 143, 218–226.

Phimphanthavong, H. (2012). Economic reform and regional development of Laos. *Modern Economy* 3(2), 179–186.

Pholsena, V. (2005) Laos in 2004: Towards Subregional Integration – 10 Years on. *Southeast Asian Affairs* 1, 171–188.

Rosser, A. (2006) Lao People's Democratic Republic. *IDS Bulletin* 37(2), 25–39.

Smith, H., Kanowski, P., Keenan, R.J. and Phimmavong, S. (2021). Lao plantation policy: prospects for change. *Forests* 12, 1132.

Time Magazine (1961). *Laos: The three Princes*. Time Magazine, New York, USA.

Wilcox, P. (2022). Laos in 2021:(im) mobility with infrastructure and COVID-19. *Southeast Asian Affairs 2022, Vol.* (1), 177–191.

World Bank (1983). The Lao People's Democratic Republic. A Country Economic Memorandum. Report No. 4125-LA. 189 p. The World Bank. Washington, DC.

World Bank (2024). *Lao Economic Monitor*. World Bank.

World Bank (2025). World Bank Open Data. Laos. https://data.worldbank.org/country/lao-pdr (accessed 05 March 2026).

Worner, W. (1989) Economic reform and structural change in Laos. *Southeast Asian Affairs* 16, 187–208.

Websites

https://data.worldbank.org/indicator/AG.LND.FRST.ZS?locations=LA

https://data.worldbank.org/country/lao-pdr.

https://en.wikipedia.org/wiki/Economy_of_Laos.

https://doi.org/10.3390/f12081132.

http://FAO.faostat/fao.org.

"Laos: The Three Princes". TIME Magazine. December 15, 1961. Archived from the original on February 5, 2011. Retrieved 2011-10-16.

World Bank online database 'World Development Indicators', available at http://databank.worldbank.org/data/home.aspx. * Asian Development Bank (ADB) online database 'Key Indicators for Asia and the Pacific 2012', available at www.adb.org/statistics

https://www.giz.de/en/worldwide/371.html.

[1] [PDF] appendix 1 past five-year national development plans. https://openjicareport.jica.go.jp/pdf/11707338_07.pdf

[2] [PDF] A short history of agricultural extension in Lao PDR. (n.d.). https://www.laofab.org/document/download/1629

[3] [PDF] Lao PDR: Governance Issues in Agriculture and Natural Resources. (n.d.). https://www.adb.org/sites/default/files/evaluation-document/35925/files/governance-issues-lao_6.pdf

[4] BBC News. (2018). Laos profile - timeline - BBC News. https://www.bbc.com/news/world-asia-pacific-15355605

[5] [PDF] Lao Agricultural Development Strategy 2020. (2010). https://www.gafspfund.org/sites/default/files/inline-files/5.%20LaoPDR_strategy.pdf

[6] [PDF] Agriculture Development Strategy to 2025 and Vision to the Year 2030. (n.d.). http://www.maf.gov.la/wp-content/uploads/2016/01/MDS-2025-and-Vision-to-2030-Eng.pdf

[7] Country Brief - The global food and economic crisis' impact on food ... (2024). https://reliefweb.int/report/lao-peoples-democratic-republic/country-brief-global-food-and-economic-crisis-impact-food-system-resilience-lao-peoples-democratic-republic-november-2023

[8] [PDF] Maximizing benefits from agricultural exports for Lao smallholder ... (2022). https://openknowledge.fao.org/server/api/core/bitstreams/73131b35-5683-454d-9bde-7a8931f14ad8/content

[9] UN Food and Agriculture Organization – Global Information and Early Warning System (GIEWS). (2001). Special Report: Lao PDR - 15 March 2001. https://www.fao.org/4/x9978e/x9978e00.htm

12 The Role of Biotechnology in Chinese Agriculture with a Case Study of Bt Cotton in Shandong Province

Shi Min, Wu Lifeng and Diemuth Pemsl

Abstract

This chapter provides a comprehensive review of biotechnology and genetically modified (GM) crops in China. It is based on an extensive literature review and the authors' own in-depth research in Shandong province on Bt cotton. The chapter begins with an overview of the regulatory framework and the implementation policies for genetically modified crops, particularly in the context of China's food security challenges. Next, it examines the adoption, diffusion and impact of Bt cotton, drawing upon a synthesis of a substantial body of existing literature. Overwhelmingly, Bt cotton is considered as a big success story in China. Hence, the chapter undertakes an assessment of the methodological challenges of benefit measurement. The third section presents the approach and findings of an in-depth case study in Linqing county in the province of Shandong. The results of the case study provide an interesting counterfactual to the large-scale impact studies by Chinese and international scientists.

The chapter concludes with an outlook for biotechnology in Chinese agriculture and some general lessons for the application of genetically modified and gene edited plants in Asia.

History of Biotechnology in China

Around the turn of the 21st century major international organizations and scientific institutions strongly encouraged the development and the commercialization of genetically modified organisms (GMO) for developing-country agriculture. The expectation was that GMOs would advance the Green Revolution of the 1960s and 1970s into a Gene Revolution, while avoiding the negative environmental effects of pesticide-fertilizer technologies of the former. This expectation was not unreasonable as the most prominent GMOs, ready for commercialization, were crop protection agents, namely insect-resistant and herbicide-tolerant crop varieties with the potential to reduce excessive applications of insecticides and herbicides. Great expectations were raised for regions which the Green Revolution had bypassed, such as sub-Saharan Africa, despite different ecological conditions as compared with the areas with widespread adoption of Green Revolution technologies, as well as institutional and geographic challenges existing there (e.g. Paarlberg 2000).

China invested early and substantively in the development of biotechnology. It now has the largest R&D capacity in this field outside the USA. Xiao and Kerr (2022) report that public R&D investment in biotechnology increased from US$26 million in 1986 to US$99 million in 2005 and reached a cumulative US$3.8 billion

 Asian Agriculture in a Climate Change-driven and Risky World (H. Waibel, S. Liebenehm and Ernah eds)
DOI: 10.1079/9781800627338.0012

over the period 2008–2020. China was also the first developing country to release GMO varieties, namely a virus-resistant tobacco variety in 1988, which, however, was later withdrawn.

Broader efforts in the research and development of GM crops started in 1986 when the Chinese government launched the 863 Program to support progress in high-tech research fields, including biotechnology (Huang and Wang, 2003). Supported by these special funds, research on insect-resistant cotton began in 1991 and led to the development of the *Guokang* variety by scientists at the Chinese Academy of Agricultural Sciences (CAAS) that was patented and approved for commercial use in four Chinese provinces in 1997. Around the same time, the USA-based company Monsanto received commercial approval for its Bollgard® cotton for Hebei province through a joint venture named JiDai established between the Provincial Government Seed Company, and Monsanto and Delta & Pine Land (Song, 1999).

Cotton was a priority crop for biotechnology R&D because of (i) its large production area and the role of the textile industry in trade and its historic importance in obtaining foreign exchange (Fang and Babcock, 2003); and (ii) the large amounts of pesticides used in growing the crop which resulted in high production costs for (mainly small-scale) farmers (Huang and Wang, 2002) and the resulting negative externalities to human health and the environment. Moreover, target pests had developed resistance against the main active ingredients of frequently used chemical pesticides (Wu et al., 1997; Huang and Wang, 2002; Wu et al., 2005).

Bt cotton was approved for commercial use in 1997. It is now the main GMO product in China and is widely regarded as a success story. In spite of that, China followed a cautious approach in regulating agricultural biotechnology, especially with food/feed crops such as rice, and to a lesser extent with maize and soybeans as animal feed.

The political motivation to develop, test, approve and release genetically modified gene edited plants is to improve domestic agricultural productivity and the level of self-sufficiency in view of rising levels of imports for maize and especially soybeans, as feed in the animal industry.

The Asia Society Policy Institute (2024) observed that China is facing a "genetically modified dilemma." On the one hand, China wants to curb the growing imports of feedstuff for its continuously expanding animal industry in view of increasingly volatile world markets and political uncertainty in trade. This is particularly the case for soybeans, where over 80% of domestic demand is imported from countries such as the USA, Brazil and other Latin American countries. Most imported soybeans are genetically modified products. On the other hand, also in China, there are consumer concerns about the safety of GMO products and therefore soybean products for direct human consumption such as tofu, soy sauce and milk are from domestically produced non-GMO varieties. In rice, consumer concerns are particularly pronounced, as found in a study with a survey of 564 consumers in eight provinces and cities by Huo and Liu (2022). Their results showed that consumer trust in scientific research institutions, food companies and the government is low, in the face of reports about loopholes in the supervision of GM rice, illegal cultivation and transaction of GM rice in the market, which had occurred repeatedly in the past, although neither Bt-rice nor vitamin B-enriched ("golden") rice has yet been approved for commercial planting. However, in December 2023, the Ministry of Agriculture and Rural Affairs (MARA, 2024) approved 37 genetically modified corn varieties and 14 GM soybean varieties. In the next section the argument of positive productivity, output increasing, self-sufficiency and food security effects of GMO crops will be examined.

GMOs, Agricultural Productivity and Food Security

The argument that GMO crop varieties can significantly contribute to the increase in output, increase productivity and improve food security in China must be carefully examined. A cautious scrutiny of this claim is especially needed, because currently approved and commercially planted GMOs are crop protection agents, i.e. insect resistant and herbicide tolerant varietal traits. Some proponents of GMOs (Qaim and Zilberman, 2003) claimed that the positive yield effects are substantial. This has prompted some researchers (Jin et al., 2024) to demand speedier regulatory and approval processes for GMOs, with the argument of huge welfare losses from delayed commercialization. More critical voices (Pretty, 2001), however, question positive yield effects and rather point to the numerous environmental and health risks of GMOs.

In the following the arguments of increasing yields through GMOs versus alternative ways of output expansion are discussed based on recent literature.

Starting out with rice, where the economic pressure to increase domestic output is less relevant, a study by Deng et al. (2019) has analysed the exploitable yield gap of Chinese rice farmers, i.e. comparing current farm yields with the potentially possible yield. They found that the yield gap could be substantially reduced in both single and double crop rice production systems with improved crop management, through investment in human capital and better monitoring techniques of crop growth. In this way, more judicious and precise input applications could be made with tactical modifications to field conditions during the growing season, in response to weather conditions and expected yield levels. The authors conclude, without any mention of GM rice, that China does have the ability to maintain its current level of self-sufficiency in rice for years to come (Deng et al., 2019).

In maize, only recently, Bt maize has been released for commercial use. Currently, the area is still small, less than 10% of the potential. There have been reports (Reuters, 2025) of "poorer-than-expected" performance of Bt maize in terms of yields. This may have cautioned the government to carry on with its roll-out programs. Once again, it is worthwhile to review findings about other ways to increase domestic maize output. For example, Luo et al. (2023), using data from some 400 crops experiment stations across China, applied a combined data-driven and a machine learning approach and showed that maize yields in China could be doubled if optimal plant density and improved management of soil organic matter would be implemented by farmers. The authors conclude that China could become self-sufficient in maize by 2030. Once again, GMO maize is not mentioned in this study, as a means to increase output.

In soybean, the self-sufficiency rate (SSR) in China is very low, i.e. in the order of 20%. As shown by a study of Wang et al. (2023), the possibilities to increase soybean output are limited. The major options are area expansion, and, to lesser extent, closing the yield gap. Both options are costly and would not solve the problem of dependency on imports. Again, their paper, published in an agronomy journal, does not make mention of GMO soybean as a possibility to raise domestic production.

As regards the field performance and economic impact of GMO maize and soybean, press reports from the Ministry of Agriculture and Rural Affairs (MARA, 2023) claim that in both crops, the resistance traits were found to be effective, and yields were increased by 5.6–11.6%. The information was based on experiment station results and pilot programs in 20 counties in the provinces of Hebei, Inner Mongolia, Jilin, Sichuan and Yunnan.

Surprisingly, unlike in cotton, no scientifically rigorous and independent impact studies have been undertaken for maize and soybean. This is unfortunate because there was the possibility to implement proper experimental designs with baseline surveys before the introduction of GM maize and soybean, and randomized, with-and- without comparisons, thereafter, allowing to perform difference-in-difference models.

In cotton, which has been the flagship GMO in China, numerous impact studies on Bt cotton have been carried out. These will be reviewed in the next section of the chapter. While Bt cotton undoubtedly can have positive yield-effects in bollworm-infested area, when chemical pesticides are no longer effective, improved pest control technologies are not the only means to raise cotton productivity. In a review paper, Dai and Dong (2014) submit that the current approach to cotton production in China should be profoundly changed. They recommend to replace the current intensive technology with a more sustainable system, including measures to reduce soil pollution from plastic and chemicals, through better management and mechanization. The authors acknowledge that pests are a major constraint in cotton, while blanket applications of pesticides are ineffective, aside from causing significant negative externalities. However, they also point out that Bt cotton is only effective against a small range of cotton pests and therefore can just be one of several components of a sustainable cotton management system (Dai and Dong, 2014).

The discussion on the productivity effects of currently available genetically modified crop varieties has shown that their contribution to increase crop yields in China must be assessed against the background of other non-GM technologies. It is important to emphasize that currently available, genetically modified and gene edited products are mostly crop protection agents. As clearly pointed out in a seminal paper (Lichtenberg and Zilberman, 1986), crop protection agents do not directly increase yields, but they reduce losses from pests. Insect-resistant GMOs, to some extent, can substitute pesticides and can also reduce crop losses if other control tactics fail. Similarly, depending on existing weed management regimes, herbicide-tolerant varieties allow farmers a more targeted control of weeds with minimum herbicide damage to crops. In this way, they can have positive yield effects.

In summary, GMO varieties with pest control traits must be assessed in the context of a "damage control world," i.e. the potential damage determines the benefit. If crops can be managed following an integrated crop and pest management approach, the probability of pests reaching significant economic levels is small and hence the necessity for chemical pesticides and respective GMOs is low. Ignoring integrated crop management principles, however, leads to overuse of pesticides, of which there are many examples in China (e.g. Xu et al., 2008; Zhang and Guanming, 2015; Yang et al., 2019; Sun et al., 2020; Cai et al., 2021). This has been the case in many cropping systems in China and other countries and has led to a dependence on pesticides and/or GMOs, keeping farmers in a kind of treadmill.

Impact Studies of Bt Cotton

The first internationally published impact study of Bt cotton in China was by Pray et al. (2001) in *World Development*. The study was based on a farm survey in 1999 with a sample of 283 cotton farmers in Northern China. Some 70% were growing Bt cotton and the remainder, i.e. 79 farmers planted non-Bt cotton. At that time Bt adoption was still in its early phase with about 10% of the total cotton area in China. The study found that Bt adopters significantly reduced pesticide use against the cotton bollworm, without any negative effect on yields. The study also reported fewer pesticide poisonings of Bt cotton farmers as compared with those planting conventional cotton.

In a follow-up study, published one year later, Pray et al. (2002) used survey data from cotton farmers in northern China, initially in the provinces of Hebei and Shandong in 1999, adding Henan in 2000 and Anhui and Jiangsu in 2001. The study confirmed the earlier findings as regards pesticide reduction in all three survey years. As can be derived from their descriptive statistics (Pray et al., 2002, table 2, p. 427), average pesticide use across years varied considerably, i.e. ranging from 11.9–32.9 kg/ha for Bt adopters and from 48.5–87.5 kg/ha for conventional cotton. In addition, the study found increased yields, labor savings and fewer pesticide poisoning cases for Bt adopters.

In an in-depth study, using data from Bt and non-Bt cotton farmers in five counties of two provinces, namely Shandong and Hebei, Huang et al. (2003) aimed to answer the question, to what extent genetically engineered (GE) crops in China can lead to reductions of pesticide use. The authors performed a descriptive and an econometric analysis, developing a pesticide use model. The data referred to the 1999 cropping season based on 282 farmers with 382 plots out of which 45 plots were non-Bt and the remaining majority were plots planted to different Bt varieties. In their descriptive statistics, the authors found large differences in pesticide use between Bt and non-Bt plots (Huang et al., 2003, table 5, p. 61). In Bt plots, the average number of pesticide applications per cropping season was 6.6, the average quantity was 11.8 kg/ha and expenditures for pesticides amounted to $32/ha on an average of 337 Bt plots. The corresponding figures for the 45 non-Bt plots were a stunning 19.8 applications, 60.7 kg/ha and $178/ha pesticide expenditures. The authors also reported (Huang et al., 2003, table 6, p. 62) that the net revenue in the non-Bt plots was negative on average, while it was consistently positive for all the Bt variants. The econometric analysis with three variants of pesticide use models (frequency, quantity, expenditures) showed that the coefficients for Bt varieties were significant and negative, thus confirming the hypothesis of a pesticide reduction effect of Bt varieties (Huang et al., 2003, table 7, p. 64). The authors acknowledge that aside from genetically modified, insect-resistant varieties, pesticide use can also be reduced by programs on integrated pest management (IPM). However, according to the paper, such programs have failed because extension agents charged with promoting IPM, earn part of their income from selling pesticides to farmers and therefore have no incentive to promote pesticide reduction.

A less optimistic view as regards the capacity of Bt cotton to sustainably solve the pest and pesticide problem in cotton production in China, has been portrayed in the paper of Wang et al. (2008). The authors had collected additional data in 2004 and combined these with earlier survey data available from 1999 onwards. They found that the initially large benefits from Bt cotton had eroded, mainly due to increasing pesticide use against secondary pests. The authors warned not to rely on genetically modified bollworm resistant cotton varieties as a sole solution.

Similar concerns were raised again later by Qiao and Yao (2015). Based on national statistics and empirical studies it was hypothesized that economic benefit of Bt cotton were decreasing when adoption had almost reached 100%, linking this to the decline in cotton area in China. However, the study found other factors responsible for this decline, such as rising costs of labor and lower cotton prices. Similarly, Qiao (2015), using nationally representative long panel data between 1997–2012, maintained that economic benefits continue, even 15 years after the commercialization of Bt cotton. In a further paper on the topic of long-term benefits, Qiao et al. (2016) acknowledged that benefits of Bt cotton, 10 years after commercialization, had declined relative to the early adoption years, due to increased pesticide use against bollworm. At the same time, the authors claimed that the technology had generated spillover benefits to non-adopters.

In response to the reality that although insect-resistant Bt cotton is praised for its ability to reduce pesticide use, Bt cotton farmers in China are still overusing pesticides. Liu and Huang (2013) carried out a survey and an artifactual field experiment in four cotton growing provinces in China. They found that the use of pesticides is related to farmers' risk preferences. Farmers who exhibit a higher degree of risk aversion tend to use larger quantities of pesticides, whereas those who are more loss averse tend to use smaller quantities.

A paper by 10 expert authors in the prestigious journal *Proceedings of National Academy of Science USA* (PNAS) (Zhang et al., 2018) showed that the large-scale expansion of Bt cotton in China on the one hand has played a critical role in regulating cotton bollworm but on the other hand has also led to effects on other cotton pests. Interestingly, the paper found that farmers continue to spray against bollworm at non-damaging levels in addition to insecticide applications against secondary pests, such as aphids. The paper also points out that other factors such as land use patterns and climate change influence the pest situation in cotton as well as in other crops.

The paper of Wang and Fok (2018) brought new aspects to the approach of impact assessment of Bt cotton in China. The authors took into account the farming background and asked farmer respondents about their level of satisfaction with Bt cotton. They found that 75% of farmers expressed dissatisfaction with Bt varieties. The authors distinguished between full-time and part-time farmers. The former group sprays less pesticides, as a result of more intensive field monitoring. Farmers who carry out farming as a side activity apply pesticides routinely in disregard of the pest situation in their cotton fields and therefore spray more often.

To conclude the review of impact studies on Bt cotton in China, Qiao and Huang (2020) scaled up benefits to all of China based on their 7-year, imbalanced panel data, between 1999 and 2012 in selected provinces. The total number of farmers in their sample was 1837 with 4127 plots, out of which 93% were Bt plots. As shown by their Bt adoption data, there were only few non-Bt plots left after 2007. The authors calculated US$8.46 billion savings on pesticides over the period from 1997 to 2012. They stated that this number will more than double if the benefits of increased yield and decreased labor use are considered. Furthermore, they calculated savings in the quantity of pesticides used with 2.19 million tons nationally, during the same period.

While calculations like this are useful exercises to stimulate discussion, at the same time, they can raise questions as regards the underlying methodology and data used.

Using table 2 of the Qiao and Huang (2020) paper, a graph was produced (Figure 12.1) to

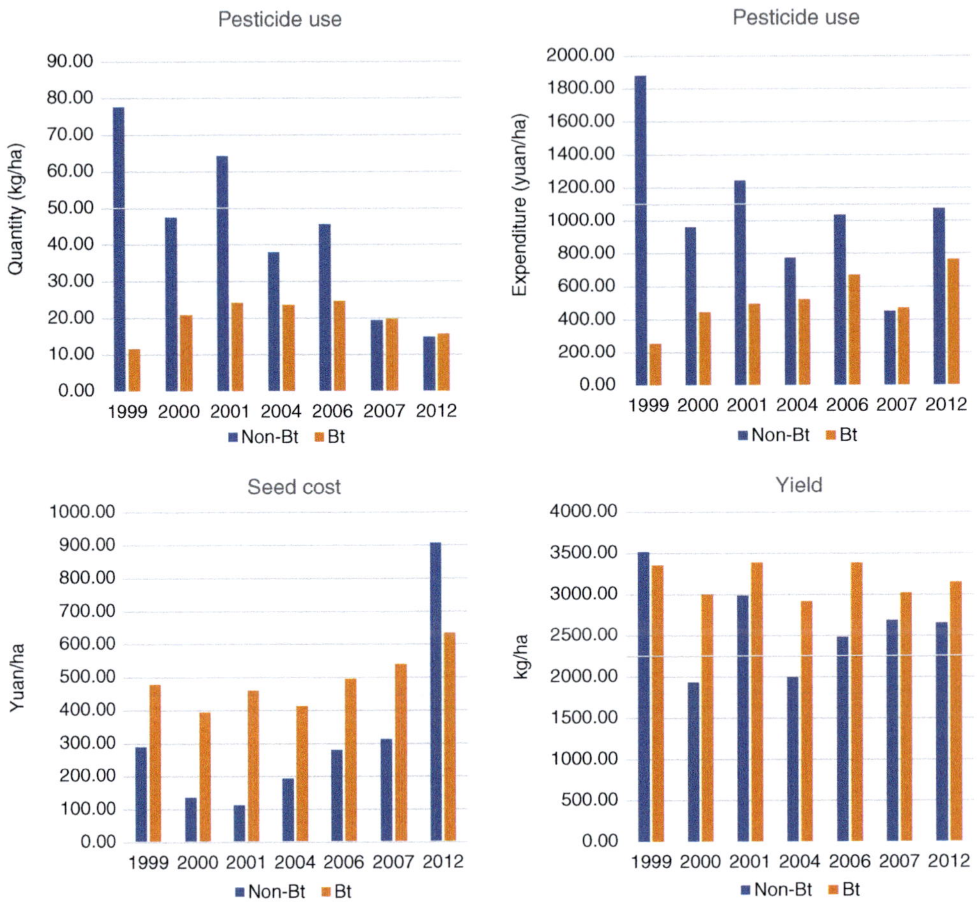

Figure 12.1. *Pesticide use, seed cost and yield in Bt and non-Bt fields, 1999–2012.* Data Source: Qiao and Huang (2020).

show quantity (kg/ha) and expenditures (Yuan/ha), costs of cotton seeds (Yuan/ha) and cotton lint yield (kg/ha). It is interesting, that, for example, in 1999, yield of non-Bt was higher than for Bt, in 2007 and in 2012, pesticide use was higher for Bt than for non-Bt and in 2012, seed costs for non-Bt were about 30% higher than for Bt varieties.

The most notable point about the study of Qiao and Huang (2020) is the fact that pesticide use on average, over the 7-year period, is still around 20 kg/ha, despite the use of genetically modified insect-resistant varieties. This is consistent with the results reported in earlier papers that pesticides continue to be overused in Bt cotton. However, studies under the FAO-EU IPM Program for Asia showed that much lower levels of pesticide use in cotton in China are possible. For example, Pananurak (2010) analyzed survey data from 535 cotton farmers in the provinces of Anhui, Hubei and Shandong in 2000 and 2002. Farmers who participated in IPM training were able to reduce the quantity of insecticides from 16.12 kg/ha/season to 8.46 kg/ha/season. The average number of insecticide applications per season was reduced from 14.54 to 6.56 while at the same time increasing cotton yields by almost 20% (see Pananurak, P. 2010, table 5.1, p. 70). Wu (2010) in a study under the same program, used an even larger sample of

1577 farmers in nine counties in the provinces of Shandong, Anhui and Hubei. He applied a difference-in-difference method with baseline data collected in 2000 and post-training data collected in 2001 and 2004. Post-training data were based on actual monitoring of pesticide use instead of recall. In addition, Wu (2010) applied various economic approaches, including a damage control function approach. He found that in addition to Bt (94% of farmers in the sample had used Bt), IPM training reduced pesticide use and increased yields. Unfortunately, the cotton IPM program had been discontinued after international funding by FAO had ended.

Naturally, the series of impact studies on Bt cotton in China (and other countries) caused reactions from the international scientific community, discussing the validity of the studies and the trustworthiness of the results. For example, Glover (2010) found several shortcomings, such as lack of acknowledgement of year-to-year variation in pest pressure and the lack of convincing causality. Smale et al. (2006) assessed 47 peer-reviewed articles that applied stated economics methods to measure the farm-level impacts of Bt cotton in developing agriculture since 1996. In general, the authors found empirical evidence of their success, at the same time, they observe methodological limitations, such as the studies ignoring the institutional and political context, the limitations of the dominantly partial budgeting nature of the studies, data quality problems and memory biases in recall data, as well as lack of a convincing causality-proof methodology. They also pointed out the outstanding importance of institutional arrangements in seed markets to assure the quality of the product. In a more recent study, Smale (2017) submits that while GM crops can help to avoid damage from pests and diseases and thus are useful for small-scale farmers, they alone cannot reduce poverty and foster rural development. Conducive institutions and enhancement of farmer knowledge are necessary conditions in this regard. In a theoretical paper in *Ecological Economics*, Pemsl et al. (2008) cautioned to draw conclusions from simple "with and without" comparisons in crop protection, when ecosystems are highly disrupted. When the ecology has been severely disturbed by prior excessive use of pesticides, it is not surprising that one can observe pesticide reduction and positive yield effects of Bt varieties in the short term. However, this is no proof, whatsoever, that the problem can be solved by a single technology such as Bt cotton varieties. Pemsl et al. (2008) suggested studies to apply a difference-in-difference approach to measure impact. However, the speed of Bt adoption in China has made this approach difficult to implement.

To conclude, the studies on the impact of Bt cotton in China have provided a wealth of empirical information and scientifically valid experiences with different methodological approaches. It can be concluded that assessing the impact of genetically modified crop varieties will remain a challenge, just as will be the case for any other technologies. It has also become clear from this review that Bt varieties are not a "silver bullet" that can solve all pest problems or save the world from famine and food insecurity.

Further insights about Bt cotton in China will be presented in the next section with an in-depth case study carried out over a period of 12 years in Linqing County in Shadong province.

Case Study: Bt Cotton in Shandong Province

The Institute of Development and Agricultural Economics of Leibniz University Hannover, Germany, in cooperation with partners in China and internationally, has conducted a number of studies coined as PhD and MSc research on cotton pest management in China. The two PhD researches of Piyatat Pananurak and Wu Lifeng on the impact of Bt and Farmer Field School in cooperation with the FAO were mentioned above. A long-term research project over a period of 12 years on Bt cotton in several villages in Linqing County, Shandong Province, over a period of 12 years (2001–2013) was carried out. Shandong province was an important cotton production area in China, while Linqing County is the location of the experimental station of the provincial cotton research center. The aim of this in-depth case study was to provide additional insights about the impact of Bt cotton complementary to the large-scale impact studies reviewed in the previous section.

The results of the case study which consists of a serious of sequential studies can help to improve the understanding of the underlying mechanisms of the impact of Bt cotton in China and may answer some of the remaining questions from the review of impact studies presented in the previous section. In Table 12.1 an overview of the research activities and the corresponding final research products are presented.

The research in Linqing County started in 2001 with the PhD research of Diemuth Pemsl (Pemsl, 2006). In 2002, she initiated a socio-economic survey among 150 cotton farmers in Linqin County, Shandong Province, including season-long monitoring of pesticide use and other inputs in order to minimize measurement and recall errors. She also collected leaf samples from cotton varieties in farmers' fields for laboratory analysis of Bt toxin content of the cotton varieties planted by farmers.

In 2005, follow-up PhD research was carried out by Wu Lifeng, a staff member of NATESC, the National Agro-tech Extension and Service Centre of China. He re-surveyed the farmers of the first study and added more respondents with the aim to assess the impact of the FAO-initiated IPM program using the Farmer Field School approach.

The panel data set from both studies was then used by Mark Völker in his Master thesis of 2006, conducting a panel data analysis of the impact of Bt cotton on pesticide use and productivity.

Six years later, Nina Dohmeier, a Master student of Hannover, went to Linqing County and re-surveyed the 150 case study farmers in 2012 and again in 2013, with the aim to find out why many Bt cotton farmers had stopped growing cotton. Based on the data collected in 2012 and 2013, Shi Min, who had received his

Table 12.1. Empirical research on Bt cotton in Shandong Province by Leibniz University Hannover, Germany. (Authors' own compilation).

Year	Researcher	Research Activity	Reference/Publication
2002	Diemuth Pemsl	Case study with survey of 150 cotton farmers in Linqing County, Shandong Province, including season-long monitoring of pesticide use and leaf sampling for Bt toxin analysis of cotton varieties planted by farmers	Pemsl, D.E. (2006). Economics of agricultural biotechnology in crop protection in developing countries: The case of Bt-cotton in Shandong Province, China. Pesticide Policy Publication Series. Special Issue No. 11. Leibniz University Hannover, Germany
2005	Lifeng Wu	Re-survey of 150 case study farmers plus additional farmers to assess impact of Farmer Field School project	Wu, L. (2010). Farmer field school and Bt cotton in China: An economic analysis. Pesticide Policy Publication Series. Special Issue No. 15. Leibniz University Hannover, Germany
2006	Marc Völker	Panel Data Analysis of 2002 and 2005 data from 150 of case study farmers for Bt cotton impact assessment	Völker, M. (2006) Economic impact of genetically modified cotton varieties in China: a panel data analysis. MSc Thesis Leibniz University Hannover
2012	Nina Dohmeier	Re-survey of 150 case study farmers from 2002 in 2012 and 2013; cotton dis-adoption model	Dohmeier, N. (2012) Plant Biotechnology and Rural Development: A Case Study from Shandong Province, China. MSc Thesis. Leibniz University Hannover
2013	Shi Min	Analysis of data collected in 2012 and 2013; modeling cotton dis-adoption and rural transformation	Min, S. (2017). The reduction of cotton cultivation in Shandong Province of China: Does off-farm employment matter? *China Agricultural Economic Review*, *9*(4), 607–622.

PhD degree from Leibniz University in 2016, wrote a paper to answer this question.

The data from this case study in Shandong province have resulted in numerous publications which are briefly summarized below, for drawing a number of interesting conclusions.

The first scientific paper from the Shandong case study was presented at the 7th International Consortium on Applied Bioeconomy Research (ICABR) Conference in Ravello, Italy (Pemsl et al., 2003). The analysis was based on the survey of 150 case study farmers. Estimation of a damage control function allowed to compute the marginal productivity of the two damage abatement agents, Bt trait and chemical pesticides. Unlike in other papers (e.g. Qaim and Zilberman, 2003) who incorporated Bt as a dummy variable, the Bt toxin measurements was incorporated as a continuous variable, allowing more precise measurements. Results showed that both the pesticide and the Bt variables were insignificant.

A follow-up paper (Pemsl et al., 2005), submitted an explanation why Bt cotton farmers in Shandong continue to spray against cotton bollworm. They found that a major reason was the highly variable effectiveness of the Bt resistance trait due to the often low toxin levels. The paper pointed out the need to better regulate and monitor cotton seed markets.

In the paper published in *Agricultural Systems* in 2007 (Pemsl and Waibel, 2007), a bio-economic model was developed, making use of the data collected in Linqing County. The paper compares the economic performance of different crop protection strategies in cotton including the use of transgenic varieties in Shandong Province, China. Applying a Monte Carlo simulation model with three strategies, namely conventional insecticide strategies, planting of bollworm-resistant Bt varieties and the combination of both. Results showed that a combination of Bt with judicious, need-based application of pesticides is the superior strategy. The authors concluded that agroecosystems variables, i.e. pest pressure and population levels of beneficial organisms need to be considered in impact assessment of Bt cotton.

A paper presented at the 34th German Development Economics Conference (Pemsl et al., 2008) used panel data from 2002 and 2005, collected from the case study farmers in Linqin County. The descriptive analysis revealed that pesticide use is increasing while quality of Bt varieties decreased. In the damage control function, the Bt-variable was consistently insignificant in various model variants, raising concerns as regards the effectiveness of the resistance trait and offering an explanation why farmers continue to spray frequently against bollworm.

This aspect was further explored in a paper, published in the *American Journal of Agricultural Economics* (Kuosmanen et al., 2008) using the 2002 data. The authors developed an advanced form of the damage control function framework, applying a two-stage, semiparametric approach. On the basis of the model, a plausible explanation for the continuous high levels of pesticide applications in Bt cotton was provided. The authors concluded that farmers' decisions can be explained in the light of decreasing effectiveness of Bt varieties and increasing pest populations, although based on the marginal productivity analysis, farmers tend to overuse pesticides.

Wu (2010) used the panel data from Linqing County and compared three groups of farmers: (1) trained on IPM in farmer field schools, (2) neighboring (exposed) farmers but non-participants of IPM training and (3) untrained and unexposed farmers in other villages. He found lower pesticide use and higher yields for trained over untrained farmers. He also observed some positive spillover effects for pesticide reduction but no yield effects on exposed farmers.

Applying the damage control function method and using the Linqing panel data of 2002 and 2005 (Pemsl et al., 2011), confirmed the findings of decreasing effectiveness of Bt cotton and more pesticide applications in response to secondary pests. The paper also found that farmers with more knowledge in cotton crop and pest management used lower levels of pesticides.

In 2012, a Master's student from the Leibniz University Hannover, Nina Dohmeier, in cooperation with Chinese partners re-surveyed the 150 case study farmers in Linqing County. She found that meanwhile about two-thirds of the farmers had stopped growing cotton. Using an endogenous switching model, she showed that the decision to dis-adopt cotton was economically superior. A major reason for this was that many farmers increasingly engaged themselves in complementary off-farm wage employment, effectively making them part-time farmers. In 2013, she carried out another survey of the case study farmers in Linqin County.

Professor Shi Min, now at Huazhong Agricultural University in Wuhan, who graduated from Leibniz University in 2016, further examined the impact of off-farm employment on the reduction of cotton cultivation using the data collected in 2012 and 2013. By that time, there were only 144 farmers left as six households had given up farming altogether. Based on the panel data, he explored possible causes of the reduction of cotton cultivation in Shandong Province, particularly examining the role of off-farm employment in the light of expanding rural labor markets. He developed a simultaneous equations model, using a three-stage least squares approach. His results showed that the emergence of off-farm labor markets was a major reason for the reduction of cotton acreage in Shandong Province. His paper is important in relation to the assessment of Bt cotton, as it shows that a crop/pest management technology, no matter how good (or bad) it is, cannot determine the path of rural transformation and economic development.

It is believed that overall, the findings of this case study in Linqing County, Shandong province in China make an important contribution to the discussion on the impact of Bt cotton in China. It must of course be acknowledged that it is just a case study of one single county. The bounty of findings, well documented in scientific papers and published in respectable journals, underlines the importance of this case study.

Conclusions

This review of biotechnology development in China aimed to draw lessons that can be learned from the assessment of a technology that is of high concern not just for scientists but for the wider public as well. In this regard, the numerous studies that were conducted in China are highly educational, for Asia and the rest of the world. While in many countries the debate is stuck in biosafety concerns, China has been progressive in pushing forward in what was often lacking, namely cost-benefit analyses of the technology, in this case, Bt cotton. These studies made the arguments transparent and have helped the discussion to become more rational.

Unfortunately, such studies are absent for other GMO crops such as maize and soybean. This is ill-fated, because here, the possibility would exist, to apply a proper difference-in-difference approach, with surveys before and after introduction, in treatment and control groups. The lack or absence of such studies was exactly the fear raised by Zadoks and Waibel (2000), who compared the introduction of genetically modified plants in world agriculture with the introduction of chemical pesticides (e.g. DDT) and weed killers, (e.g. 2,4 D) some 50 years earlier. For both technologies, there was initially great enthusiasm and a strong belief of having a new "*silver bullet technology*" to solve the world's food security problems. The authors predicted that for GMOs, the debate would be dominated by safety concerns and regulatory issues and therefore would suffer from a lack of scientifically rigorous cost-benefit assessments, prior to the release and commercialization of GMOs.

Another lesson from GMO cotton in China is that one cannot praise genetically modified insect-resistant crop varieties as a key-turn technology for increasing productivity. After all, as mentioned before, such products are pest control agents that reduce damage from pests but cannot increase yields directly. Also, they can have negative externalities, such as secondary pests.

There are GMO products with other traits, such as drought resistance, fertilizer efficiency or improved nutritional contents of plants, now in the research and development pipeline. However, the expectations that were raised in 2010 in a *Science* paper (Godfray et al., 2010) of what would be out in the field after 5, 10 or 15 years, i.e. in 2015, 2020 and 2025, did not materialize. As noted by Noack et al. (2024) in another paper in *Science*, "large-scale GM crop adoption has been limited to just two traits: herbicide tolerance and insect resistance". And, "GM and gene-edited crops and traits should not be seen as a substitute for good agronomic practices but should be integrated smartly into sustainable production systems," This statement can be underlined through the findings of the agronomic research studies on rice, maize, soybean and cotton, carried out in China and cited above. Improved crop management, using modern scientific tools, are indispensable for growth in productivity and sustainable agriculture.

References

Asia Society Policy Institute (2024). China's Genetically Modified Dilemma; https://asiasociety.org/policy-institute/chinas-genetically-modified-dilemma. Access date 30.08.2025.

Cai, J., Xiong, J., Hong, Y., and Hu, R. (2021). Pesticide overuse in apple production and its socioeconomic determinants: Evidence from Shaanxi and Shandong provinces, China. *Journal of Cleaner Production*, 315, 128179.

Dai, J., and Dong, H. (2014). Intensive cotton farming technologies in China: Achievements, challenges and countermeasures. *Field Crops Research*, 155, 99-110.

Deng, N., Grassini, P., Yang, H., Huang, J., Cassman, K. G., and Peng, S. (2019). Closing yield gaps for rice self-sufficiency in China. *Nature Communications*, 10(1), 1725.

Dohmeier, N. (2012) Plant Biotechnology and Rural Development: A Case Study from Shandong Province, China. MSc Thesis. Leibniz University Hannover.

Fang, C., and Babcock, B. (2003). China's Cotton Policy and the Impact of China's WTO Accession and Bt Cotton Adoption on the Chinese and U.S. Cotton Sectors. Food and Agricultural Policy Research Institute (FAPRI) Publications (archive only) 03-wp322, Center for Agricultural and Rural Development (CARD) at Iowa State University.

Glover, D. (2010). Is Bt Cotton a pro-poor technology? A review and critique of the empirical record. *Journal of Agrarian Change*, 10(4), 482–509.

Godfray, H. C. J., Beddington, J. R., Crute, I. R., Haddad, L., Lawrence, D., Muir, J. F., ... and Toulmin, C. (2010). Food security: the challenge of feeding 9 billion people. *Science*, 327(5967), 812–818.

Huang, J., and Wang, Q. (2002). Agricultural biotechnology development and policy in China. *AgBioForum*, 5(4), 122–135.

Huang, J. and Wang, Q. (2003) Biotechnology policy and regulation in China. Working paper series, 195. Brighton: IDS.

Huang, J., Hu, R., Pray, C., Qiao, F., and Rozelle, S. (2003). Biotechnology as an alternative to chemical pesticides: a case study of Bt cotton in China. *Agricultural Economics*, 29(1), 55–67.

Huo, L., and Liu, Y. (2022). Genetically modified rice: Do Chinese consumers support or go against it? *Frontiers in Psychology*, 13, 813862.

Jin, Y., Kristkova, Z. S., Kardung, M., and Wesseler, J. (2024). Impacts of accelerating agricultural R&D transfer on global food security. *GM Crops & Food*, 15(1), 1–12.

Kuosmanen, T., Pemsl, D., and Wesseler, J. (2006). Specification and estimation of production functions involving damage control inputs: A two-stage, semiparametric approach. *American Journal of Agricultural Economics*, 88(2), 499–511.

Lichtenberg, E., and Zilberman, D. (1986). The econometrics of damage control: why specification matters. *American Journal of Agricultural Economics*, 68(2), 261–273.

Liu, E. M., and Huang, J.K. (2013). Risk preferences and pesticide use by cotton farmers in China. *Journal of Development Economics*, 103, 202–215.

Luo, N., Meng, Q., Feng, P., Qu, Z., Yu, Y., Liu, D. L., ... and Wang, P. (2023). China can be self-sufficient in maize production by 2030 with optimal crop management. *Nature Communications*, 14(1), 2637.

Min, S. (2017). The reduction of cotton cultivation in Shandong Province of China: Does off-farm employment matter? *China Agricultural Economic Review*, 9(4), 607–622.

Ministry of Agriculture and Rural Affairs – MARA (2023). https://www.moa.gov.cn/ztzl/ymksn/xhsbd/202308/t20230825_6435073.htm

Ministry of Agriculture and Rural Affairs (2024). http://www.moa.gov.cn/govpublic/nybzzj1/202312/t20231207_6442285.htm (in Chinese)

Noack, F., Engist, D., Gantois, J., Gaur, V., Hyjazie, B. F., Larsen, A., ... and Kremen, C. (2024). Environmental impacts of genetically modified crops. *Science*, 385(6712), eado9340.

Paarlberg, R. (2000) The Global Food Fight. *Foreign Affairs*, 79(3), 24–38.

Pananurak, P. (2010). Impact assessment of farmer field schools in cotton production in China, India and Pakistan. Pesticide Policy Project Publication Series, Special Issue No. 14; Inst. of Development and Agricultural Economics, Leibniz University Hannover.

Pemsl, D. E., Waibel, H., and Gutierrez, A. P. (2003). Productivity analysis of Bt cotton: a modelling approach. In *Proceedings: Productivity, Public Goods and Public Policy: Agricultural Biotechnology Potentials.* 7th International ICABR Conference, Ravello, Italy.

Pemsl, D., Waibel, H., and Gutierrez, A. P. (2005) 'Why Do Some Bt-Cotton Farmers in China Continue to Use High Levels of Pesticides? *International Journal of Agricultural Sustainability*, 3(1), 44–56.

Pemsl, D. E. (2006). Economics of agricultural biotechnology in crop protection in developing countries: The case of Bt-cotton in Shandong Province, China. Pesticide Policy Publication Series. Special Issue No. 11. Leibniz University Hannover, Germany.

Pemsl, D., and Waibel, H. (2007). Assessing the profitability of different crop protection strategies in cotton: case study results from Shandong Province, China. *Agricultural Systems*, 95(1–3), 28–36.

Pemsl, D. E., Völker, M., Wu, L., and Waibel, H. (2008). Challenging the miracle of transgenic crops in China: findings from a case study in cotton using panel data. *Proceedings of the German Development Economics Conference*, Zürich, No. 34.

Pemsl, D. E., Gutierrez, A. P., and Waibel, H. (2008). The economics of biotechnology under ecosystem disruption. *Ecological Economics*, 66(1), 177–183.

Pemsl, D. E., Voelker, M., Wu, L., and Waibel, H. (2011). Long-term impact of Bt cotton: findings from a case study in China using panel data. *International Journal of Agricultural Sustainability*, 9(4), 508–521.

Pray, C., Ma, D., Huang, J., and Qiao, F. (2001). Impact of Bt cotton in China. *World Development*, 29(5), 813–825.

Pray, C. E., Huang, J., Hu, R., and Rozelle, S. (2002). Five years of Bt cotton in China–the benefits continue. *The Plant Journal*, 31(4), 423–430.

Pretty, J. (2001). The rapid emergence of genetic modification in world agriculture: contested risks and benefits. *Environmental Conservation*, 28(3), 248–262.

Qaim, M. and Zilberman, D. (2003) Yield effects of genetically modified crops in developing countries. *Science*, (299), 900–902.

Qiao, F., and Yao, Y. (2015). Is the economic benefit of Bt cotton dying away in China. *China Agricultural Economic Review*, 7(2), 322–336.

Qiao, F. (2015). Fifteen years of Bt cotton in China: the economic impact and its dynamics. *World Development*, 70, 177-185.

Qiao, F., and Huang, J. (2020). Sustainability of the economic benefit of Bt cotton in China: Results from household surveys. *Journal of Development Studies*, 56(11), 2045–2060.

Qiao, F., Huang, J., and Zhang, C. (2016). The sustainability of the farm-level impact of Bt cotton in China. *Journal of Agricultural Economics*, 67(3), 602–618.

Reuters, (2025). https://www.reuters.com/world/china/china-ramps-up-gm-corn-planting-state-controls-patchy-trials-keep-industry-2025-04-30/. Access date: 30_08_2025.

Smale, M. (2017). GMOs and poverty: the relationship between improved seeds and rural transformations. *Canadian Journal of Development Studies/Revue canadienne d'études du développement*, 38(1), 139–148.

Smale, M., Zambrano, P., and Cartel, M (2006). Bales and balance: a review of the methods used to assess the economic impact of Bt cotton on farmers in developing economies. *AgBioForum*, 9(3), 195–212.

Song Y. (1999). Introduction of transgenic cotton in China. *Biotechnology and Development Monitor*, 37, 14–17.

Sun, S., Zhang, C., and Hu, R. (2020). Determinants and overuse of pesticides in grain production: a comparison of rice, maize and wheat in China. *China Agricultural Economic Review*, 12(2), 367–379.

Völker, M. (2006) Economic Impact of genetically modified cotton varieties in China: a panel data analysis. MSc Thesis Leibniz University Hannover, Germany.

Wang, G., and Fok, M. (2018). Managing pests after 15 years of Bt cotton: Farmers' practices, performance and opinions in northern China. *Crop Protection*, 110, 251–260.

Wang, S., Just, D. R., and Pinstrup-Andersen, P. (2008). Bt-cotton and secondary pests. *International Journal of Biotechnology*, 10(2–3), 113–121.

Wang, Y., Ling, X., Ma, C., Liu, C., Zhang, W., Huang, J., ... and Deng, N. (2023). Can China get out of soy dilemma? A yield gap analysis of soybean in China. *Agronomy for Sustainable Development*, 43(4), 47.

Wu, K., Liang, G., and Guo, Y. (1997). Phoxim resistance in *Helicoverpa armigera* (Lepidoptera: Noctuidae) in China. *Journal of Economic Entomology*, 90(4), 868–872.

Wu, K., Mu, W., Liang, G., and Guo, Y. (2005). Regional reversion of insecticide resistance in *Helicoverpa armigera* (Lepidoptera: Noctuidae) is associated with the use of Bt cotton in northern China. *Pest Management Science: formerly Pesticide Science*, 61(5), 491–498.

Wu, L. (2010). Farmer field school and Bt cotton in China: An economic analysis. Pesticide Policy Publication Series. Special Issue No. 15. Leibniz University Hannover, Germany.

Xiao, Z., and Kerr, W. A. (2022). Biotechnology in China–regulation, investment, and delayed commercialization. *GM Crops & Food*, 13(1), 86–96.

Xu, R., Kuang, R., Pay, E., Dou, H., and de Snoo, G. R. (2008). Factors contributing to overuse of pesticides in western China. *Environmental Sciences*, 5(4), 235–249.

Yang, M., Zhao, X., and Meng, T. (2019). What are the driving factors of pesticide overuse in vegetable production? Evidence from Chinese farmers. *China Agricultural Economic Review*, 11(4), 672–687.

Zadoks, J. C., and Waibel, H. (2000). From pesticides to genetically modified plants: history, economics and politics. *NJAS: Wageningen Journal of Life Sciences*, 48(2), 125–149.

Zhang, C., and Guanming, S. (2015). Productivity effect and overuse of pesticide in crop production in China. *Journal of Integrative Agriculture*, 14(9), 1903–1910.

Zhang, X., Tang, H., Cao, D., (2003). Problems and countermeasures in Chinese pesticide market. *Pesticide Science and Administration*, 4, 31–34 (in Chinese).

Zhang, W., Lu, Y., van der Werf, W., Huang, J., Wu, F., Zhou, K., ... and Rosegrant, M. W. (2018). Multidecadal, county-level analysis of the effects of land use, Bt cotton, and weather on cotton pests in China. *Proceedings of the National Academy of Sciences USA*, 115(33), E7700–E7709.

Zhao, J. H., Ho, P., and Azadi, H. (2011). Benefits of Bt cotton counterbalanced by secondary pests? Perceptions of ecological change in China. *Environmental Monitoring and Assessment*, 173(1), 985–994.

13 The Mekong River and Village Development in Thailand

Suputtra Boriwut, Thanaporn Krasuaythong, Chomponuch Nantajit and Somkid Naprom

Abstract

This chapter looks at the development of rural villages in Thailand. The focus is on villages located along the Mekong River. The empirical basis is household and village level survey data. As regards village data, a 2023 survey of 110 villages, along the Mekong River, out of which 26 were on the Thai side of the river has been administered. The household data are from the TVSEP household panel using the 2007 and 2017 panel wave. In addition, a special Covid-19 survey was conducted in 2021 with semi-formal interviews with village heads. The analysis is largely based on descriptive statistics. Results indicate that reliance of the villages on the river as a source of livelihood was reduced during the past two decades. Specifically, the rural people living in Mekong villages are paying the price of hydropower development; they suffer losses in natural resources and health shocks such as Covid-19 have exposed the weakness of rural economies and make it harder to cope with the effects of climate change, thus impeding the safety net function of the river. The chapter also provides a point of departure for further research of village development in the Mekong River Basin.

Introduction

Development in rural Thailand continues to lag behind the urban areas. This has led to a significant rural-urban divide, as already pointed out in Chapter 4. Less developed physical and social infrastructure and imperfect insurance systems make the people living in rural areas more vulnerable to idiosyncratic and co-variate shocks. Climate change, with more extreme weather events, increases the risk of drought and floods with detrimental effects for agricultural productivity (Liebenehm *et al.*, 2023). Villages located in the Mekong River Basin are particularly exposed to such risks because of the side effects of major development interventions such as hydropower electricity dams and extraction of sand and gravel from the main river and its tributaries. Hydropower dams on the main stem of the Mekong River were first established by China but have been spreading to Laos, with the completion of the Xayburi Dam in 2019 and more such projects in the planning or construction phase. Soukhaphon *et al.* (2021) studied river communities in the Mekong Basin, including the Mun River, a tributary of the Mekong in north-eastern Thailand. The study found that hydropower dams are affecting fish migrations, river hydrology and sediment transfers, often negatively impacting riparian communities, even up to 1000 km away.

The second major outputs of the river are sand and gravel. These are natural resources, increasingly in demand and needed as construction material for the rapid expansion of urban

 Asian Agriculture in a Climate Change-driven and Risky World (H. Waibel, S. Liebenehm and Ernah eds)

DOI: 10.1079/9781800627338.0013

developments in Asia. As pointed out by Bendixen *et al.* (2019), on a global level, sand and gravel are being extracted faster than they can be replaced. Continuous sand mining from rivers has cascading effects on the physical environment, such as riverbed widening and lowering, on the biological environment through reduced biodiversity, including the aquatic and shoreline flora and fauna, on the chemical environment, affecting air and soil quality through pollution and last but not least, on the anthropogenic environment with damaged infrastructure, reduced access to water and agricultural losses (e.g. Rentier and Cammeraat, 2022).

While the Mekong is a transboundary natural resource, cross-country relationships and externalities are a major topic. However, as pointed out by Kuenzer *et al.* (2013) the even bigger issue is perhaps in terms of urban versus rural. Hydropower dams and sand extraction sites have visibly changed the river's appearance with potentially negative effects for the livelihoods of the rural population. This is significant, considering the high proportion of the 190 million rural people of the Mekong Basin being dependent on the natural resources for their livelihoods (Morton and Olson, 2018).

Almost 20% of the 4900 km of the Mekong flows through Thailand, forming most of its border with Laos. The majority of settlements in the river basin are rural villages, while only a few of them are provincial capital towns, such as Nong Khai, Nakhon Phanom or Mukdahan. Development projects at the Mekong and its tributaries have mainly benefited the urban population in Thailand with cheap electricity and good quality river sand for the construction industry. Attributable negative externalities include irregularities in water flows, as a results of unfavorable water management regimes of the dams, leading to occasional heavy floods and droughts. Furthermore, loss of livelihoods, from loss of river bank land, loss of crop production, loss of capture fisheries, and other aquatic plants and organisms, environmental and water pollution, as well as higher risks of floods and droughts. Also, the presence of sand companies causes dirt and noise pollution and an increased risk of accidents when heavy trucks pass through the narrow village roads.

The prevailing conditions of the Mekong villages are one of the motivations to include a chapter about the situation of rural villages in the Mekong Basin in this book. The chapter makes use of three sources of data. First, a qualitative study of 10 villages in the north-eastern province of Nakhon Phanom, located at the Mekong, over a length of about 100 km, was carried out in 2020. Second, in 2023, a historical survey of 31 villages in the eight provinces of Thailand, connected to the Mekong River was administered. The latter was part of a Mekong-wide survey of 110 villages, in all six Mekong counties. Third, a panel data set from the Thailand Vietnam Socioeconomic Panel (Klasen and Waibel, 2013) of 54 villages in the Mekong Basin in the provinces Nakhon Phanom and Ubon Ratchathani, from 2007 and 2017 (Waibel *et al.*, 2021) was used.

The main hypothesis of this paper is that to date, villages in the Mekong River Basin are more vulnerable than in the past. They have undertaken coping measures, which however are insufficient to fully mitigate the negative effects from hydropower dams and sand extraction at the river.

As shown by the analysis of the qualitative and quantitative data used in this study, villages along and in the vicinity of the Mekong River have changed dramatically in the past two decades.

Methodology and Data

In this paper, a qualitative and quantitative research approach is combined. First, in-depth discussions with village heads, of randomly selected villages, were carried out. The villages can be divided into two groups, those who were in close vicinity of the Mekong River, i.e. 5 km or less and those at a distance of up to 40 km maximum away from the Mekong, yet within the river basin. A set of 10 guiding questions were used, prompting respondents to compare current and past livelihood situations of the village including major shocks and disasters during the past 20 years, observed changes in the river and its environment and possible reasons for these changes as perceived by the respondent, and, finally, expected changes in the Mekong Basin and their possible effects on the village as well as requests for government support. Since the interviews were carried out during the Covid-19 pandemic, related issues were discussed as well.

The second assessment of 31 Mekong villages in Thailand was part of a larger assessment of economic and environmental changes in the Mekong River Basin. The study covered a total of 110 villages in all six Mekong counties, ranging from Southern China to the Mekong Delta in Vietnam and was carried out in 2023. The aim of this project was to investigate changes in the livelihoods over the past two decades, including changes in the sources of income over time such agriculture, river fisheries, aquaculture, products from ecosystems services such as algae and river weeds, wage income, income from small- and medium-sized businesses, remittances and public transfers of the Mekong villages. Furthermore, observed and perceived changes in the river and its environment were collected along with satellite data on changes in rainfall, land use and vegetation cover. Villages were selected by means of a sampling frame which was based on equal distances within the range of the river covered in the study, i.e. villages were 25–30 km apart from each other. All villages included in the survey are located in proximity of the river, i.e. with direct access to the river, not further away than 1.5 km. Personal interviews were undertaken with village representatives such as village head and village committee members. Results for the 31 Thai villages are presented as descriptive statistics comparing three points in time, i.e. 2003, 2013 and 2023.

The third empirical source used in this study is panel data that comes from the still ongoing, long-term household panel "TVSEP," the Thailand Vietnam Socio Economic Panel (https://www.tvsep.de/en/) carried out in Thailand and Vietnam since 2007. For Thailand, the data that include 2200 households in 220 villages in the provinces of Buri Ram, Ubon Ratchathani and Nakhon Phanom; the latter two provinces are riverine to the Mekong. From the panel database 54 villages were identified as located within the Mekong River Basin. Therefore, by using the 2007 and 2017 panel wave, it is possible to compare the development and their environments, based on socioeconomic household data. Analysis of the data is carried out by presenting average per village based on the up to 10 sample households per village.

It is believed that on the basis of these three data sets, credible empirical evidence for the impact of changes in the Mekong Basin can be shown. This will provide a good foundation for assessing the consequences for the constraints, opportunities and needs of the villages for village development. At the same time, this can be an avenue for further research. Also, making use and interpreting the testimonials of the 10 village heads, we can undertake some ground testing of the findings in the literature. In the last step we delve more deeply into the changes experienced by Mekong villages using a 10-year panel data set from two Mekong provinces.

Results

In the following, we first present the results of the semi-formal interviews with the 10 households in Nakhon Phanom province. This analysis can help to further underline some of the findings of the various literatures about the Mekong River Basin and give some guidance in the interpretation of the two subsequent data sets.

The second result is the presentation of the historical review of changes in village conditions and development in the 31 Mekong villages, spread over seven of the eight provinces bordering the Mekong River, namely, Chiang Rai, Loei, Nong Khai, Bueng Kan, Nakhon Phanom, Mukdahan and Ubon Ratchathani. Hereby, descriptive statistics have been used. Finally, the panel data set of 54 villages in Nakhon Phanom and Ubon Ratchathani, was analyzed by means of descriptive statistics, comparing 2007 with 2017.

Qualitative Case Study

In Table 13.1 the results of the in-depth interviews with village heads in 10 Mekong villages in Nakhon Phanom are presented. In the seven columns the main issues of the villages are summarized. The first column indicates the location of the village in relation to the river. Villages 1–6 are within 5 km away from the river indicated by "n." The remaining four villages are more than 5 km away but within a range of 40 km, indicated by "a" in column 1.

Table 13.1. Results of in-depth interviews with village heads in 10 Mekong villages in Nakhon Phanom Province. (Authors' own table based on interviews by Somkid Naprom).

Village No./ River Distance	Current Conditions	Major Past Shocks	Causes of Shocks	Changes in River	Expected Changes in River	Village Prospects
1 (n)	*Positive*: people do more trading *Negative*: high expenditures	Flooding in rice fields	Don't know	Less water; less fish; inconsistent water flow because of "*China's dam*"; forest loss, become rubber plantation	More drought and water shortage	Economic difficulties
2 (n)	*Positive*: None *Negative*: Price of rice is too low; household debts; drug problems	Flood	Nong Han and Mekong River are the cause of flooding	Irregular water flow, can no longer predict; forecast like "*China's dam*: "*they open and close as they want*"	"*Some day the Mekong will be just a sandbar*"	Lack of water for agriculture
3(n)	*Positive*: off-farm employment increased (government jobs, trade, tourism) *Negative*: Poor irrigation system	Flood	More people *China's dam*	Low and irregular water level; cannot forecast unlike before; flood, affects vegetable and aquaculture at river bank	More dams in China and Laos; more drought and more water disputes	Lack of water for agriculture
4 (n)	*Positive* :successful investment in rubber plantations, some migrant workers in Singapore and Taiwan *Negative*: unemployment	none	Not applicable	less water; more sandbanks because of *China's dams*	River become shallower and more sandbanks	Lack of water for agriculture
5 (n)	*Positive*: infrastructure improved *Negative*: "*nature is gone; climate becomes drier*"	Drought	Negative effect of rubber plantation on microclimate and water table	Less water; the flow of water is irregular because of China's dams	More drought	Lack of water for agriculture
6 (n)	*Positive*: people rely more on commercial crops (rubber and vegetables in a dome); irrigation system improved *Negative*: prices of inputs going up while product prices do not; lack of long-term prospects in villages	Drought	Forest destruction, unreliable water supply from Mekong	Less water, China's dam and Lao's dam	More drought and more flood	Lack of water for agriculture

Continued

Table 13.1. Continued.

Village No./ River Distance	Current Conditions	Major Past Shocks	Causes of Shocks	Changes in River	Expected Changes in River	Village Prospects
7 (a)	*Positive*: two rice crops per year due to irrigation system *Negative*: household debt is going up	Sometimes flooding	Insufficient drainage system	River gets dry and more sandbanks; Thai shore above Laos side, making river shore in Thai side drier	More droughts	No effect on the village because of long distance from village to Mekong
8 (a)	*Positive*: most people have off-farm occupations; hire machinery services to cultivate their farm land *Negative*: Lack of good jobs with long-term prospects	Flood	Poor drainage system	River becomes narrower and shallower; many shops along Mekong Shore extend the area to Mekong	Will become further narrow and shallow	No significant effect for village
9 (a)	*Positive*: None *Negative*: Welfare decreased and HH debt increased	Flood and drought	Natural resources destruction, especially forest, monetization of village life	There is more construction along Mekong shore, a landmark and more local attractive trees; development project of the province is changing nature	Mekong will be dirtier	No answer
10 (a)	*Positive*: agricultural system has become diversified (rubber, oil palm, etc.); welfare of village much increased (most HH now have a car) *Negative*: infrastructure (roads, irrigation, electricity) still insufficient	No mention	Village located on elevated land	more shops along Mekong; more concrete construction because of trading, province's project, people have more income	Will be more beautiful	No answer

Column 2 displays the assessment of current village conditions as positive and negative. In Village 2 (a) and Village 7 (n), no positive points were mentioned. In the remaining eight villages, structural changes in agriculture and the evolution of off-farm job opportunities were mentioned as being positive for the current situation.

As regards negative conditions, high input and low output prices, household debt and other social problems were frequently mentioned. Major shocks mentioned are flooding and droughts. When asked for the reasons, respondents in Mekong-near villages sometimes blamed changes in the Mekong, particularly the hydropower dams in China. The spread of rubber plantations and forest destruction are seen as other causes.

When asked for changes in the Mekong River, again, the Chinese dams were frequently mentioned, also by villages further away. Interestingly, no one mentioned the Xayaburi Dam in Laos which is much nearer than the dams in China. The main issue as regards the river is the irregularity of water flows which villagers believe is due to the management of the dams. Also, there is no prior information when the dams release water. That the river is getting shallower and drier is mentioned also by the respondents from the villages further away (see Table 13.1). When asked for the future of the river, most respondents expect the river to become drier, including a statement, "*someday, the river will just be a sandbar.*" Regarding the question how changes in the Mekong will affect the prospects for village development, near-Mekong villages dominantly mention more water shortages for agriculture. This is different for the further-away villages where respondents did not see such effects.

In summary, this qualitative case study revealed that people living in villages in the Mekong River Basin are aware of the ecological changes in the river and generally have plausible explanations of the impact of village development.

Mekong Village Survey

In this section, results of the Mekong-wide survey of 110 villages for the 31 villages in Thailand. These villages were spread out throughout the Mekong provinces of Thailand, namely 4 in Chiang Rai, 5 in Loei, 6 in Nong Kai, 5 in Bueng Kan, 5 in Mukdahan and 6 in Ubon Ratchathani.

Table 13.2 presents some basic village characteristics such as distance from town, villages with sand extraction, population, awareness of the location of nearest hydropower dams, village size in terms of area, population and number of households as well as migration.

As shown in Table 13.2, on average the villages are not far away from district towns. Also, most villages are aware of the existence of hydropower dams, even when further way. About 20% of the villages in the sample had a sand extraction company within the village area. On average, the administrative area of villages was 400 ha with a standard deviation of 300. Average village population was 752 with an average of 219 households or about 3.5 members

Table 13.2. Basic statistics of 31 Mekong Villages in Thailand.

	Mean	Standard Deviation	Share/(%)
Distance to town (km)	11.61	9.82	
Respondents aware of hydropower dams	–	–	20/31
Villages with sand extraction companies			6/31
Village area (ha)	400	300	
Village population (2023)	752	376	
No. of households (2023)	219	124	
Households with migrants			
Annual change in village population 2013 vs 2003	–	–	(+ 2.44)
Annual change in village population 2023 vs 2013	–	–	(+ 3.0)

Source: Mekong Village Survey 2023.

per household. Interestingly, village population has grown considerably in both observation periods, with a compound annual growth rate (CARG) of 2.44 and 3% respectively, based on the formula $(EV/BV)^{1/n} - 1$, where EV is ending value, BV is beginning value and n is the number of years. The data suggest that Mekong villages are active settlements who manage to cope with the potential adversities from the changes in the river.

Table 13.3 shows changes in land use over the 20-year observation period. Land use is broken down into forest, land for cropping and grazing, for built environment, vacant land and other land use (roads, sports fields, etc.). Notably forests have declined dramatically from about 40% in 2003 to about 20% in 2023. Agriculture had already been the dominant land use since 2003 and is expanding by replacing forest land. Land for the built environment, i.e. for residential houses, other buildings and other physical infrastructures has increased remarkably. The share of land allocated for houses including residential lots and other buildings has increased. This is corresponding with population growth in the villages as mentioned above and suggests that there has been no major out-migration in rural villages along the Mekong River.

Table 13.4 presents income sources of the Mekong Villages over three points in time, i.e. 2023, 10 years and 20 years previously. Village representatives were asked for the percentage of households who pursue a certain source of income.

Most households derive income from crops, and this did not change a lot during between 2003 and 2023. However, households engaged in fishing declined from just over 40% in 2003 to little over 15% 20 years later. Natural resources were as a source of income for almost one-third of the households in 2003 and went down to below 10% in 2023. Off-farm wage employment has remained fairly constant over time while self-employment and other income sources increased considerably between 2003 and 2013 but only little in the decade after.

Respondents were also asked to rank the three major sources of income during the three reference points. As shown in Table 13.5, income from crops has been the first income source in all three time periods. However, different income sources can have the same rank. For example, in 2003, wage employment was ranked second and third. Note that livestock was ranked higher than wage employment in 2023 which underlines that the villages are still dominantly agricultural. Note, that river fisheries even 20 years ago were not among the three major sources of income.

Table 13.3. Land use in Mekong Villages over time (%).

Land Use	2003	2013	2023
Forest	36.3	27.6	22.3
Agriculture	46.8	52.6	56.5
Buildings	10.1	13.3	15.4
Vacant land	2.7	1.7	1.2
Other	4.1	4.8	4.6

Source: Village Survey 2023.

A Decade of Development in Mekong Villages

In the last section, a further look is taken at the development of 54 Mekong villages in two provinces, namely Nakhon Phanom and Ubon Ratchathani, using a 10-year panel data set from the TVSEP panel.

Table 13.4. Household income by source over time, in % of households.

Source	2003	2013	2023
Crop	80.2	75.3	73.5
Livestock	36.7	24.1	17.8
Fishing	40.8	28.4	15.4
Off-farm wage employment	30.7	31.0	26.9
Other (incl self-employment)	14.1	20.5	23.0
Natural resources	27.4	15.6	8.1
Others			

Source: Village Survey 2023.

Table 13.5. Priority sources of income, 2003, 2013 and 2023.

	2003	2013	2023
Crop	First	First	First
Livestock	(not listed)	Second	Second
Wage employment	Second/Third	Second	Third

Source: Village Survey 2023.

Table 13.6 shows some crucial development parameters, comparing 2007 and 2017 as two points in time. In each village, 10 households had been selected, based on systematic random sampling.

The first interesting result is that the village population, both in terms of households per village and total village population, had increased between 2007 and 2017 by 24% and 13% respectively (see Table 13.6). In general, this confirms the findings in the previous section, i.e. Mekong villages seem to be attractive living places, despite the ecological challenges arising from the changes in the Mekong River. As is the case in all villages in Thailand (Rigg *et al.*, 2012), a large share of the village population are migrants working outside the village. However, migration has declined during the observation period by about 5.5 percentage points. The decline in migrants is also reflected in the share of remittances in household income, going down from 20% to 16%.

Household income has almost doubled during the decade, corresponding with a daily per capita income of about PPP $3 in 2003 and about PPP $7 in 2017, well above the poverty threshold. For the interpretation of the income data, it is important to look at the composition of household income. Corresponding with the large share of migrant workers, wage employment was the major source of income in 2007 and still in 2027. However, the share has gone down from 34% to 23%. The second major source is income from crops. This has increased only slightly from 17% to 18%. However, when income from livestock is included, agriculture has become the main source of income with 29% in 2017. Note that livestock includes aquaculture. While, on average per village, there were 33 households who operated a fish pond (or a cage in the river) in 2007, this was reduced by more than 50% to only 17% 10 years later. The most dramatic income reduction occurred in natural resource extraction which includes income from fishing. In 2017 only 1% of household income was what can also be called environmental income while it was still 7% 10 years before. This is reflected in the households who undertook fishing. While in 2007 slightly over half the households were engaged in fishing these were only 30% 10 years later.

Small-scale businesses in the village increased moderately by 5 percentage points making it the third most important income source in 2017 after agriculture and wage employment.

A crucial factor is household debt. Apparently, the growth in income has come along with

Table 13.6. A decade of development in 54 Mekong villages in two Mekong provinces.

Parameter[1)]	Unit	2007	2017	% Difference
Households per village	No./village	134	158	+ 24
Village population	Persons/village	589	612	+13
Migrants	%	50.1	44.6	-5.5
Average no. of migrants	No./HH	2.02	1.76	-0.26
Average household income	PPP $/HH	5012	9899	+97.5
Annual debt repayment	PPP $/HH	663	3246	+489.6
Over indebtedness (DSR >40)	% of HHs	12.5	25.4	+12.9
Income shares:				
Crop production	%	17	18	+1
Livestock and aquaculture	%	3	11	+8
Natural resource extraction	%	7	1	-6
Remittances	%	20	16	-4
Off-farm wage employment	%	34	23	-11
Non-farm self-employment	%	15	20	+5
Public transfers and others	%	4	11	+7
Households with fish ponds/cage culture	No.	33	17	-51.5
Households engaged in fishing	No.	276	141	-51.5

Source: Own calculation by C. Nantajit based on TVSEP panel data. Note: The number of households interviewed was 537 in 2007 and 469 in 2017; there are 10 households per village in the panel, however attrition has occurred.

an enormous increase in household debt. This has increased by a factor of almost 5, leading to an increase in the debt-service ratio (DSR) by 12.9 percentage points. Hence, over 25% of households were classified as being over-indebted in 2017. These figures correspond well with the findings of Chichaibelu and Waibel (2017) who studied borrowing behavior of people in rural Thailand.

In general, the analysis shows that while welfare in Mekong villages had increased, there are at least two downsides. One is the increasing household debt problem and secondly the dramatic weakening of the safety net traditionally provided by the natural environment. Most remarkable, however, is the reduction in the income share of natural resource extraction which includes fishing in the Mekong but also the harvesting of timber and non-timber forest products (e.g. mushroom and honey collection, game hunting). Hence it is doubtful whether the loss in environmental income is compensated by higher increases in agricultural productivity as predicted in a simulation study by Orr *et al.* (2012).

These long-term panel data, independently collected for research purposes, support the testimonials of the village heads as reported in the first section of the paper. By and large, the natural safety nets of rural people in the Mekong Basin are gone. While this may have only a small effect on average income, it can weigh heavily, given the risky economic and natural conditions to be expected in the future. A slow-down in the global economy will reduce the pull-effect of labor migration with negative effects for the household debt situation.

In summary, while the panel data support the hypothesis that development and exploitation of the Mekong water resources has helped rural villages in the basin to gain more income, on the other hand, several factors suggest increasing vulnerability to external shocks, both co-variate and idiosyncratic, and a weakening of the resilience of rural areas, mainly because of natural resources destruction.

Summary and Conclusions

This analysis of development of villages in the Mekong Basis has revealed a number of issues which are important for designing rural structural transformation policies and sustainable development.

Firstly, as shown by the qualitative study changes, people living in Mekong villages are well aware of the environmental changes caused by major development interventions in the river. Oftentimes, village representatives blame China's hydropower dams for the irregularities in water flow with an increased risk for flood and drought. While the general economic development has also benefited from economic growth of the country the outlook for the river is perceived as bleak, exacerbated by the effects of climate change.

The second study reinforced the indications from the qualitative interviews with village heads in more concrete empirical terms. The decline in forest and the increase in crop land can be highlighted. However, most of that land is rain-fed which makes agricultural production riskier. The expectation that irrigated agriculture will receive a push from the many hydropower dams at the river or its tributaries cannot be confirmed by the results of this study. Another indicator of the changes in the environment is the decline in natural resources as a source of income. To date, Mekong villages no longer can rely on food collected from river banks and river environments as they did 20 years ago or earlier. Capture fisheries have declined in all Mekong villages and these were never mentioned as even a third important source of income. The primary source of income has always been agriculture, i.e. crops and livestock.

On the other hand, village economies are active. Generally, village populations have increased during the 20-year reporting period. Also, village economies appear to be fairly robust as most household members are village residents. Villages have generally benefited from economic growth of the country and have diversified their village economies with a growing population. This observation is confirmed by the analysis of the 10-year panel data set of some 500 households in 54 villages in the Mekong Basin. While household income has increased the safety-net function of the Mekong River basically is lost. In addition, household debt has risen dramatically as is the case all over rural Thailand.

Pretty *et al.* (2010) in a global analysis on the sustainability of agriculture came forward with a hundred questions as regards the future of agriculture in the world. From the analysis of Mekong villages, we conclude that many of these questions apply to the future of the 31 Mekong villages in Thailand. Hydropower dams and extensive sand mining (Saiyarod, 2023) endanger their future because the additional risks emerging from the Mekong converted the river from being a safety net to become a threat (Campbell, 2009).

References

Bendixen, M., Best, J., Hackney, C. andIversen, L.L. (2019) Time is running out for sand. *Nature* 571(7763), 29–31.

Campbell, O. (2009) Development scenarios and Mekong river flows. pp. 389–402, In: The Mekong Biophysical Environment of an International River Basin; Academic Press, Cambridge, Massachusetts, USA.

Chichaibelu, B.B. and Waibel, H. (2017). Borrowing from “pui” to pay “pom”: Multiple borrowing and over-indebtedness in rural Thailand. *World Development* 98, 338–350.

Chilamkurthy, K., Marckson, A.V., Chopperla, S.T. and Santhanam, M., (2016) A statistical overview of sand demand in Asia and Europe. In: *International Conference UKIERE CTMC, Vol. 16*.

Klasen, S. and Waibel, H. (2013). Introduction and key messages. In: *Vulnerability to Poverty: Theory, Measurement and Determinants, with Case Studies from Thailand and Vietnam*. Palgrave Macmillan, London. pp. 1–14.

Kuenzer, C., Campbell, I., Roch, M., Leinenkugel, P., Tuan, V.Q. *et al.* (2013) Understanding the impact of hydropower developments in the context of upstream–downstream relations in the Mekong River basin. *Sustainability Science* 8, 565–584.

Liebenehm, S., Jaretzky, H., & Waibel, H. (2023). Extreme weather and agricultural management decisions among smallholder farmers in rural Thailand and Vietnam. *Agricultural Economics* 54(5), 593–622.

Mekong River Commission. (2011) *Assessment of Basin-wide Development Scenarios – Basin Development Plan Programme, Phase 2. Vientiane, Lao PDR*. Mekong River Commission, Vientiane, Laos.

Mekong River Commission. (2021) *Basin Development Strategy 2021–2030 and MRC Strategic Plan 2021–2025*. MRC Secretariat, Vientiane, Laos.

Mekong River Commission. (2024) Fisheries, Vientiane, Laos. Available at: http://www.mrcmekong.org/fisheries (accessed 1 August 2024).

Morton, L.W. and Olson, K.R. (2018) The pulses of the Mekong River Basin: rivers and the livelihoods of farmers and fishers. *Journal of Environmental Protection* 9(4), 431–459.

Orr, S., Pittock, J., Chapagain, A. and Dumaresq, D., (2012) Dams on the Mekong River: lost fish protein and the implications for land and water resources. *Global Environmental Change* 22(4), 925–932.

Pretty, J., Sutherland, W.J., Ashby, J., Auburn, J., Baulcombe, D. *et al.* (2010) The top 100 questions of importance to the future of global agriculture. *International Journal of Agricultural Sustainability* 8(4), 219–236.

Rigg, J., Salamanca, A. and Parnwell, M. (2012). Joining the dots of agrarian change in Asia: a 25 year view from Thailand. *World Development* 40(7), 1469–1481.

Saiyarod, P. (2023) Retrofitting the Mekong: Community-based environmental responses to Chinese transnational infrastructure in a Thai border town. *Journal of Southeast Asian Studies* 54(3), 480–501.

Rentier, E.S. and Cammeraat, L.H. (2022). The environmental impacts of river sand mining. *Science of the Total Environment* 838, 155877.

Soukhaphon, A., Baird, I.G. and Hogan, Z.S. (2021). The impacts of hydropower dams in the Mekong River Basin: a review. *Water* 13(3), 265.

Waibel, H., Nantajit, C., Phami, P. and Naprom, S. (2021). *COVID-19 and Rural Development in the Mekong River Region: Case Studies from Thailand and Laos*. Tokyo, pp. 222–254.

14 Exploring Differences in Rural Household Debt between Thailand and Vietnam: Economic Environment versus Household Characteristics

Bezawit Beyene Chichaibelu

Abstract

This study compares rural household debt dynamics in Thailand and Vietnam using harmonized microdata from the "Vulnerability in Southeast Asia" project. Differences in credit market participation, levels of household debt, and over-indebtedness are analyzed using decomposition methods to distinguish the roles of household characteristics versus structural economic environments. Results indicate that rural Thai households carry markedly heavier debt burdens and face a higher likelihood of over-indebtedness than their Vietnamese counterparts. These gaps arise primarily from Thailand's more permissive institutional and market environment, characterized by broader credit availability and weaker constraints on household borrowing. The findings underscore that while expanding rural credit access can promote investment, safeguards are needed to prevent excessive debt, with context-specific policies that strengthen lending standards, consumer protections, and financial literacy.

Introduction

Over the past two decades, rapid financial development in Asia has brought about expanded credit access and deeper financial inclusion. However, this progress has also coincided with a sharp rise in household debt, raising questions about the sustainability of debt and the long-term welfare implications for vulnerable populations. In countries like Thailand, household debt has grown at an alarming pace, with the debt-to-GDP ratio exceeding 90% in recent years (World Bank Group, 2024) and rural households are increasingly entangled in persistent cycles of borrowing and repayment distress (Chantarat *et al.*, 2023). Evidence shows that for many Thai agricultural households, reliance on credit to smooth consumption amid volatile income has led to a pattern of revolving debt, where loan repayments are frequently financed by new borrowing, eroding long-term financial resilience and compounding poverty risks (Chichaibelu and Waibel, 2017; Chantarat *et al.*, 2023).

While rising household debt is observed across Asia, there are substantial differences in both the scale and structure of household debt across countries (ADB, 2015; Organization for Economic Co-operation and Development (OECD), 2016), particularly between Thailand and Vietnam. Thailand's rural credit environment is characterized by high penetration of formal and semi-formal lenders, such as the Bank for Agriculture and Agricultural Cooperatives (BAAC) and Village Funds, with relatively lenient lending practices and widespread use of debt moratoriums (Chantarat *et al.*, 2023). In contrast, Vietnam has maintained more conservative credit policies, despite significant efforts to improve financial

 Asian Agriculture in a Climate Change-driven and Risky World (H. Waibel, S. Liebenehm and Ernah eds)
DOI: 10.1079/9781800627338.0014

inclusion through the Vietnam Bank for Social Policies and cooperative credit models (Varangis *et al.*, 2019). These differences in institutional environments are likely to shape the extent and nature of household indebtedness, yet systematic micro-level comparative evidence remains scarce.

Aggregate indicators of household debt can also obscure substantial heterogeneity in debt burdens and financial vulnerability across population groups (Aminudin and Tissot, 2015). Recent studies highlight the need to shift focus from macro-level credit growth to household-level dynamics, particularly in rural areas where informal borrowing persists, financial literacy is limited, and access to risk mitigation tools remains uneven (Chichaibelu and Waibel, 2018; Nguyen *et al.*, 2019). In both Thailand and Vietnam, rural households face the dual burden of income insecurity and limited insurance coverage, but the institutional mechanisms through which credit markets interact with these risks differ markedly. Understanding how these factors jointly shape debt holding and over-indebtedness is critical for designing policies that promote both financial inclusion and financial stability.

This study contributes to the literature on household finance and rural household indebtedness by examining the cross-country differences in credit market participation, levels of debt holding, and over-indebtedness among rural households in Thailand and Vietnam. A set of decomposition techniques is applied to harmonized household survey data from the "Vulnerability in Southeast Asia" project to isolate the contribution of household characteristics (e.g., income, education, assets) from structural factors embedded in the economic and institutional environment. This approach follows the empirical strategies employed in European and US contexts (Christelis *et al.*, 2013; Christelis *et al.*, 2021; Bover *et al.*, 2016) and adapts them to the rural Asian setting, where informal financial systems and state-led credit programs play a central role.

By identifying how much of the cross-country variation in rural household indebtedness stems from differences in borrower profiles versus differences in credit market structure and regulation, this study provides insights relevant for policy design. The findings suggest that policies solely focused on expanding credit access may be insufficient or even harmful if they fail to account for borrower capacity and the broader institutional environment. In contexts like Thailand, where broader credit supply coexists with weak debt screening and limited financial literacy, the risk of chronic over-indebtedness may outweigh the benefits of financial inclusion. Conversely, Vietnam's more cautious financial ecosystem may offer lessons for balancing access with sustainability.

Rural Credit Markets in Thailand and Vietnam

Rural credit markets in Thailand and Vietnam have undergone significant expansion, supported by government-led initiatives to promote financial inclusion and rural development. Despite similarities in goals, enhancing access to finance for underserved rural populations, both countries have pursued distinct institutional and regulatory paths. These differences have shaped the structure of rural credit markets, household-level borrowing behavior, credit access, and the extent of household indebtedness in each context.

Thailand's rural credit market is characterized by a system in which formal, semi-formal, and informal lenders coexist. The Bank for Agriculture and Agricultural Cooperatives (BAAC) is the country's dominant rural financial institution, accounting for over 90% of formal agricultural loans. Since its establishment in 1966, BAAC has expanded its coverage nationwide, providing credit not only for agricultural production but also for consumption, small enterprise development, and rural infrastructure. Complementing BAAC is the Village and Urban Community Fund, a large-scale government initiative launched in 2001 that channels revolving funds to local communities for micro-lending at the village level (ADB, 2014). Together, these institutions have dramatically expanded financial access in rural areas.

However, Thailand's credit expansion has been accompanied by rising concerns over debt sustainability. Recent studies indicate that many rural households are engaged in multiple borrowing, often from overlapping formal and informal sources, and may use new loans to service existing debt rather than to invest in productive assets (Chichaibelu and Waibel, 2017; Chantarat *et al.*, 2023). In politically sensitive periods, credit policies have often relaxed eligibility requirements or included debt moratoria, reinforcing expectations of periodic loan forgiveness and undermining credit discipline. As a result, household debt-to-income ratios have increased significantly in rural regions, with a growing share of households considered over-indebted (Chantarat *et al.*, 2023).

In contrast, Vietnam's rural credit market reflects a more centralized and socially embedded structure. The Vietnam Bank for Agriculture and Rural Development (VBARD) and the Vietnam Bank for Social Policies (VBSP) are the two principal institutions providing rural credit. While VBARD operates on commercial terms, VBSP offers subsidized credit to poor and near-poor households through an extensive network that engages mass organizations such as the Women's Union and Farmers' Union as intermediaries. These groups are responsible not only for borrower selection but also for loan monitoring and repayment support (Nguyen *et al.*, 2019; Ta, 2022). This community-based system leverages social capital and peer accountability to mitigate default risk and encourage responsible borrowing (Chichaibelu and Waibel, 2018).

Vietnam's approach emphasizes financial discipline and targeted social policy. Loans from VBSP and VBARD are often conditional on income level and purpose-specific, with lending caps, staggered disbursements, and mandated uses tied to productive activity or education. Group-based lending and social oversight have been effective in keeping delinquency rates low, and repayment rates for VBSP loans exceed 98% (Varangis *et al.*, 2019). This model also appears to minimize the risk of over-indebtedness, as borrowers face tighter eligibility criteria and social pressure to maintain financial commitments (Nguyen *et al.*, 2019).

Nevertheless, Vietnam's rural credit markets face their own challenges. Despite substantial progress, Vietnam's rural credit landscape still falls short of supporting the country's accelerating agricultural transformation. Formal lenders focus largely on short-term, poverty-reduction loans aimed at bottom-of-the-pyramid clients, leaving many non-farm rural enterprises and growth-oriented agribusinesses without adequate financing (Varangis *et al.*, 2019; Nguyen *et al.*, 2019). Consequently, informal lenders continue to dominate rural areas, particularly in the Mekong Delta and remote mountainous areas, often at steep cost (Nguyen *et al.*, 2019).

Data and Methods

Data

The 2008 data of rural household in Thailand and Vietnam collected by the project "Vulnerability in Southeast Asia", a long-term research project funded by the German Research Foundation (DFG), is used. The survey collected data from around 4200 rural households from six provinces in Northeastern Thailand and the North Central Coast and Central Highland of Vietnam. The six provinces, namely Buriram, Ubon Ratchathani and Nakhon Phanom from Thailand, Ha Tinh, Thua Thien Hue, and Dac Lac from Vietnam, were first purposively selected targeting rural households that are either poor or face risk of falling into poverty to meet the general objective of the project (Hardeweg *et al.*, 2012). Then, three-stage cluster sampling design was implemented to select the sample of households. Firstly, sub-districts were randomly selected from strata with probability that is proportional to the population density in each province. After selecting the sub-districts, two villages were again selected randomly with probability proportional to size. Finally, 10 households were sampled in each village from a household list ordered by household size using a systematic random sampling technique that gave each household equal chance of being selected. Hence, the households in our sample are representative of rural households in Northeastern Thailand and the North Central Coast and Central Highland of Vietnam.

The data contain detailed information on households' borrowing, loan defaults, and arrears along with a full set of household level data such as households' demographics, social and economic characteristics and special modules on risks and shocks. This detailed data on the financial situation of households allows us to examine rural households' borrowing behavior in the two countries and decompose the differences into their separate underlying factors.

Measures of Household Debt Outcomes

To analyze differences in household debt outcomes across rural Thailand and Vietnam, a range of dependent variables capturing both the incidence and intensity of borrowing is used. These include a binary indicator for debt holding, the natural logarithm of outstanding household debt among borrowers, and a set of commonly used objective indicators of over-indebtedness. Specifically, we examine the debt service ratio (DSR), defined as the share of annual income spent on debt repayments; the debt-to-income ratio (DIR); the debt-to-asset ratio (DAR); and an indicator for default, based on whether households reported late or missed repayments. These measures reflect short- and long-term repayment burdens and are widely employed in studies of household financial distress. While each captures a different dimension of over-indebtedness, they are well-suited to cross-country comparison, avoiding the subjectivity inherent in self-assessed financial hardship.

Figure 14.1 shows the log debt distribution by country in 2008. The distribution for Vietnam lies to the left and is denser at the lower tail, indicating that rural households tend to hold smaller debts. In contrast, Thailand's distribution is shifted to the right and thicker at the upper tail, reflecting higher levels of debt across the distribution, particularly at the top end.

Table 14.1 compares the prevalence and conditional amounts of rural household debt in Thailand and Vietnam, measured in PPP-adjusted 2005 USD. In both 2008 and 2010, Thai rural households were more likely to hold debt, with over 80% indebted in 2008 compared with 66% in Vietnam. Conditional on borrowing, Thai households also held larger debts across the distribution. At the median and 90th percentile in 2008, debt levels in Thailand were about twice and four times those in Vietnam, respectively. While Thai household debt declined slightly after the 2008 crisis, likely due to increased repayments supported by government stimulus (Khandker *et al.*, 2012), it rebounded rapidly, surpassing pre-crisis levels and even those in the USA by 2014 (ADB, 2015; OECD, 2016). Meanwhile, debt among Vietnamese rural households rose modestly over the same period, narrowing the cross-country gap slightly.

Table 14.1 further presents the four indicators of over-indebtedness to compare debt burdens among rural households in Thailand and Vietnam. The debt service ratio (DSR), the share of annual income spent on debt repayments, shows that 43% of Thai rural households were over-indebted in 2008 (using a 40% threshold),

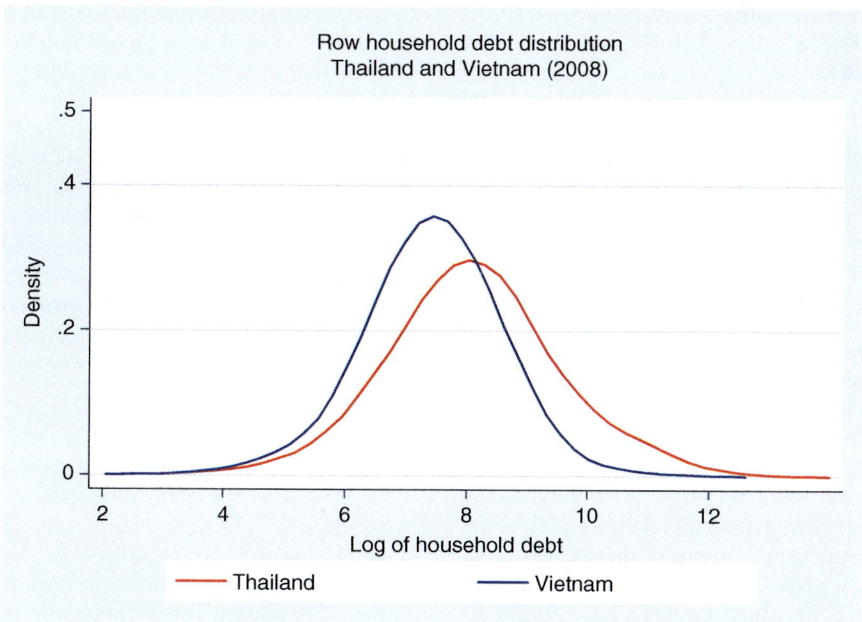

Figure 14.1. Row distribution of rural households' debt in Thailand and Vietnam in 2008. (Author's own image).

Table 14.1. Summary statistics for debt holdings (prevalence and conditional amounts), DSR DIR, DAR and default in 2008 and 2010. (Author's own table).

		Thailand		Vietnam	
		2008	2010	2008	2010
Outstanding debt	Prevalence	0.82	0.74	0.66	0.69
	10th percentile	617	580	471	568
	50th percentile	3205	2926	1812	2270
	90th percentile	20294	16389	6196	8614
DSR	Prevalence>40%	0.43	0.14	0.11	0.10
	10th percentile	0.00	0.00	0.00	0.00
	50th percentile	0.31	0.00	0.02	0.01
	90th percentile	1.63	0.54	0.46	0.40
DIR	Prevalence>50%	0.54	0.36	0.32	0.38
	10th percentile	0.05	0.06	0.00	0.05
	50th percentile	0.71	0.41	0.37	0.48
	90th percentile	4.59	2.50	1.83	2.42
DAR	Prevalence>50%	0.09	0.05	0.03	0.04
	10th percentile	0.01	0.01	0.00	0.01
	50th percentile	0.08	0.07	0.07	0.08
	90th percentile	0.50	0.40	0.28	0.33
Default	Prevalence	0.11	0.05	0.05	0.07

compared with just 11% in Vietnam. Among the most burdened, the top 10% of Thai households spent nearly twice their annual income servicing debt, while their Vietnamese counterparts spent around half. The debt burden gap widens along the DSR distribution. Two other measures, i.e. debt-to-income (DIR) and debt-to-asset (DAR) ratios, show a similar pattern, indicating that Thai households face greater short- and long-term repayment constraints. Lastly, default rates also suggest greater financial stress in Thailand, where a larger share of rural borrowers reported late or missed payments.

Explanatory Variables

The explanatory variables used in the decomposition analysis capture a range of socio-economic and demographic characteristics that influence household participation in the debt market and the extent of their indebtedness. These variables are selected based on established literature on household debt in both developed and developing contexts (e.g. Brown and Taylor, 2008; Disney *et al.*, 2008; D'Alessio and Iezzi, 2013; Schicks, 2014; Christelis *et al.*, 2021; Wu *et al.*, 2015). The analysis controls for household head's age, gender, education level, marital status, household size, main occupation, income and wealth quintiles, exposure to shocks, financial expectations and risk attitudes. Risk attitudes and financial outlooks were constructed from survey questions on future expectations and willingness to take risks. Most of these variables are operationalized through categorical groupings, with specified base categories, to allow meaningful interpretation of their effects in the analysis.

Table 14.2 compares the average characteristics of rural households across countries. Thai households tend to be wealthier and more concentrated in higher income and asset quintiles, suggesting greater borrowing capacity. In contrast, Vietnamese households are younger and better educated, pointing to higher future earning potential but also show greater risk aversion, which may limit their willingness to borrow heavily.

Empirical Methodology

To analyze cross-country differences in rural household debt dynamics between Thailand and Vietnam, we employ three complementary decomposition techniques: the non-linear Oaxaca–Blinder method (Fairlie, 1999), the standard

Table 14.2. Average household characteristics by country in 2008 and 2010. (Author's own table).

	Vietnam	Thailand
Age of HH head <39	0.214	0.067
Age of HH head 40–49	0.306	0.241
Age of HH head 50–59	0.235	0.268
Age of HH head ≥60	0.245	0.424
Female HH head	0.209	0.335
Married HH head	0.852	0.782
Household size	5.276	5.490
Illiterate	0.113	0.045
Primary education	0.317	0.843
Secondary education	0.513	0.0936
Higher education	0.057	0.0184
Agricultural HH	0.665	0.606
Self-employed HH	0.0751	0.082
Off-farm employed HH	0.209	0.184
Inactive HH	0.0509	0.128
Income quintile 1	0.307	0.200
Income quintile 2	0.234	0.200
Income quintile 3	0.183	0.200
Income quintile 4	0.169	0.200
Income quintile 5	0.107	0.200
Financial wealth quintile 1	0.663	0.215
Financial wealth quintile 2	0.015	0.185
Financial wealth quintile 3	0.0687	0.199
Financial wealth quintile 4	0.0993	0.201
Financial wealth quintile 5	0.154	0.200
Real wealth quintile 1	0.286	0.200
Real wealth quintile 2	0.287	0.200
Real wealth quintile 3	0.180	0.200
Real wealth quintile 4	0.119	0.200
Real wealth quintile 5	0.128	0.200
Income fluctuation (t–1)	0.613	0.681
Unexpected shocks to expense	0.454	0.372
Expected shocks to expense	0.119	0.0966
Unexpected shocks to income	0.616	0.406
Better future financial expectation	0.506	0.513
Same future financial expectation	0.421	0.375
Worst future financial expectation	0.073	0.112
Risk averse	0.523	0.352
Risk neutral	0.197	0.416
Risk taker	0.28	0.232
Observations	2024	2091

Oaxaca–Blinder decomposition (Blinder, 1973; Oaxaca, 1973), and the Recentred Influence Function (RIF) regression approach (Firpo *et al.*, 2009). These approaches allow observed differences to be decomposed into composition effects arising from variation in household characteristics and structural effects reflecting differences in how characteristics translate into outcomes across institutional, economic, and cultural contexts.

All three decomposition methods rely on counterfactual distributions to estimate how household debt outcomes would change if one group had the same characteristics or operated under the same structural environment as the other. In this analysis, Thai rural households serve as the reference group, allowing us to assess how Vietnamese households would perform if they shared the observable characteristics or faced the debt-generating structure of their Thai counterparts. The debt outcome gap is consistently defined as the difference between Thai and Vietnamese households. A positive coefficient effect indicates that Thailand's institutional or structural environment is more favorable to borrowing and indebtedness, while a positive covariate effect suggests that Thai households have characteristics more conducive to accessing and managing debt. The identification strategy follows Fortin *et al.* (2011), relying on the assumptions of common support and conditional independence to distinguish between endowment and structural effects.

To explain differences in binary outcomes like debt holding and over-indebtedness, pooled logit models are estimated and Fairlie's extension of the Oaxaca–Blinder decomposition for non-linear settings (Fairlie, 1999) is applied. The method expresses the gap in predicted probabilities as the sum of (i) changes due to substituting the Thai distribution of covariates into the Vietnamese sample and (ii) changes in the response of those covariates to country-specific environments. For continuous outcomes like log debt levels, DSR, DIR, and DAR, the standard linear Oaxaca–Blinder decomposition at the mean (Blinder, 1973; Oaxaca, 1973) is used, based on a pooled linear regression model with normalization to address the choice of base categories in categorical variables and ensure invariance to the omitted base (Yun, 2005). Finally, to uncover heterogeneous effects across the debt distribution, the re-centred influence function (RIF) regression approach of Firpo *et al.* (2009) is employed. Unconditional quantile regressions are estimated at selected quantiles (e.g. P25, median, P75). Decomposition at each quantile parallels the mean case, revealing whether determinants such as income, wealth, or risk attitudes matter more or the structural factors in the tails than at the center of the distribution.

Results

This section presents the results of the decomposition analysis examining cross-country differences in rural household debt outcomes between Thailand and Vietnam. We report findings from three complementary methods, focusing on debt participation, the level of indebtedness, and indicators of over-indebtedness. Together, the results provide insight into the extent to which observed differences are driven by household characteristics versus structural factors.[1]

Decomposing the Prevalence of Debt and Over-indebtedness

Table 14.3 presents the decomposition of cross-country differences in debt prevalence and over-indebtedness between rural households in Thailand and Vietnam into covariate and coefficient effects. The results from the aggregate decomposition show that the observed differences are primarily driven by coefficient effects, consistently favoring Thai households. This suggests that Thailand's institutional, cultural, and economic context is significantly more conducive to borrowing and higher debt burdens than Vietnam's. If Vietnamese households operated in a similar environment, their debt participation and over-indebtedness rates would closely resemble those of Thai households.

The detailed decomposition highlights that the coefficient effect is largely driven by the constant term, representing the most economically disadvantaged group, implying that even poorer Thai households are more likely to access credit and become over-indebted than their Vietnamese counterparts. This aligns with evidence of widespread debt among low-income groups in Thailand (ADB, 2013). Income also contributes notably to the coefficient effect in DSR-based over-indebtedness, suggesting that at comparable income levels, Thai households face higher debt service burdens. While adverse shocks and risk preferences explain much of the cross-country gap in default rates.

Conversely, covariate effects play a smaller and generally negative role, particularly for debt prevalence. This suggests that Thai households would be even more likely to borrow if they had the same characteristics as their Vietnamese counterparts, especially given that Vietnamese rural households tend to be younger and more educated, traits typically associated with higher credit demand and repayment potential. However, these favorable characteristics are outweighed by the strong influence of Thailand's structural environment. Differences in exposure to adverse shocks and income fluctuations also contribute to the covariate effect, though in opposing directions due to their differing incidence across countries (see Table 14.2).

Decomposing the Amount of Debt Holdings

Table 14.4 presents the Oaxaca–Blinder decomposition results at the mean of the conditional debt distribution. The findings show that most of the observed difference in the average log of household debt between rural households in Thailand and Vietnam is driven by the coefficient effect, suggesting that Thailand's economic environment is more supportive of higher debt holding. Specifically, 71% of the debt gap (about 0.497 points) is attributed to structural differences (top panel, Table 14.4), with financial wealth and financial expectations contributing most strongly. This implies that for any given level of financial wealth or expectations, Thai households are more likely to hold higher debt.

The remaining 29% of the debt gap is explained by covariate differences. Thai rural

[1] The discussion focuses exclusively on the decomposition results, without detailing the first-level regression estimates that form their basis. Although these preliminary regressions provide the coefficients and model fit statistics essential for constructing the decompositions, they are omitted here to maintain conciseness and keep attention on the main analytical findings. The complete set of first-level regression outputs, including all variable estimates, significance levels, and model diagnostics, can be provided upon request.

Table 14.3. Decomposition of differences in the prevalence of debt and over indebtedness in 2008. (Author's own table).

	Debt	DSR >40%	Default
Overall			
Thailand	0.817***	0.427***	0.111***
Vietnam	0.662***	0.113***	0.056***
Total difference	0.155***	0.314***	0.055***
Covariate effect	−0.033***	−0.009	−0.004
Coefficient effect	0.188***	0.324***	0.059***
	(0.01)	(0.02)	(0.01)
Covariate effect			
Female	−0.001	0.003	−0.002
Age	−0.015***	−0.015**	−0.003
Education	−0.010*	−0.021	0.017
Married	−0.001	−0.001	−0.000
HH size	0.003***	0.002**	0.004
Occupation	−0.001	0.002	0.003
Income	−0.001	0.004	0.000
wealth	−0.003	0.023***	−0.016
Financial expectation	−0.001	0.001	0.003
Adverse shocks	−0.007***	−0.008*	−0.011
Income fluctuation (t−1)	0.006***	0.004*	0.009
Risk preference	−0.001	−0.003	−0.008
Coefficient effect			
Female	−0.020**	−0.015*	0.001
Age	−0.032	0.017	−0.042**
Education	−0.020	−0.043	0.029
Married	−0.040	−0.059	−0.007
HH size	−0.074*	0.022	−0.012
Occupation	0.000	−0.007	0.004
Income	0.023	0.096***	0.023
Wealth	0.030	−0.026	−0.003
Financial expectation	−0.064	−0.051	−0.019
Adverse shocks	−0.042**	0.010	0.023*
Income fluctuation (t−1)	0.002	−0.024	0.018
Risk preference	−0.038	−0.098***	0.029*
Constant	0.462***	0.501***	0.013
Observations	4211	4211	4211

Notes: Results are from decomposition analyses that compare the prevalence of debt and over-indebtedness among rural households in Vietnam to those in Thailand using coefficients from pooled logit regression models. Results are based on the Oaxaca–Blinder Decomposition Method. Standard errors are available upon request. *, **, & *** represent statistical significance at the 10%, 5%, & 1% level respectively.

households tend to have higher income, real, and financial wealth, as well as greater exposure to income fluctuations, which are all factors associated with greater debt holding. However, education contributes negatively to the covariate effect, reflecting that Vietnamese households are generally more educated, and education is positively correlated with debt holding. Overall, both better endowments and a more conducive economic environment explain why Thai households carry more debt on average.

To explore how these effects vary across the debt distribution, Table 14.5 presents results from the RIF-regression decomposition. The debt gap between countries widens at higher percentiles, and the relative importance of covariate and coefficient effects shifts along the distribution. Up to the median, the covariate effect is positive and significant, indicating that Thai households' stronger endowments explain much of their higher debt levels. Beyond the median, however, covariate effects become negligible, and the gap

Table 14.4. Decomposition of differences in average log of debt, DSR DIR and DAR in 2008. (Author's own table).

	Amount of Debt	DSR	DIR	DAR
Overall				
Thailand	8.133***	48.346***	106.224***	17.956***
Vietnam	7.436***	15.590***	63.779***	11.861***
Total difference	0.697***	32.755***	42.445***	6.095***
Covariate effect	0.200***	12.303***	10.881**	−2.805**
Coefficient effect	0.497***	20.452***	31.564***	8.900***
Covariate effect				
Female	0.001	0.244	−0.764	−0.135
Age	0.016	−0.199	−0.798	0.311
Education	−0.171***	−1.838**	−5.655***	−2.490***
Married	−0.010*	−0.383**	−0.159	−0.083
HH size	0.001	0.074	−0.014	0.015
Occupation	0.012	0.109	0.793	0.206
Income	0.021***	−0.691	−1.005	0.184**
Financial wealth	0.164***	12.115***	10.967**	1.697*
Real wealth	0.166***	3.006***	6.341***	−2.979***
Financial expectation	−0.001	−0.043	1.716*	0.049
Adverse shocks	−0.013	−0.459	−2.839***	−0.094
Income fluctuation (t−1)	0.011**	0.285	0.513*	0.170**
Risk preference	0.003	0.081	1.787*	0.343
Coefficient effect				
Female	−0.008	−0.917	−2.949	−1.120**
Age	−0.045	1.457	−0.982	−0.641
Education	0.092	−1.946	7.409	2.331
Married	0.055	−1.164	−0.753	−1.949
HH size	0.103	−0.590	−1.014	4.366*
Occupation	0.046	−0.502	1.102	0.758
Income	−0.121	−17.249***	−6.301	0.297
Financial wealth	0.179***	7.400***	10.766***	2.167***
Real wealth	−0.093	4.436	−8.637	−7.298***
Financial expectation	0.248*	−3.278	14.408	3.060
Adverse shocks	−0.097	2.604	−1.871	−1.515
Income fluctuation (t−1)	−0.012	0.127	−2.637	0.319
Risk preference	−0.118	−12.082***	−4.558	−1.524
Constant	0.267	42.155***	27.581	9.649*
Observations	3117	3975	3141	3397

Notes: Results are from decomposition analyses that compare the average amount of debt, DSR, DIR and DAR of rural households in Vietnam to those in Thailand using coefficients from pooled linear regression models. Outstanding amount of debt is conditional on participation in debt markets. Results are based on the Oaxaca–Blinder decomposition method. Standard errors are available upon request. *, **, & *** represent statistical significance at the 10%, 5%, & 1% level respectively.

is fully explained by the coefficient effect. This shift suggests that structural and institutional factors increasingly account for the debt gap at higher levels of debt.

Detailed decomposition confirms that at the lower end of the distribution, income, wealth, and income volatility drive the covariate effect. At the upper end, the same variables contribute to the coefficient effect, underscoring that Thai households with similar endowments to their Vietnamese counterparts are still more indebted, likely due to a more enabling credit environment.

Overall, the RIF-regression results reinforce the mean-based findings while offering a richer

Table 14.5. Decomposition of differences in log of debt at the 10th, 25th, 50th, 75th and 90th percentiles in 2008. (Author's own table).

	Log of debt				
	10th	25th	50th	75th	90th
Overall					
Thailand	6.509***	7.188***	8.064***	8.905***	9.880***
Vietnam	6.138***	6.896***	7.518***	8.154***	8.707***
Total difference	0.371***	0.292***	0.546***	0.751***	1.173***
Covariate effect	0.280*	0.209**	0.221***	0.130	-0.048
Coefficient effect	0.090	0.083	0.325***	0.621***	1.221***
	(0.18)	(0.11)	(0.09)	(0.10)	(0.16)
Covariate effect					
Female	−0.023	−0.002	−0.000	0.008	0.006
Age	0.046	0.013	0.021	0.003	−0.014
Education	−0.173***	−0.161***	−0.128***	−0.122***	−0.266***
Married	−0.008	−0.017*	−0.008	−0.013*	−0.012
HH size	0.001	0.001	0.001	0.000	0.000
Occupation	0.009	0.010	0.018**	0.023**	0.016
Income	0.018**	0.014**	0.015**	0.022**	0.029**
Financial wealth	0.323***	0.215***	0.143**	0.026	−0.081
Real wealth	0.112***	0.118***	0.147***	0.182***	0.243***
Financial expectation	−0.005	0.004	0.001	−0.003	−0.007
Adverse shocks	−0.054**	−0.011	−0.009	−0.010	0.014
Income fluctuation (t−1)	0.017**	0.011*	0.007	0.011*	0.014
Risk preference	0.015	0.015	0.013	0.002	0.009
Coefficient effect					
Female	−0.027	−0.012	−0.046	0.012	0.061
Age	−0.085	−0.041	0.074	−0.008	−0.234
Education	−0.156	0.052	−0.143	0.122	0.443
Married	0.036	0.235	−0.119	0.158	0.302
HH size	−0.221	−0.006	0.078	−0.007	0.336
Occupation	0.008	−0.007	0.039	0.070	0.227***
Income	0.091	−0.005	0.097	0.144	1.540*
Financial wealth	0.347***	0.251***	0.121**	0.137**	0.058
Real wealth	−0.528	0.534	0.149	0.577	2.747**
Financial expectation	0.313	0.302*	0.439***	0.196	0.403
Adverse shocks	−0.053	−0.038	−0.064	−0.101	0.035
Income fluctuation (t−1)	0.047	−0.016	0.066	−0.109	−0.050
Risk preference	−0.167	−0.181*	0.007	−0.195*	−0.164
Constant	0.484	−0.985	−0.372	−0.373	−4.484***
Observations	3005	3005	3005	3005	3005

Notes: Results are from decomposition analyses that compare the distribution of amount of debt of rural households in Vietnam to those in Thailand using coefficients from pooled linear regression models. Outstanding amount of debt is conditional on participation in debt markets. Results are based on the RIF-Regression decomposition method. Standard errors are available upon request. *, **, & *** represent statistical significance at the 10%, 5%, & 1% level respectively.

view of how drivers of the debt gap differ across the distribution. At lower debt levels, better household endowments in Thailand explain most of the difference; at higher debt levels, structural factors dominate, highlighting the role of Thailand's economic environment in enabling greater borrowing.

Decomposing the Indebtedness Indicators

The RIF-regression decomposition results indicate that the higher levels of debt observed among Thai rural households are partly linked to stronger endowments such as greater income and wealth which may enhance their ability to service debt.

However, when turning to measures of debt burden (DSR, DIR, and DAR), Thai households still exhibit significantly higher indebtedness on average, despite their economic advantage (Table 14.4). This debt burden gap is predominantly driven by coefficient effects, pointing to a more permissive economic and institutional environment in Thailand.

Detailed decomposition shows that financial wealth consistently contributes to the coefficient effect across all three indicators (Table 14.4). One likely explanation is the widespread use of savings-secured and group-based loans in Thailand, where accumulated savings signal creditworthiness and determine borrowing limits (Coleman, 1999). Furthermore, for DSR and DAR, the large positive constant term highlights that even economically disadvantaged households in Thailand face a more permissive lending environment than their peers in Vietnam.

Covariate effects also explain part of the debt burden gap, i.e. around 37% for DSR and 25% for DIR, driven largely by higher financial and real wealth in Thailand (Table 14.4). Conversely, education contributes negatively, as Vietnamese households tend to be more educated, and education correlates positively with higher debt burdens. Interestingly, risk tolerance and pessimistic financial expectations also play a role in explaining higher debt burdens among Thai households, particularly for DIR. For DAR, the negative covariate effect indicates that rural households in Vietnam have characteristics associated with higher debt burdens; however, Thailand's more favorable economic environment, which is more tolerant of higher rural debt, offsets this effect.

The RIF decomposition results across the distribution of DAR, DSR, and DIR highlight that the debt burden gap widens at higher percentiles and is increasingly explained by coefficient effects. For DAR, coefficient effects fully account for the observed gap, even overriding a negative covariate effect (Table 14.6). Income, financial wealth, occupation, and household size are key drivers of the coefficient effect. Notably, the economic environment is less favorable to disadvantaged Thai households at the bottom of the DAR distribution but becomes more enabling at the top. For DSR, financial and real wealth dominate both coefficient and covariate effects, while the economic environment again appears especially tolerant of high debt among poorer households. For DIR, financial wealth drives the gap at the lower end, while real wealth and the occupation of the household head, particularly self-employment, explain the large difference at the upper end, suggesting that Thailand's economic context is especially accommodating of higher debt burdens for self-employed rural households[2].

Covariate effects vary across indicators. For DSR and DIR, Thai households' characteristics such as wealth, financial expectations, and risk preferences positively contribute to the debt burden gap at the lower end of the distribution. However, variables like age, education, and income exert a negative effect, favoring Vietnamese households. For DAR, the covariate effect is generally negative, particularly at the top percentiles, again driven mainly by education (Table 14.6).

Overall, Thai rural households consistently face higher levels of indebtedness across all indicators. While differences in household characteristics explain part of the gap, the dominant factor is Thailand's economic and institutional environment, which appears more tolerant of, and even conducive to, high levels of rural household debt.

Conclusion and Policy Implications

This study examined cross-country differences in rural household debt between Thailand and Vietnam, using decomposition methods to disentangle the roles of household characteristics and broader structural conditions. Across all indicators, including debt prevalence, outstanding debt, and multiple measures of over-indebtedness, Thai households consistently exhibit higher levels of debt and debt burden compared with their Vietnamese counterparts. As debt burdens rise, particularly at higher levels of indebtedness, structural factors captured by the coefficient effect play an increasingly prominent role.

[2] For brevity, detailed RIF regression results for the two indebtedness indicators DSR and DIR are not presented here; these results can be provided upon request.

Table 14.6. Decomposition of differences in DAR distribution at the 25th, 50th, 75th and 90th percentiles in 2008. (Authors' own table).

	Debt to asset ratio			
	25th	50th	75th	90th
Overall				
Thailand	2.963***	8.229***	19.669***	44.793***
Vietnam	2.349***	6.425***	13.832***	27.891***
Total difference	0.614**	1.803***	5.837***	16.902***
Covariate effect	−0.428	−1.325*	−4.797***	−17.612***
Coefficient effect	1.042**	3.128***	10.635***	34.514***
Covariate effect				
Female	−0.028	−0.041	−0.392**	−0.805
Age	−0.008	−0.101	−0.076	−0.044
Education	−0.526***	−1.149***	−2.947***	−8.531***
Married	−0.016	−0.074	0.006	0.008
HH size	0.007	0.009	0.008	0.051
Occupation	0.014	0.209**	0.346*	0.178
Income	0.020	0.063	0.126	0.362
Financial wealth	0.870***	1.959***	2.207**	1.187
Real wealth	−0.871***	−2.388***	−4.431***	−11.024***
Financial expectation	−0.025	0.244*	0.044	0.508
Adverse shocks	−0.054	−0.113	0.070	−0.883
Income fluctuation (t−1)	0.042*	0.095*	0.162*	0.350
Risk preference	0.146**	−0.039	0.080	1.033
Coefficient effect				
Female	−0.011	−0.242	−1.235**	−2.141
Age	−0.190	0.364	−1.983	−0.106
Education	0.232	4.001**	2.211	1.295
	(0.89)	(1.88)	(3.21)	(11.49)
Married	0.075	0.697	−0.750	−7.621
HH size	−1.280*	0.847	−0.864	19.667**
Occupation	−0.043	0.609*	1.916***	3.227
Income	5.547***	6.291	17.488**	66.716**
Financial wealth	0.855***	1.732***	2.506***	4.379*
Real wealth	2.907	−18.707***	−19.689	−188.845***
Financial expectation	0.665	0.679	6.078**	8.915
Adverse shocks	−0.686*	−1.092	−0.449	3.676
Income fluctuation (t−1)	0.272	−0.057	−0.910	0.448
Risk preference	−0.123	−2.048**	−1.938	−0.321
Constant	−7.177**	10.053	8.253	125.224**
Observation	3288	3288	3288	3288

Notes: Results are from decomposition analyses that compare the distribution of amount of outstanding debt to asset ratio of rural households in Vietnam to those in Thailand using coefficients from pooled RIF-regression models. Results are based on the RIF-Regression decomposition method. Standard errors are available upon request. *, **, & *** represent statistical significance at the 10%, 5%, & 1% level respectively.

Specifically, Thailand's institutional and economic environment, which likely includes more accessible credit markets, less restrictive lending criteria, and debt-tolerant culture, enables greater household indebtedness, even among economically disadvantaged rural households. At the lower end of the debt distribution, household characteristics contribute positively in Thailand mainly through higher incomes, greater asset ownership, and stronger financial expectations. Meanwhile, Vietnam's younger and more educated rural population may have greater credit demand and repayment capacity, yet remains limited by systemic borrowing constraints.

From a development perspective, these findings carry two policy implications. First, while expanding rural credit access can support productive investment, inadequate safeguards risk trapping vulnerable households in cycles of over-indebtedness, especially where low financial literacy and informal lending expose the poor to financial distress. Second, country-specific responses are needed: Vietnam could safely broaden credit access for better-off, young, and educated households by strengthening collateral frameworks and diversifying financial products, provided prudent lending standards are maintained. Thailand, by contrast, must curb excessive borrowing through stronger consumer protections, credit-risk regulation, and targeted financial literacy efforts, particularly for the most economically disadvantaged households who bear the greatest debt burdens.

References

Aminudin, U. and Tissot, B. (2015) Assessing Household Financial Positions-An Asian Perspective. IFC Bulletins Chapters, in: Bank for International Settlements (ed.), Assessing household financial positions in Asia, volume 40, Bank for International Settlements, Hong Kong. Available at: https://www.bis.org/ifc/publ/ifcb40.htm.

Asian Development Bank (2013) *Thailand Financial Inclusion Synthesis Assessment Report. Technical Assistance Consultant's Report. TA7998-THA*. Asian Development Bank, Manila.

Asian Development Bank (2014) *Financial Inclusion in Asia: Country Surveys*. Asian Development Bank Institute, Tokyo, Japan.

Asian Development Bank (2015) *Asian Development Outlook 2015: Financing Asia's Future Growth*. Asian Development Bank, Manila, Philippines.

Blinder, A.S. (1973) Wage discrimination: Reduced form and structural estimates. *Journal of Human Resources* 8(4), 436–55.

Bover, O., Casado, J.M., Costa, S., Caju, P.D., McCarthy, Y. *et al.* (2016) The distribution of debt across euro area countries: The role of individual characteristics, institutions and credit conditions. *International Journal of Central Banking* 2(12), 71–128.

Brown, S. and Taylor, K. (2008) Household debt and financial assets: Evidence from Germany, Great Britain and the USA. *Journal of the Royal Statistical Society: Series A (Statistics in Society)* 171(3), 615–643.

Chantarat, S., Chawanote, C., Ratanavararak, L., Rittinon, C., Sa-ngimnet, B. *et al.* (2023) *Financial Lives and the Vicious Cycle of Debt among Thai Agricultural Households*. The Puey Ungphakorn Institute for Economic Research, Discussion Paper NO. 204. Puey Ungphakorn Institute for Economic Research, Bangkok.

Chichaibelu, B.B. and Waibel, H. (2017) Borrowing from "pui" to pay "pom": Multiple borrowing and over-indebtedness in rural Thailand. *World Development* 98, 338–350.

Chichaibelu, B.B. and Waibel, H. (2018) Over-indebtedness and its persistence in rural households in Thailand and Vietnam. *Journal of Asian Economics* 56, 1–23.

Christelis, D., Ehrmann, M. and Georgarakos, D. (2021) Exploring differences in household debt across the United States and euro area countries. *Journal of Money, Credit and Banking* 53(2–3), 477–501. http://10.1111/jmcb.12769.

Christelis, D., Georgarakos, D. and Haliassos, M. (2013) Differences in portfolios across countries: Economic environment versus household characteristics. *Review of Economics and Statistics* 95(1), 220–236.

Coleman, B.E. (1999) The impact of group lending in Northeast Thailand. *Journal of Development Economics* 60(1), 105–141.

D'Alessio, G. and Iezzi, S. (2013) *Household Over-Indebtedness: Definition and Measurement with Italian Data (No. 149)*. Bank of Italy, Roma. Available at: 10.2139/ssrn.2243578.

Disney, R., Bridges, S. and Gathergood, J. (2008) *Drivers of Over-Indebtedness: Report to the Department for Business, Enterprise and Regulatory Reform*. Center for Policy Evaluation, University of Nottingham, UK.

Fairlie, R.W. (1999) The absence of the African-American owned business: An analysis of the dynamics of self-employment. *Journal of Labor Economics* 17(1), 80–108.

Firpo, S., Fortin, N. and Lemieux, T. (2009) Unconditional quantile regressions. *Econometrica* 77(3), 953–73.

Fortin, N.M., Lemieux, T. and Firpo, S. (2011) Decomposition methods in economics. In *Handbook of Labor Economics, Vol. 4*, Ashenfelter, O. and Card, D. (eds); Elsevier, Amsterdam, pp. 1–102.

Hardeweg, B., Klasen, S. and Waibel, H. (2012) Establishing a database for vulnerability assessment. In: Klasen, S. and Waibel, H. (eds.) *Vulnerability to Poverty-Theory, Measurement, and Determinants*. Palgrave Macmillan, Basingstoke, Hampshire.

Khandker, S.R., Koolwal, G.B, Haughton, J. and Jitsuchon, S. (2012) *Household Coping and Response to Government Stimulus in an Economic Crisis: Evidence from Thailand*. Policy Research Working Paper 6016. World Bank, Washington, DC.

Nguyen, H.-K.T., Nguyen, V.-H.T., Vuong, T.-T., Ho, M-T. and Vuong, Q-H. (2019) The new politics of debt in the transition economy of Vietnam. *Austrian Journal of South-East Asian Studies* 12(1), 91–110.

Oaxaca, R.L. (1973) Male-female wage differentials in urban labor markets. *International Economic Review* 14(3): 693–709.

Organization for Economic Co-operation and Development (2016) *Economic Outlook for Southeast Asia, China and India 2016: Enhancing Regional Ties*. OECD Publishing, Paris, France.

Schicks, J. (2014) Over-indebtedness in microfinance – an empirical analysis of related factors on the borrower level. *World Development* 54, 301–324.

Ta, N.L. (2022) *Factors Affecting Credit Accessibility of Farm Households in Rural Areas of Vietnam: A Case Study in Haiphong City*. Universite de Liege, Belgium.

Varangis, P., Buchenau, J. and Chen, R. (2019) Vietnam – Agriculture Finance Diagnostic Report: Financial Inclusion Support Framework – Country Support Program. World Bank Group Washington, D.C., Available at: http://documents.worldbank.org/curated/en/884321587357455934.

World Bank Group. (2024) *Thailand Economic Monitor: Unlocking the Growth Potential of Secondary Cities*. World Bank, Bangkok.

Wu, W., Fasianos, A. and Kinsella, S. (2015) *Differences in Borrowing Behaviour Between Core and Peripheral Economies – Economic Environment Versus Financial Perceptions*. Working Papers No 201516, Geary Institute, University College Dublin, Ireland.

Yun, M.-S. (2005) A simple solution to the identification problem in detailed wage decompositions. *Economic Inquiry* 43 (4), 766–72.

15 Food Self-sufficiency and Building-integrated Urban Agriculture: Lessons from Singapore

Tomoki Fujii, Christoph Waibel, Xinyi Du and Zhongming Shi

Abstract

Singapore aims to achieve 30% food self-sufficiency by 2030, known as the 30-by-30 target, and this chapter reviews Singapore's changing landscape for urban agriculture and its recent progress towards the 30-by-30 goal, highlighting key challenges such as land constraints, high production costs, and resource limitations. Building-integrated agriculture (BIA), which utilizes building surfaces such as rooftops, façades, and balconies for food production within urban environments, is examined as a potential way to increase self-sufficiency through a simulation. Despite the BIA's potential, practical issues – including regulatory concerns and infrastructure limitations – must be addressed before it can be implemented at scale. Insights from Singapore's experience offer valuable lessons for other dense, land-scarce cities around the globe in seeking to enhance local food and sustainable production and strengthen resilience against supply shocks and disruptions.

Introduction

As a small and densely populated island city-state, ensuring food security has always been an essential concern for the government. Singapore has taken steps to diversify source countries for food, grow food overseas, and increase the stockpile to be prepared for unexpected supply shocks. In recent years, Singapore has reoriented its agricultural policy towards increasing local production to raise food self-sufficiency substantially. This chapter first reviews the changes in food self-sufficiency in Singapore and the 30-by-30 target – a target to locally produce 30% of Singapore's nutritional needs by 2030. The chapter then discusses the critical economic and geographic factors that affect the promotion of urban agriculture. Based on this discussion, we examine the concept of building-integrated agriculture (BIA) as a potential way to utilize the urban built environment to address food self-sufficiency and undertake a simulation-based, simple cost-benefit analysis on a city scale. Finally, we discuss challenges and opportunities in urban farming and draw lessons that are relevant to other cities around the world.

Food Self-sufficiency in Singapore

From the time of independence in 1965 until the 1980s, family farms played a critical role in Singapore's food security. The agricultural sector was sizable, with 9% of the total population reportedly engaged in agricultural activities or indirectly dependent on farming and fishing for a livelihood. The *Annual Report of the Primary Production Department* published in 1970

DOI: 10.1079/9781800627338.0015

suggests that poultry, eggs, and livestock were entirely locally produced, and 50% and 30% of the needs for vegetables and fish were satisfied respectively from local produce (Singapore Primary Production Department, 1970). With rapid economic changes, Singapore witnessed a significant decline in the number of farms and the area of agricultural land in the 1970s. Nevertheless, Singapore shifted towards more capital-intensive farming and maintained a high self-sufficiency rate in some key food items. For instance, in 1982, Singapore had 100% self-sufficiency for hen's eggs, 95% for pigs, 70% for chickens and ducks, and 25% for fish and vegetables (Gamboa and Chew, 1984).

However, due to the competing needs for land and concerns for pollution, Singapore's agricultural policy was reoriented towards promoting agribusinesses aimed at high value produce for export or specialized local markets, and family farms were phased out in the 1980s (Chow, 2014). Among the most prominent changes in the agricultural policy was pig farming. In 1984, the government decided to phase out pig farming by 1989 despite rosy production figures (Choo, 2022). Singapore's agricultural sector continued to decline. The land allocated for food production decreased to around 600 hectares, down from about 15,000 hectares in the 1960s (Montesclaros and Teng, 2019). The shares of agriculture, forestry, and fishing in value-added and employment in 2022 were, respectively, 0.03% and 0.11%, down from 0.25% and 1.22% in 1991 (World Bank, 2024). Unsurprisingly, coupled with Singapore's much larger current population, the self-sufficiency rates today are much lower than they were in the 1970s. The proportion of local production as a percentage of total food production has been around 30% for hen eggs, 7% for seafood, and 3% for vegetables in the fourth quarter of 2023 (SFA, 2024).

The 30-by-30 Target

In recent years, Singapore has started to reconsider its food policy because Singapore's food supply can be affected by the disruption in food production elsewhere due to climate change (Montesclaros and Teng, 2019). Singapore is also vulnerable to disruptions in food supply chains, such as pandemics, trade policy shocks, and external conflicts and wars, as it is a small, open economy. The relevance of such vulnerability was apparent during the Covid-19 pandemic, when many countries imposed export restrictions or bans on some essential products, including food, personal protection equipment, medicines, and hand sanitizers.

The announcement of the creation of the SFA and 30-by-30 target on March 7, 2019 by Mr. Masagos Zulkifli, the then Minister for the Environment and Water Resources, was thus timely (Ministry of Sustainability and the Environment, 2019). Shortly after the announcement, the SFA was incorporated to have regulatory oversight across the entire food chain, transform the agri-food industry, and pursue three broad strategies to strengthen Singapore's food security: Diversify Import Sources, Grow Local, and Grow Overseas. Each of these "three food baskets" will be critical for the Singapore Food Story that it aspires to write as with the "four national taps" – water from local catchment areas, imported water, reused water, and desalinated water – in the "Singapore Water Story" to move towards water self-sufficiency (Irvine *et al.*, 2014).

Achieving the 30-by-30 target is a key target under the Grow Local strategy. The announcement of the 30-by-30 target also marks a significant turning point for Singapore's agricultural policy. In Mr. Zulkifli's speech, he discussed various measures to support existing egg and fish farms and promote state-of-the-art indoor and rooftop farms. The government has also invested in local farms to promote capacity-building and innovation through schemes such as the Agricultural Productivity Fund and the Startup SG Equity Scheme for agri-technology (Tortajada and Lim, 2021). The government also established the Agri-Food Innovation Park to catalyze innovation in the agri-tech ecosystem in the Greater Sungei Kadu area, which forms part of a larger Northern Agri-Tech and Food Corridor with food-related industries and enables the industry's expansion in the future (Ministry of Trade and Industry, 2019).

At the onset of the Covid-19 pandemic, at around the time Singapore went into lockdown (locally known as "circuit breaker") period, SFA announced the 30×30 Express grant to ramp up local production (Ministry of Sustainability and the Environment, 2020). Efforts were also made to expand community gardens, empower individuals to grow their own food, and promote awareness of local produce.

However, local produce's affordability and high labor and land costs remain among the key constraints to scaling up local production (Tortajada and Lim, 2021), and it has been difficult to foster commercially viable urban farms. TransZPlus Engineering, which is the holding company of the local indoor farm I.F.F.I. – a recipient of the 30×30 Express grant and said to have the capacity to produce 300 tonnes of leafy greens annually – entered liquidation in November 2023 and closed the farm. Sky Gardens, which had touted its patented vertical farming systems as an innovative and environmentally friendly way of growing vegetables, was also reported to have scaled down its operations (Tan, 2024). Vegetable farm, VertiVegies, is also reported to have scrapped plans to build what would have been one of Singapore's largest indoor farms in Lim Chu Kang. The fish farming business also faces similar issues (The Business Times, 2024).

Correspondingly, the progress towards achieving the 30-by-30 target has been mixed at best. On the one hand, there is a marked increase in the local production of hen eggs from 25.5% to 34.9% from 2019 to 2023. In contrast, the corresponding figure for seafood hovered around 7%, and that for vegetable sectors has marginally declined from around 4% to 3% (Figure 15.1). These data, however, may not capture the long-term trends because they only cover five years, and most of the years were affected by the Covid-19 pandemic.

It is a challenge to assess a longer-term trend of food self-sufficiency in Singapore due to the lack of publicly available data. For example, a commonly used measure of food self-sufficiency, the food self-sufficiency ratio (SSR) – the percentage of the local production in the total local food supply, which is the sum of the local production and net imports (Food and Agriculture Organization, 2001) – is neither readily available nor computable from the publicly available sources to the best of our knowledge. Hence, we derive

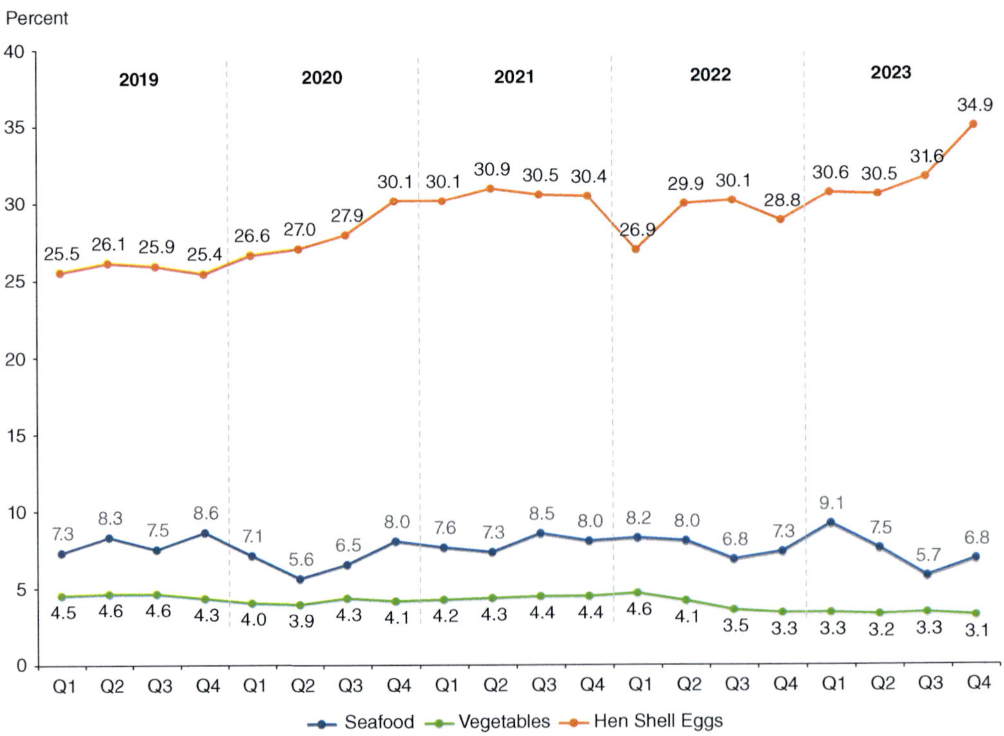

Figure 15.1. Local Production as a Percentage of Total Food Consumption. (Source: Adapted from Singapore Food Agency (SFA), 2022 and 2024. Licensed under Singapore Open Data License (ODL)).

an alternative value-based measure, which is the percentage of the local production value out of the total food consumption,[1] because we can use the monetary values of local productions in the total food consumption as these data are available from Singapore Department of Statistics (2024). Note that the value-based measures may not necessarily exhibit the same trend as the self-sufficiency ratio, because the value-based figures reflect both the quantity and price. Nevertheless, as Figure 15.2 shows, we find that the local production share of hen eggs has gone up substantially between 2012 and 2023, whereas those for seafood and vegetables remained almost the same during this period.

While locally produced vegetables, eggs, and seafood are all viewed to be fresher and safer than imported counterparts, the self-sufficiency ratio has only visibly increased for eggs, and not for vegetables and seafood. This contrast may come from the differences in the competition environment and the maturity of the industry. For example, Mr. Koh Yeow Koon, who is a managing director of local egg farm Seng Choon Farm, explains in an interview that the egg industry is a unique player in the market with a single product – hen shell eggs, whereas vegetables and seafood have diverse varieties and various sources, making it difficult to achieve economies of scale (Tham, 2024).

Therefore, local producers would need to exert extra efforts to stay competitive, and there are some potentially promising cases. For example, Sustenir – which grows microgreens, edible flowers, and vegetables in an indoor farm – has been able to produce more nutritious and palatable vegetables while diversifying its products. Sustenir's revenue increased fivefold over the past five years despite the headwinds from rising energy costs and falling private investment, among others (Feng, 2024).

Nevertheless, significant challenges are clearly present in progressing towards the 30-by-30 target. Even though urban agriculture (UA) is a significant source of food production, providing an estimated 15–20% of global food output (Smit *et al.*, 1996), UA is traditionally based on the utilization of vacant land within or at the fringe

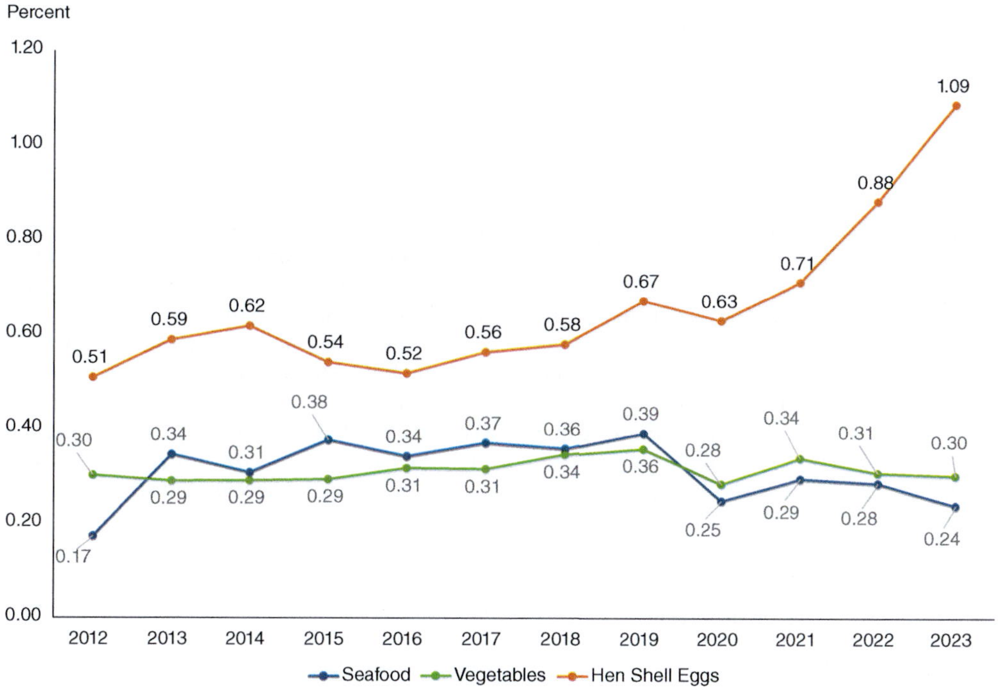

Figure 15.2. Local Production Share in Food Consumption. (Source: Adapted from Singapore Department of Statistics, 2024).

of a city (Redwood, 2012) and has addressed food security primarily in developing countries and cities with large informal settlements (Redwood, 2012). Given that Singapore is a highly developed city-state with virtually no informal settlements, a new, different UA model is needed for Singapore. In the rest of this chapter, we offer a discussion on UA using the technologies for building integrated agriculture (BIA). While these technologies are still not yet fully mature, they are highly relevant to Singapore as argued below. These technologies may help Singapore to advance towards the 30-by-30 target and, with the accumulation of expertise and advancement of technology, Singapore may be able to foster a high value-added industry that could essentially replicate Singapore's food story elsewhere and contribute towards a carbon-free world economy and attainment of the Sustainable Development Goals (SDGs).

Technologies for Building-integrated Agriculture

In Singapore, small-scale UA can be observed throughout the city. Private crop growth can be found in community gardens and rooftops and on the many balconies and access corridors of the multi-story public housing buildings supplied by the Housing and Development Board (HDB), where 80% of the population resides (HDB, 2024). Considering that Singapore's surface area is majorly urbanized, the prevalent land scarcity would require the expansion of agricultural activity into the urban fabric, if Singapore aims to achieve its 30-by-30 goals (see Diehl, 2024, for changes in forested, cultivated, and urban areas in Singapore between 1915–2018). Governmental bodies such as NParks incentivize home-grown produce, for example, by information campaigns and by distributing free seeds to HDB inhabitants (Sajan, 2020). As a recent study by Kosorić *et al.* (2019) shows, growing crops on balconies, rooftops, and decks of parking garages is socially accepted. This community-oriented UA model contrasts with business-oriented vertical (indoor) farming, where crops are grown in controlled and sealed environments with artificial light and water supply. However, the architectures of buildings on which existing community-oriented UA is based are mostly not designed to promote urban farming. By designing the architecture of buildings to use façades, roofs, and balconies, different architectural designs are required than for conventional open-space UA or high-tech indoor vertical farming. Thus, the agricultural output could be increased by adopting BIA technologies.

The topic of BIA has gained interest in recent years, especially in South-East Asia. Sengodan (2022) assessed the potential of vertical farming and BIA for Malaysian high-rise buildings, and differentiated between roof, balcony, façade, and interior as well as underground farming, and finds that different cultivation methods (soil-based, hydroponics, aeroponics, aquaponics) as well as business models are required for each form of farming.[2] The Singapore Food Agency (SFA) considers BIA in their classifications for different types of urban farming methods into land-based agricultural farms, indoor agricultural farms, and rooftop agricultural farms (SFA, 2025).

As a nascent field, BIA does not have a singular technology, clear architectural typology, or straightforward farming method. Table 15.1 exemplifies the diversity of approaches for buildings-related urban farming in Singapore, showing which consequences each of the respective concepts have on operation and adoption. The list includes classical concepts such as greenhouses on rooftops and community gardens on vacant urban land, but also unconventional ideas such as utilizing façade areas on buildings. Examples of urban farming related to buildings are also presented in Song *et al.* (2022a) and Tablada and Kosoric (2024).

To stimulate the development of new BIA solutions, it helps to examine some specific existing implementations: Eriksson (2024) designed alternative architectural solutions for a window-integrated system, as shown in Figure 15.3(a). The motivation is to develop a system that can be easily integrated into existing buildings and operated from inside through the windows. Different variants are proposed based on sliding mechanisms, rotating panels, or panels attached on hinges, such that plants can be faced outwards towards the sun for growing and accessed easily later for harvesting and maintenance. These concepts can work with soil, but hydroponics systems are also possible. Figure 15.3(b)

Table 15.1. Overview of Urban Agricultural Systems in Singapore. (Authors' own table).

Type and Description	Examples
Indoor (vertical) farming: Crop cultivation in indoor spaces. Profitable for high-value crops (micro-greens and leafy vegetables) and in high-density regions (with challenging outdoor climates) such as Singapore, Tokyo, and the Middle East, but also in Europe and the USA. No need for pesticides due to controlled environments. **Adoption**: Commercial; established industry; typical for high-density cities. **Initial investment**: Very high (artificial lighting, irrigation, cooling systems). **Challenges**: High energy costs especially for lighting; Only profitable for high-value crops.	GroGrace, GoFarm, and VertiVegies, Singapore indoor and vertical farming companies.
Rooftop greenhouse: Installed on vacant building rooftops (car parks, industrial buildings). **Adoption**: More common for commercial ventures due to high investment cost, but can also be found in community initiatives. **Initial investment**: High. **Challenges**: Structural requirements on rooftop need to be checked.	ComCrop, commercial rooftop greenhouse farm.CitySprouts, community-driven initiative.
Rooftop garden: Open air farming on carparks, shopping malls, or residential buildings (no greenhouse). The Singapore Food Authority (SFA) is promoting this farming type with a tender for HDB car park rooftops. **Adoption**: Commercial or community-driven. **Initial investment**: Medium to low. **Challenges**: Structural requirements on rooftop need to be checked.	Ang Mo Kio car park rooftop, commercial rooftop farm.HDB Jurong car park rooftop, community-driven rooftop garden.
Façade: Vertical surfaces on buildings, which are usually unused otherwise, except for windows. Underexplored opportunity with technical and logistic challenges, thus rather fitting to commercial tracks, for example by renting out façade surface area to businesses. **Adoption**: No significant adoption so far. Commercial more likely. **Initial investment**: High to very high. **Challenges**: Structural and safety requirements need to be checked. Logistics challenging (how to access facades). Business model unclear. Will impact aesthetics of the building.	Vertical farm on HDB building block 146 in Tampines, reported in Temasekfoundation.org.sg
Balcony: Growing crops on balconies. **Adoption**: Private, since balconies are usually within apartments (challenge to access). **Initial investment**: Low. Does not require extra infrastructure other than the farming system. **Challenges**: Upscaling difficult and relies on inhabitants' motivation.	To be found in many apartments across Singapore.How-To-Grow online guides are widely available. Training courses are also offered. Private initiatives also documented online for balconies and for window-farming.

Windows: Behind (i.e. inside the building) or in front of (i.e. outside) windows. Constructively, outside placed approaches require mechanisms such as sliding racks or extensions to house the crops. Inside the building, plants can simply be placed without special infrastructure. **Adoption**: Mostly private (accessing), but possible also in e.g. office spaces. **Initial investment**: Low (inside the building) to high (outside the building). **Challenges**: If outside, requires regulatory approval (safety), and aesthetic considerations. Upscaling difficult due to logistic challenges.	
Corridors: Utilizing the many semi-public corridor spaces in HDB buildings. **Adoption**: Private, but since corridors are semi-public, business-approach in principle possible. **Initial investment**: Low. **Challenges**: Upscaling is difficult, requires occupants' motivation. In case of a commercial approach, occupants' privacy needs to be respected and nuisance avoided.	
Vacant urban lots: Unused infrastructure area (sides of roads, median strips or traffic islands), and soil-grown approaches can be directly used. Public spaces, thus requires municipal approval. **Adoption**: Mostly private and municipal, but commercial possible. **Initial investment**: Very low, no extra infrastructure needed. **Challenges**: Traffic impairment and safety risk. Limited vacant lots.	Information and map of allotment gardens in Singapore by NParks GardeningSG.
Community garden: Public spaces, such as parks or green spaces. **Adoption**: Private and commercial, but limited area. **Initial investment**: Very low. **Challenges**: Low scaling potential due to limited parks. Conflict of interest to utilize park as either recreational space or farming space.	Information and map of community gardens in Singapore by NParks.EdibleGardenCity, social enterprise.
Home garden/private property: Production mostly limited to home consumption or neighbor sharing. Sometimes, also for hen eggs and broiler. **Adoption**: Private, low potential due to Singapore's high urban density. **Initial investment**: Very low. **Challenges**: Increasing urban density; conflicting utility to use garden as recreational space.	Private gardens relatively rare due to Singapore's density.

Figure 15.3. Examples of recent BIA solutions. (a) Window-integrated system (Source: Eriksson, 2024. Used with permission.) (b) Façade agrivoltaics system (Source: Adapted from Tablada *et al.*, 2018. Licensed under CC BY 4.0.).

provides an example of an experimental façade farming system with photovoltaic panels to maximize surface utilization by both farming and electricity generation. Further generalized BIA designs are presented in Figure 11.3 of Tablada and Kosorić (2022), including options for in front of a façade wall, on balconies, behind or in front of windows, and within a double-skin façade.

Expanding the concept of BIA, green facades can also be combined with solar photovoltaic[3] (PV) panels due to potential mutual benefits by bringing both systems together: partial shading from semi-transparent PV can improve crop yield, especially in hot and sunny climates, while evapotranspiration of plants cools down PV panels and improves electrical performance. This concept is known as Agrivoltaics and, again, is originally intended for open-space land. However, newer developments already experimented with Agrivoltaics on building facades, such as the "Vertikka" project (Hartmann *et al.*, 2024), or the experimental set-up at the Tropical-Technologies lab of the National University of Singapore, as Figure 15.3(b), taken from Tablada *et al.* (2018) shows.

The most important challenge for BIA technologies remains in high initial and operating costs, preventing them from gaining larger popularity. Especially for indoor vertical farming, high-power consumption for automation and lighting is a significant cost contributor. On the other hand, Sengodan (2022) empirically demonstrates that rooftop greenhouses on high-rise buildings might become a significant contributor to urban food production. Furthermore, the direct utilization of natural sunlight and rainwater for plant growth, waste heat from building air-conditioning, and renewable electricity generation from solar photovoltaic power are recommended to be integrated to improve the profitability of BIA and vertical farming.

Building-Integrated Agriculture (BIA) in Singapore: City-scale Production Potential and Simple Break-even Cost Analysis

As shown in the previous section, there is already a strong base for UA in Singapore, and the new BIA technologies discussed above could help improve food self-sufficiency in Singapore. To assess whether utilizing buildings and building surfaces such as façades and roofs can have a significant impact on food security, this section presents a simple break-even cost analysis based on physics-based yield simulations. In this analysis, we use the estimated yield and market price of output to derive the annual revenue from UA per m^2, which provides the break-even cost per year for the UA to be financially sustainable. Note that we do not provide cost estimates, because the implementation costs vary greatly, depending on the condition. However, in vertical farming, lighting, racking growing systems, and building costs can comprise 80–85% of the capital expenditure (Baumont de Oliverira, 2022), whereas operational expenditure typically consists of labor, energy use, maintenance, depreciation, and inputs with the first two constituting largest shares (Monzini and Orsini, 2025).

The current analysis relates to some previous works. For example, Song *et al.* (2022a) conducted a feasibility study of BIA in Singapore (for basil, mustard, okra, and peanuts), applied to one residential block. A computational framework using mathematical optimization to optimally allocate surfaces on building façades and roofs to either photovoltaics panels for renewable electricity generation or for BIA urban farming is shown in Waibel and Shi (2023) and applied to an HDB block. Astee and Kishnani (2010) examine the potential of BIA utilizing rooftops. Song *et al.* (2022b) compare the potential quantity of leafy vegetables produced under conventional soil-based farming, vertical farming with natural lighting, and indoor vertical farming.

In this section, we utilize a bottom-up physics-based simulation program to first assess solar irradiation on all building surfaces (roofs and facades) in Singapore in a high spatio-temporal resolution (monthly irradiation values on a 0.7m × 0.7m grid over the entire 3-dimensional urban fabric). The building geometry is obtained from OpenStreetMaps (OSM), and missing building height information is interpolated from adjacent buildings. Using k-medoids clustering, a representative point is identified for each district in Singapore and for each cardinal direction (roof, North, East, South, West), respectively. For these representative points, high-resolution annual hourly solar simulations are conducted. The data has been published in Caviezel *et al.* (2023). The resulting time series are used as inputs to a BIA-crop yield simulation program based on the Daily Light Integral method (Shi, 2023a, b). It translates solar irradiation on surfaces into crop yield potential, with experimentally obtained coefficients for different crop types. Figure 15.4(b) provides the Cal-Heatmap, which helps to determine the suitability of growing different crop types on a building surface, such as roof or façade, and is used to determine the yield schedules in the BIA simulation for the following crops that are potentially suitable in buildings in Singapore: Red Amaranth, Bitter Gourd, Cai Xin Hong Kong, Flat Mustard Green, Mustard Green, and Lettuce. As can be seen in the figure, high-density planning areas have higher normalized BIA-yield potential, since these areas have more building surface areas (especially more vertical façade areas).

Figure 15.5 shows the results of the yield simulation for all considered crops and separated between roof surfaces (horizontal) and façade surfaces (vertical). In Figure 15.5(a), results are normalized per person, assuming a population of 5.9 million in Singapore, and in Figure 15.5(b),

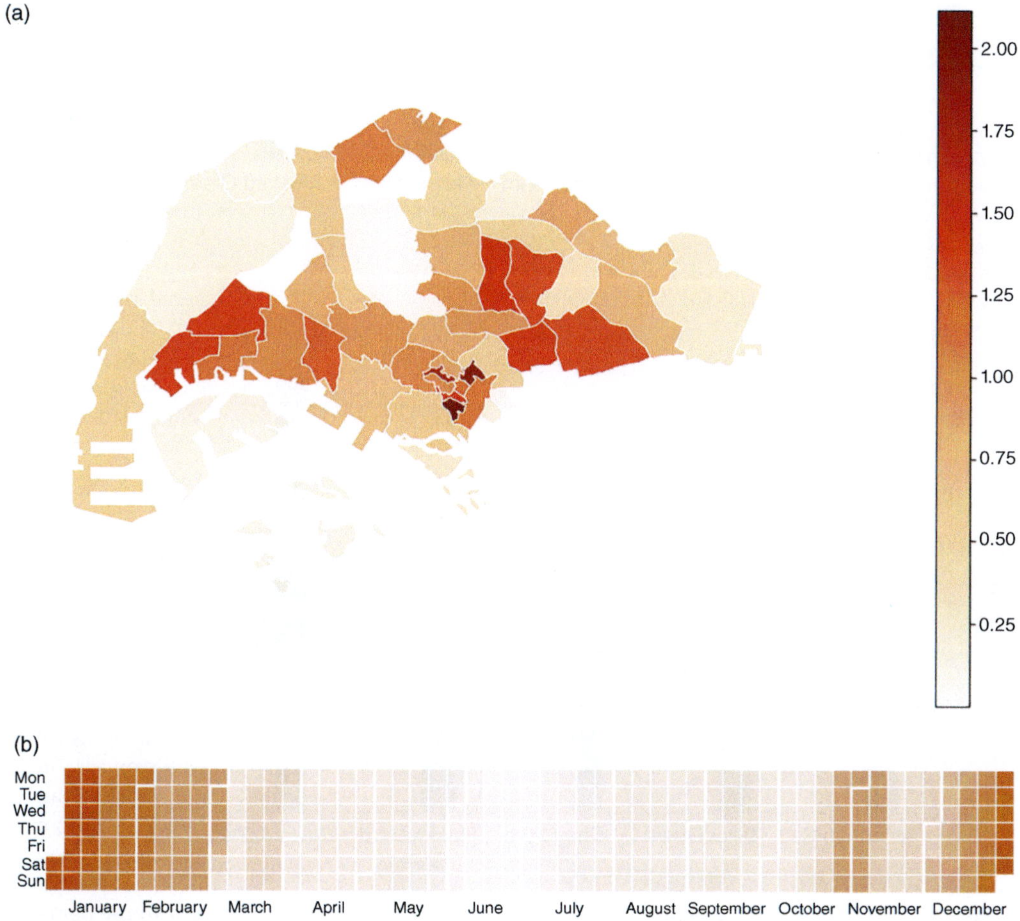

Figure 15.4. Calculation methodology for vegetable yield potential assessment, exemplified for red amaranth as crop type. (a) Annual red amaranth yield potential on building surfaces normalized by land area of each planning area of Singapore (kg/m²/year). (b) A residential building in the "Clementi" planning area. The figure shows the planning calendar to grow red amaranth on a south-facing balcony on Floor #5. (Source: Adapted from OpenStreetMap contributors, 2024. Data licensed under the Open Database License (ODbL)).

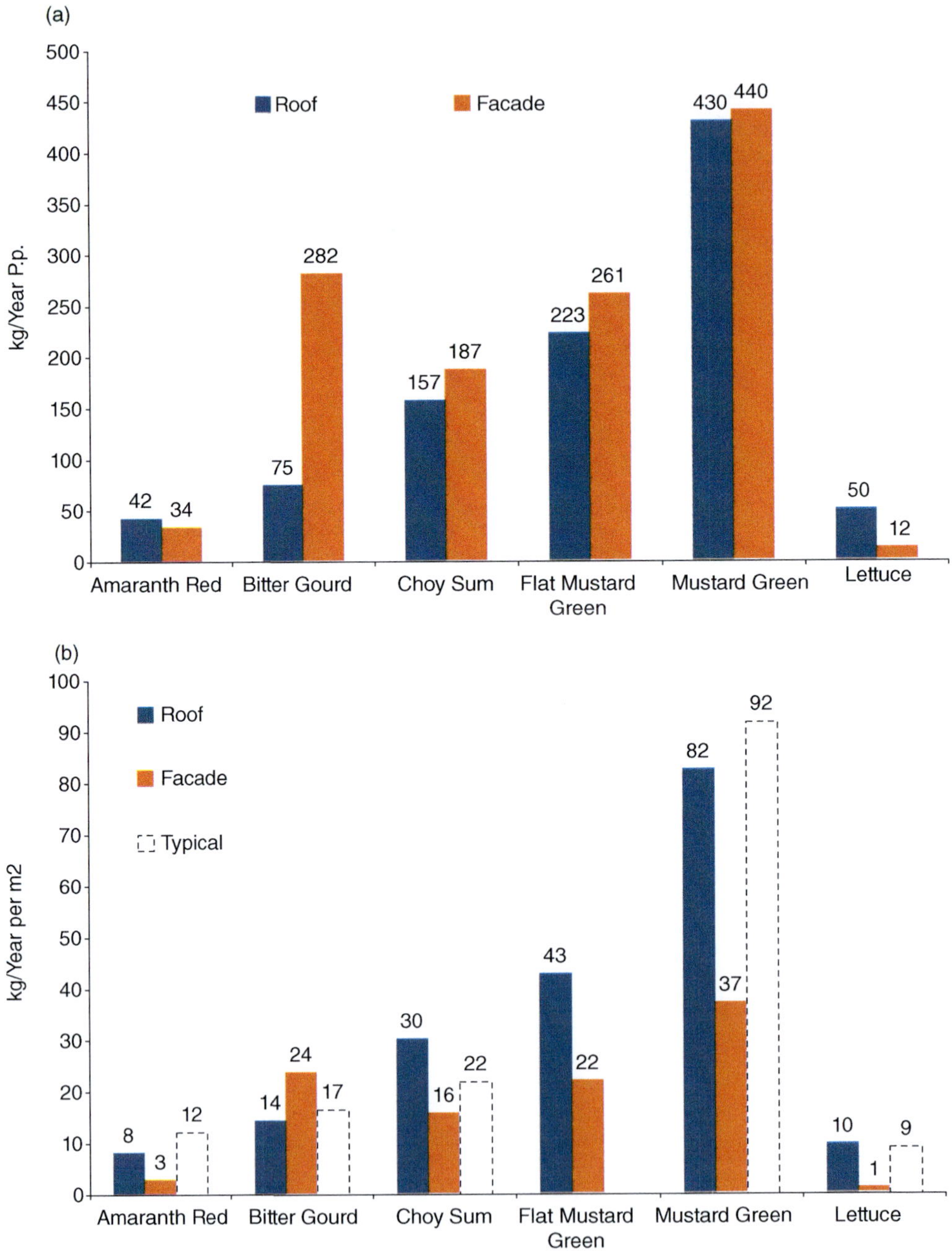

Figure 15.5. *Crop production potential on Singapore's buildings. (a) crop production in kg per person per year. (b) BIA production potential per m² of surface area and typical yield expectations for regular farming.* (Sources: Adapted from TNAU, 2013; ICAR-CCARI, 2018; TNAU, 2022; Organic Garden Info, nd; Wang and Liu, 2024).

crop yield results are shown in kg/year per m^2 of building surface area. The per m^2-results compare well with typical yield expectations using regular soil-based farming methods, indicating that our simulated results are feasible. Generally, surface normalized crop potential is higher on roof surfaces, presumably due to higher solar irradiation. Interestingly, this is not the case for Bitter Gourd, where the potential is higher on façade surfaces. Presumably, the prevalent solar irradiation patterns and intensities on Singapore facades are more favorable and roof irradiation might be too high. Considering the person-normalized results in Figure 15.5(a), high yields can be theoretically achieved in Singapore building surfaces alone. For lettuce, 50 kg/year per person can be achieved when all available roof surfaces are utilized, and additionally, 12 kg/year per person on all available façade surfaces. For mustard green, these values rise to 430 kg/year per person on roof surfaces and 440 kg/year per person on façade surfaces. While the surface-normalized yield for mustard green is lower on vertical façade surfaces than on horizontal roofs, the total yield is higher since Singapore has more vertical building surfaces in total due to its high density.

Note that the figures for mustard green are particularly high relative to the projected annual consumption of leafy vegetables per capita of 16 kg in 2030 (Teng and Montesclaros, 2019). If we ignore the composition of leafy vegetables, by achieving slightly above 1% of the production potential of mustard green would enable Singapore to achieve 30% self-sufficiency in leafy greens.

Table 15.2 shows a simple break-even cost analysis, which provides the maximal annualized total costs that BIA farming system could have to be financially sustainable for horizontal roof surfaces and vertical façade surfaces, based on the prevailing market price of each crop and production potentials reported in Figure 15.5(b). Relevant annualized BIA system costs here would include all investment and operational costs, including labor costs (in the case of a commercial approach) or opportunity costs (in the case of private and community-driven concepts).

As Table 15.2 shows, the most profitable option is growing mustard green on roofs, as the maximal annualized BIA system cost can be up to 890 SGD per m^2, which means that growing mustard green is most likely to be financially sustainable. When grown on facades, flat mustard green has the highest annualized break-even cost. Overall, Cai Xin Hongre Kong, flat mustard green, and mustard green are more likely to be viable options than amaranth red, bitter gourd, and lettuce, whether grown on roofs or façades. We also find that growing vegetables on façades tends to have a lower break-even cost than growing on roofs. This would not be surprising as changing façades require more significant investments and operating costs.

While it is not straightforward to provide average annualized BIA system costs, suggestive cost figures could be obtained from existing research. For example, according to Monzini and Orsini (2025), operating expenses of vertical farming is typically EUR 500–1000 per m^2, but these figures are not directly applicable to Singapore. The annual total costs per m^2 implied by parameters used by Song *et al.* (2022b) are 161 SGD per m^2 for vertical farming with natural lighting and 796 SGD per m^2 for indoor vertical farming.[4] While these figures are also not directly applicable to our context, the order of magnitude

Table 15.2. Annualized break-even cost for BIA systems on facades (vertical area) and roofs (horizontal area) based on current market prices for different crops.

Crop type	Market price [SGD/kg]	Annualized break-even cost of roof system [SGD/m^2 a]	Annualized break-even cost of facade system [SGD/m^2 a]
Amaranth red	14.00	114, -	40, -
Bitter gourd	8.40	121, -	199, -
Cai Xin Hong Kong	15.00	453, -	237, -
Flat mustard green	10.80	462, -	238, -
Mustard green	10.80	891, -	401, -
Lettuce	16.50	158, -	17, -

Sources: Adapted from Quan Fa Organic Farm, 2025; MarketFresh.sg., 2025.

of these figures appear to suggest two things. First, indoor farming may still be too expensive to be financially viable. Second, BIA in Singapore could be financially viable if the location and crop are carefully chosen so as to have favorable production and cost conditions.

The calculations made above underscore the overall UA potential, exploiting building surfaces in Singapore. However, these calculations only indicate what may be possible under a variety of assumptions. Notably, our assumed available building surfaces are assessed by an approach like – but more optimistic than – a prominent report on the potential for photovoltaic utilization in Singapore by the Solar Energy Research Institute of Singapore (SERIS, 2020).[5] This leads to a theoretically high production potential as all the available building surfaces were assumed to be utilized for UA in the simulation. In practice, when BIA technologies are put to actual use, various other issues assumed away in the simulation – including regulatory questions around privacy, safety, structural requirements, and fire protection, among others – need to be assessed.

Discussion: Critical Factors for Domestic Food Production

This chapter discussed the food self-sufficiency issue and 30-by-30 target in Singapore and recent BIA technologies as a potential way to improve food self-sufficiency there. We also argued that Singapore could potentially foster a new industry and contribute to SDGs by promoting UA through the adoption of BIA technologies. Nevertheless, adopting the BIA technologies will not automatically translate into the attainment of the 30-by-30 target as it does not address the self-sufficiency of food items other than leafy vegetables. Further, there are various other potentially important factors that affect food self-sufficiency.

First, while the BIA model may help address the space constraint, it will become difficult for any single UA firm to scale up except for the case of indoor farms, which is generally costly. As food production is likely to be spread out across buildings or within a building, new management issues are likely to arise when we adopt BIA at a scale. Coupled with the high labor cost in Singapore, the BIA approach may be more suited to operate as a community-driven UA. Community-driven approach may also help foster the awareness of the food self-sufficiency issue, and this is an area that requires further research.

Second, UA needs to be supported by relevant policies, particularly at the current stage where the commercial viability of urban farming remains uncertain. Investment in research and development activities in agri-technology, subsidies for local production, and campaigns to promote local produce would help increase food self-sufficiency, and policies to nurture UA talents are also needed to ensure the sector's long-run sustainability. Hence, the "agri-food sandbox" would enable innovative agricultural technologies to be tested in a more streamlined manner and an "agriculture talent programme" would help attract and nurture agricultural technology talents. As an additional source of policy support, once a reasonable production capacity is attained, it may be reasonable to consider mandating a small fraction of local procurement for some large food consumers to help urban farms enjoy a stable demand and realize the potential benefits from economies of scale. Besides the food security perspective, policy support for UA could be justified because greening the building surface is found to be beneficial for the urban microclimate (Alsaad, 2022).

Third, there is a set of external factors directly relevant to food price and food supply. As food produced overseas will compete with the local produce, any increase in overseas food prices can increase the viability of local farms. Other potentially important factors include the freight and insurance charges, external policy environments (e.g. tariffs and geopolitical tensions), and fluctuations in the foreign exchange market.

Fourth, supply chain efficiency also matters since the value of certain perishable food items can decline rapidly. Optimizing shipment, reducing waiting times, and enhancing warehousing efficiency can lower the overall costs of delivering quality food to the end consumers. While Singapore boasts an efficient freight and warehousing system as a global logistics hub, this ironically challenges local food producers. Still, local farms have a potential advantage in freshness and lower freight and distribution costs.

Finally, consumer behavior and perception matter to promote food self-sufficiency. If the consumers see value in locally grown foods and are willing to pay for them, UA will become more promising (Redwood, 2012). Consumers can also be producers in community-drive UA. While the scale of such activities may be limited in the short run, it can potentially raise awareness about the potential benefits of local food production from the perspectives of food safety, food security, and the environment, among others. This is an important consideration, as consumers are willing to pay a higher price for local produce once they clearly recognize its environmental benefits (Nakajima, 2022).

Acknowledgments

We thank Liebenehm, Sabine, Edward Ti, and Hermann Waibel for their helpful comments. Wat Zhengyi Joel provided research assistance.

Endnotes

[1] We choose to use the total food consumption as the denominator primarily due to the data availability. Therefore, the local production share in food consumption depends on the share of the food item in food consumption.

[2] Different approaches to indoor farming exist, where the most important ones are (i) soil-based, coming closest to regular farming methods, (ii) hydroponics, where crops are grown in a non-soil surrogate medium (such as hemp fibre, coco coir, or rockwool) with controlled flow of a nutrient-rich water solution, (iii) aeroponics which is similar to hydroponics but where the roots of the plants are suspended in the air and irrigated with a nutrient-dense mist, and (iv) aquaponics, where crop growth is coupled with aquaculture for raising aquatic animals such as fish or prawns, thus benefiting from mutual benefits of both systems where the nutrient-rich effluent water from the aquaculture is streamed through a hydroponics systems for filtration.

[3] Photovoltaic (PV) is a renewable energy generation technology that converts solar energy into electricity and is considered as one of the pillars in the renewable energy transition.

[4] We use the ratio of annual total cost of 2,013 SGD per tower and 23,650 SGD per container to the footprint of 12.5 m^2 and 29.7 m^2 to arrive at these figures.

[5] In the SERIS report, only continuous surfaces of over 130 m^2 are considered as economically viable. We do not impose the same restriction in our simulations.

References

Alsaad, H., Hartmann, M., Hilbel, R. and Voelker, C. (2022) The potential of facade greening in mitigating the effects of heatwaves in Central European cities. *Building and Environment* 216, 109021. DOI: 10.1016/j.buildenv.2022.109021

Astee, L.Y. and Kishnani, N.T. (2010) Building integrated agriculture: Utilising rooftops for sustainable food crop cultivation in Singapore. *Journal of Green Building* 5(2), 105–113. DOI: 10.3992/jgb.5.2.105

Baumont de Oliveira, F. (2022) A typology review for vertical plant farms: Classifications, configurations, business models and economic analyses. *Research Gate, Preprint*. DOI: 10.13140/RG.2.2.24729.49766/2

Caviezel, D., Waibel, C., Schläpfer, M. and Schlueter, A. (2023) Vehicle-to-grid coupled photovoltaic optimization for Singapore at a district resolution. In: Blanco-Marigorta, A.M., Del Rio Gamero, B., Melián Martel, N. and El Kori, N. (eds.) *Proceedings of ECOS 2023 – The 36th International Conference on Efficiency, Cost, Optimization, Simulation and Environmental Impact of Energy Systems.* Curran, Las Palmas de Gran Canaria, Spain, pp. 3327–3338.

Choo, R. (2022) Journeys of Transformation: Singapore's Pig Farmers after Pig Farming. Available at: www.sfa.gov.sg/food-for-thought/article/detail/journeys-of-transformation-singapore's-pig-farmers-after-pig-farming (accessed 20 March 2025).

Chow, C. (2014) Agriculture and the end of farming in Singapore. In: Barnard, T.P. (ed.) *Nature Contained: Environmental Histories of Singapore*. NUS Press, Singapore, pp. 216–240.

Diehl, J.A. (2024) In pursuit of a traditional livelihood: Insights from Singapore's farmers. In: Al-Kodmany, K., Govind, M., Khan, S. and Kole, C. (eds.) *Sustainable Urban Agriculture.* CRC Press, Boca Raton, Florida, pp. 85–112.

Eriksson, J. (2024) Building Integrated Agrivoltaics. Master thesis, ETH Zurich, Zurich, Switzerland.

Feng, Z. (2024) Made-in-SG Veggies: Start-up Plans to Bring More Locally Grown Spinach, Kale to Market. Available at: www.straitstimes.com/business/made-in-sg-veggies-start-up-plans-to-bring-more-locally-grown-spinach-kale-to-market (accessed 13 August 2024).

Food and Agriculture Organization (2001) *Food Balance Sheets: A Handbook, 1st edn*. Food and Agriculture Organization, Rome, Italy.

Gamboa, E. and Chew, L. (1984) Singapore sets its sights on feeding itself. Available at: https://eresources.nlb.gov.sg/newspapers/digitised/article/straitstimes19840117-1.2.36 (accessed 21 March 2025).

Hartmann, M., Kohlhepp, G., Alsaad, H., Beier, S. and Völker, C. (2024) Planung eines Reallabors zum interdisziplinären Monitoring von Mikroklima, Hygrothermik und Grauwasseraufbereitung an Grünfassaden. In: Vogel, A., Völker, C. and Kornadt, O. (eds.) *Proceedings Bauphysiktage in Weimar 2024: Bauphysik in Forschung und Praxis*. Bauhaus-Universität Weimar, Weimar, Germany, pp. 158–165.

HDB (2024) Public Housing – A Singapore Icon. Available at: www.hdb.gov.sg/about-us/our-role/public-housing-a-singapore-icon (accessed 28 November 2024).

ICAR-Central Coastal Agricultural Research Institute (CCARI) (2018) *Red Amaranthus: Potential Leafy Vegetable of Goa – Genetic Resources, Characterization and Utilization, 1st edn*. ICAR-CCARI, Goa, India.

Irvine, K., Chua, L., and Eikass, H.S. (2014) The four national taps of Singapore. A holistic approach to water resources management from drainage to drinking water. *Journal of Water Management Modeling*, 22, C375. DOI: 10.14796/JWMM.C375

Kosorić, V., Huang, H., Tablada, A., Lau, S.K. and Tan, H.T. (2019) Survey on the social acceptance of the productive façade concept integrating photovoltaic and farming systems in high-rise public housing blocks in Singapore. *Renewable and Sustainable Energy Reviews* 111, 197–214. DOI: 10.1016/j.rser.2019.04.056

MarketFresh.sg. (2025) Vegetables. Available at: https://marketfresh.com.sg/collections/vegetables (accessed 6 February 2025).

Ministry of Sustainability and the Environment (2019) Speech at the Committee of Supply Debate by Mr Masagos Zulkifli, Minister for the Environment and Water Resources. Available at: www.mse.gov.sg/latest-news/speech-at-the-committee-of-supply-debate (accessed 20 March 2025).

Ministry of Sustainability and the Environment (2020) 30x30 Express - Ramping Up Local Production to Enhance Singapore's Food Security. Available at: www.mse.gov.sg/resource-room/category/2020-04-08-press-release-on-30-x-30-express/ (accessed 21 March 2025).

Ministry of Trade and Industry (2019) Fact Sheet: Agri-Food Innovation Park. Available at: www.mti.gov.sg/-/media/MTI/COS-2019/Factsheet_AFIP.pdf (accessed 21 March 2025).

Montesclaros, J.M.L. and Teng, P.S. (2019) Supporting Singapore's "30-By-30" Food Security Target. Available at: https://rsis.edu.sg/rsis-publication/nts/supporting-singapores-30-by-30-food-security-target-finding-the-sweet-spot-in-property-taxation/ (accessed 21 March 2025).

Monzini, J. and Orsini, F. (2025) *Understanding Commercial Urban Agriculture: An Overview, 1st edn*. FAO, Rome, Italy.

Nakajima, M. (2022) Sustainable Food Consumption: Demand for Local Produce in Singapore. *Sustainability* 14(19), 12330. DOI: 10.3390/su141912330

Organic Garden Info (nd) Growing Mustard Greens. Available at: www.organicgardeninfo.com/growing-mustard-greens.html (accessed 21 March 2025).

Quan Fa Organic Farm (2025) Organic Vegetables. Available at: https://quanfaorganic.com.sg (accessed 6 February 2025).

Redwood, M. (2012) *Agriculture in Urban Planning: Generating Livelihoods and Food Security, 1st edn*. Routledge, London, UK.

Sajan, C. (2020) NParks Seeding a Nation of Gardeners. Available at: www.straitstimes.com/lifestyle/home-design/nparks-seeding-a-nation-of-gardeners (accessed 22 March 2025).

Sengodan, P. (2022) An overview of vertical farming: Highlighting the potential in Malaysian high-rise buildings. *Pertanika Journal of Science & Technology* 30(2), 949–981. DOI: 10.47836/pjst.30.2.06

Shi, Z. (2023a) Building-Integrated Agriculture (BIA): plot your BIA calendars and visualise your assessment outcome! Available at: www.cityenergyanalyst.com/blog-posts/2022/09/20/cea-bia-plugin (accessed 25 November 2024).

Shi, Z. (2023b) The BIA Plug-In is Now Compatible With CEA v3.34.1[v2.0]. Zenodo. Available at: https://doi.org/10.5281/zenodo.8349607 (accessed 21 March 2025).

Singapore Department of Statistics (2024) Private Consumption Expenditure. Available at: https://tablebuilder.singstat.gov.sg/table/TS/M015041# (accessed 12 October 2024).

Singapore Food Agency (2022) Singapore Food Statistics 2021. Available at: https://www.sfa.gov.sg/news-publications/publications/sgfs (accessed 13 October 2024).

Singapore Food Agency (2023) Singapore Food Statistics 2022. Available at: https://www.sfa.gov.sg/news-publications/publications/sgfs (accessed 13 October 2024).

Singapore Food Agency (2024) Singapore Food Statistics 2023. Available at: https://www.sfa.gov.sg/news-publications/publications/sgfs (accessed 13 October 2024).

Singapore Food Agency (2025) Agriculture Technologies. Available at: https://www.sfa.gov.sg/food-science-and-technology/technology/agriculture-technologies (accessed 10 March 2025).

Singapore Primary Production Department (1970) Annual Report 1970. Available at: https://www.nlb.gov.sg/main/book-detail?cmsuuid=03d40cbe-bff2-48fb-9481-16f3e69f2f69 (accessed 10 October 2024).

Smit, J., Ratta, A. and Nasr, J. (1996) *Urban Agriculture: Food, Jobs and Sustainable Cities, 1st edn*. United Nations Development Programme, New York, US.

Solar Energy Research Institute of Singapore (2020) UPDATE of the Solar Photovoltaic (PV) Roadmap for Singapore. Available at: https://www.seris.nus.edu.sg/publications/technology-roadmap/ (accessed 21 March 2025).

Song, S., Cheong, J.C., Lee, J.S., Tan, J.K., Chiam, Z. *et al.* (2022a) Home gardening in Singapore: A feasibility study on the utilization of the vertical space of retrofitted high-rise public housing apartment buildings to increase urban vegetable self-sufficiency. *Urban Forestry & Urban Greening* 78, 127755. DOI: 10.1016/j.ufug.2022.127755

Song, S., Hou, Y., Lim, R.B.H., Gaw, L.Y.F., Richrds, D.R. *et al.* (2022b) Comparison of vegetable production, resource-use efficiency and environmental performance of high-technology and conventional farming systems for urban agriculture in the tropical city of Singapore. *Science of the Total Environment* 807, 150621. DOI: 10.1016/j.scitotenv.2021.150621

Tablada, A. (2024) Vertical farming systems on building envelopes. In: Al-Kodmany, K., Ng, A.K., Tablada, A. and Kole, C. (eds.) *The Vertical Farm: Scientific Advances and Technological Developments.* CRC Press, Boca Raton, Florida, pp. 38–56.

Tablada, A. and Kosorić, V. (2022) Vertical farming on facades: transforming building skins for urban food security. In: Gasparri, E., Brambilla, A., Lobaccaro, G., Goia, F., Andaloro, A. and Sangiorgio, A. (eds.) *Rethinking Building Skins: Transformative Technologies and Research Trajectories*. Woodhead Publishing, Cambridge, UK, pp. 285–311.

Tablada, A., Kosoric, V., Huang, H., Chaplin, I.K., Lau, S.-K. *et al.* (2018) Design optimization of productive façades: Integrating photovoltaic and farming systems at the tropical technologies laboratory. *Sustainability* 10(10), 3762. DOI: 10.3390/su10103762

Tamil Nadu Agricultural University (TNAU) (2013) Minor Vegetable Crops. Available at: https://agritech.tnau.ac.in/horticulture/horti_vegetables_minor%20vegetables.html (accessed 21 March 2025).

Tamil Nadu Agricultural University (TNAU) (2022) Bitter Gourd (*Momordica charantia* L.). Available at: https://agritech.tnau.ac.in/horticulture/horti_vegetables_bittergourd.html (accessed 21 March 2025).

Tan, C. (2024) Mega Indoor Veggie Farm I.F.F.I Shuts Down in Latest Blow to Local Food Production. Available at: www.straitstimes.com/singapore/mega-indoor-veggie-farm-iffi-shuts-down-in-latest-blow-to-local-food-production (accessed 21 March 2025).

Teng, P. and Montesclaros, J.M.L.P. (2019) Singapore's '30 by 30' Strategy: Can Food Self-Production Be Achieved? Available at https://dr.ntu.edu.sg/handle/10356/80923 (accessed 23 March 2025).

Tham, D. (2024) In Focus: Will Singapore Meet its 30 by 30 Food Sustainability Goal? Available at: www.channelnewsasia.com/singapore/vegetables-seafood-singapore-food-price-cost-farmers-security-4442626 (accessed 21 March 2025).

The Business Times (2024) Singapore Vegetable, Seafood Production Fell in 2023 Due To Construction Challenges, Inflation: SFA. Available at: www.businesstimes.com.sg/singapore/singapore-vegetable-seafood-production-fell-2023-due-construction-challenges-inflation-sfa (accessed 20 March 2025).

Tortajada, C. and Lim, N.S.W. (2021) Food Security and COVID-19: Impacts and Resilience in Singapore. *Frontiers in Sustainable Food Systems* 5, 740780. DOI: 10.3389/fsufs.2021.740780

Waibel, C. and Shi, Z. (2023) Veggies and PV: Optimization of building-integrated agriculture in an energy hub. In: Andersen, M., Smith, B., Schwartz, Y., Waibel, C. and Lindelöf, D. (eds.) *Journal of Physics: Conference Series*. IOP Publishing, Bristol, UK, pp. 172001.

Wang, Y. and Liu, G. (2024) Production Guide for Choy Sum – An Emerging Asian Vegetable in Florida. Available at: https://edis.ifas.ufl.edu/publication/HS1380 (accessed 21 March 2025).

World Bank (2024) World Development Indicators. Available at: https://databank.worldbank.org/source/world-development-indicators (accessed 10 October 2024).

16 Synthesis

Ernah, Sabine Liebenehm, Sirman Purba and Hermann Waibel

Abstract

In this chapter we draw the major conclusions and lessons that have emerged from the policy chapters in Part II and the special topics in Part III, following the overall theme of the book, climate change and risk. Hereby we focus on three topics. First, we look at the role that agriculture has played in the countries covered by the book. Second, we summarize the lessons from the structure of agriculture and the rural areas and the changes over time thereof. Hereby the farm size issue as it has emerged from the policy chapters is a major topic. Third, we look at technology in agriculture, including climate-smart technologies and biotechnologies, drawing mainly from the chapters in Part III. The synthesis chapter paves the way for the last chapter that presents an outlook for the future.

Agriculture and Economy

As elaborated in the country policy chapters, in all the eight countries, agriculture's share of GDP has been declining over the past decades. Most of the eight countries are now below or around 10% GDP share (see Table 16.1). Notable exceptions are Laos with 16.1% and India with 18%. Consistent with agricultural development patterns around the world, labor shares consistently are a multiple of GDP shares, demonstrating lower labor productivity in agriculture as compared with other economic sectors. Most countries have labor shares of around 30%, however, China is at 22%, India at 45% and not surprisingly, Laos is highest with close to 70%. The slowly declining labor shares have been attributed to a sluggish transformation process. However, as the discussion in the different chapters has shown, labor shares and labor productivity are a problematic issue. There are measurement problems in labor statistics, such as the definition of rural households, part-time migration and seasonal employment within and outside agriculture. Also, since home food production and consumption are often not accounted for, it is quite possible that labor productivity of the rural and farm population is higher than the statistics suggests. Nevertheless, increasing labor productivity in agriculture remains an issue in Asia and beyond.

Agriculture has been a major driver of poverty reduction in all Asian countries (Christiansen *et al.*, 2001; Timmer, 2008). This has particularly been the case in China which claimed eradication of poverty in 2020 (Table 16.1). A major role has been attributed to land reform, open trade policy and R&D investments that have raised productivity significantly. With the exception of Bangladesh and Laos, all countries covered in this book now have headcount ratios below 10%. Inequality is similar across the eight

 Asian Agriculture in a Climate Change-driven and Risky World (H. Waibel, S. Liebenehm and Ernah eds)
DOI: 10.1079/9781800627338.0016

Table 16.1. Indicators of agriculture's role in the economy by country. (Authors' own table).

Country[1]	GDP Share[2]	Labor Share[2]	Poverty Headcount[2]	Gini Coefficient[2]
Bangladesh	11%	35.3%	18.7%	0.499
China	7%	22.3%	'Zero' (since 2020)	0.467
India	18%	45.8%	5.3%	0.410
Indonesia	12.5%	28.8%	9.4%	0.361
Laos	16.1%	69.6%	16.9%	0.388
Philippines	9.4%	24.5%	15.5%	0.391
Thailand	8.6%	30%	3.4%	0.335
Vietnam	12%	33%	5.7%	0.370

[1]alphabetical ordering; [2]most data refer to 2023; World Bank and national statistics sources.

countries, ranging between 0.335 and 0.499. Bangladesh and interestingly China is next. Unfortunately, no good quantitative measure of rural–urban inequality is available. However, it is fair to say that generally, the rural–urban divide is a significant problem in these Asian countries.

Structural Change and Rural Transformation

Structural change in an economy and rural transformation are interconnected and can reinforce one another. By transferring surplus resources, agriculture can facilitate economic growth through the expansion of industry and service sectors. Economic growth provides the resources for the modernization of agriculture through investment in R&D and agricultural extension. As noted by Huang (2018), rural Asia has experienced significant transformations, albeit with large differences among the countries. This observation is confirmed by the country case studies in this book. To illustrate, we use four indicators to compare the countries, namely, range of agricultural growth between 2010 and 2020, crop productivity, using the range of average annual rice yields as proxy, the share of irrigated land in total agricultural land and the status of land reform, judged by the current degree of implementation (Table 16.2).

As shown in the country policy chapters, growth of the agricultural sector is generally behind overall economic growth in all the countries, however, there are large differences in agricultural growth. Taking Thailand for example, we observe highly negative agricultural growth and an overall growth rate of over 6% within the same decade. No other country in the list has these extreme ranges. The Thai case can be attributed to the "monster flood" in 2011 and a government policy that encourages the abonnement of rice harvests in case of natural disasters such as drought. Vietnam, The Philippines and Indonesia are also exposed to frequent weather extremes which is reflected in variable growth rates ranging between below 1% and over 4%.

On top of the list is China, with growth rates between 4% and 7.1%, followed by Bangladesh. The country has caught up impressively with its Southeast Asian neighbors, which is the result of government investments and policy reforms. Against common perceptions, Bangladesh has actually quite favorable natural conditions, since the river systems of the Ganges and the Brahmaputra deliver fertile silt and support high underground water levels, which reduces the costs of irrigation. This is also reflected in the high share of irrigated land, close to 80% (see Table 16.2). Irrigation is a major infrastructure and an indicator for structural change, facilitating agricultural growth. Unfortunately, data on irrigated land is sparse. For example, the agricultural censuses of Indonesia and Vietnam do not include information on the share of irrigated land. The Philippines, India, Thailand and China have shares between 41% and 63%, while Laos is considerably lagging behind with less than 10% share, despite massive investment in hydropower dams which often are not designed for irrigation.

In all the countries in our list, land reform has been a flagship project with the aim to reduce inequality created by the feudal systems of the past. Since the communist countries, namely China, Vietnam and Laos, stipulate collective ownership of land, their land reforms deal with

Table 16.2. Indicators of rural transformation. (Authors' own table).

Country[1]	Agric Growth (%)[1]	Productivity[2]	Share irrigated Land[3]	Status of Land Reform[4]
Bangladesh	3.4–6.2	4.0–4.5	78.9	Slow progress; resistance by elites
China	4.0–7.1	6.9–7.1	41.0	Reform of land use rights; collective ownership maintained
India	2.7–4.3	2.2–4.1	55.8	Abolition of 'zamindari system'; progress uneven across States
Indonesia	1.3–4.6	4.7–5.5	(n.d.)	Progress in land registration; redistribution is short of government targets
Laos	3.8–4.3	3.0–3.3	<10.0	Geographically highly variable; both, customary and statutory systems
Philippines	1.2–4.2	3.9–4.1	63.0	Slow progress; debt release act;
Thailand	(−)6.5 – (+)6.1	2.5–2.9	46.0	Land tenure insecurity continues
Vietnam	0.5–4.1	5.7–6.1	(45 –85)[5]	Extending long-term use rights, ownership remains collective

[1]Highest and lowest annual growth between 2010 and 2020 period; [2]highest and lowest rice yields per annum, different periods between 2010 and 2020, as data available; [3]in percent, most recent figures; [4]derived from country chapters; [5]only selected information available.
Source: Own compilation from country statistics.

the regulation and relaxation of land use rights restrictions. In the other more democratic countries, land reform aimed at transferring land ownership from large landholders to tenants and introducing ceilings for land ownership. By and large, in all the countries listed in Table 16.2, the land reform process did not progress as fast as expected. Technical and political constraints can be made responsible for this. On the other hand, ceilings for land ownership may actually hinder the rural transformation process towards larger farm units. Besides, for some commodities such as oil palm (Indonesia) and rubber (Thailand, Laos), large landholdings continue to exist and are excluded from land reform.

In addition to the indicators of structural change in agriculture, another important aspect is rural employment with rural off-farm labor markets. Traditionally, development theories focused on rural–urban migration to supply cheap rural surplus labor for industrial and service sector development in urban agglomerates. China is perhaps the best example where rural household members have moved to the cities for non-agricultural employment and sending remittances to their families in their natal villages. Recently, however, China has implemented its rural revitalization strategy encouraging university graduates to work in rural areas with start-ups to develop off-farm employment opportunities and thereby attracting laborers to return to their home towns in rural areas. A good example is Xishuangbanna in Southern Yunnan, a major rubber growing area. There, off-farm labor markets have developed in the tourist sector, for example, giving seasonal and permanent employment to farm household members of rubber farmers (Zhuang, 2022). In India, the rural non-farm sector has always played an important role, generating the largest number of jobs (Binswanger-Mkhize, 2013). Government infrastructure investments, public works programs and urban–rural spillovers, have been major drivers to help expanding the rural non-farm sector rural and generating rural employment. In Bangladesh there is also a trend towards more households with both farm and non-farm income. Sen *et al.* (2021) found that the share of rural households with both farm and non-farm jobs rose from 23% to 39% between 2000 and 2013, especially in areas near cities.

For the remaining countries covered in the book, i.e. Thailand, Vietnam, Philippines, Indonesia and Laos, rural development is part of the policy agenda. The Association of South East Nations (ASEAN) has been promoting a master plan for rural development between 2022–2026 with local participation in the context of territorial approaches, women and youth involvement. For example, Vietnam established its national rural development (NRD) policy, already in 2010. Thailand and The Philippines have similar programs promoting sustainable rural development. However, little is known about implementation and impact, so that Rigg's (2020) concern that "rural means reservoir of labour and source of food, almost a relict space and a residual concern" seems to still hold. The story in Chapter 13 about the Mekong villages has underlined that there is unused potential in the rural space that could be exploited given the right policies. Perhaps, as mentioned above, China could lead the way towards a better and more modern rural development.

The last and perhaps most pressing point about structural transformation in agriculture is average farm size. The small farm sizes in Asia have raised concerns that Asian countries might lose competitiveness in global agriculture (e.g. Devendra, 2007). The findings in the country chapters show that farms in Asia did not get larger during the past decades and in some cases are even getting smaller, while at the same time, agricultural output and productivity was going up. This observation is of particular significance, considering that over 50% of the farms worldwide are in China and India. Likewise 84% of farms, globally, are smaller than 2 ha (Lowder *et al.*, 2016). Table 16.3 shows the summary statistics about the most recently available data on average farm size, and the share of farms below 2 ha in the eight countries covered in the book. If not available from national statistics, average farm data was calculated or estimated from the policy chapters. In addition, a column showing the trend in farm size was added. Although farm size data suffer from limitations of the national statistics and definitional problems, the data portray a picture, consistent with other literatures (e.g. Yamauchi *et al.*, 2021) and confirmed by extensive rural household survey experience of the chapter authors and the editors.

As shown in Table 16.3, the country with the lowest average farm size, of just slightly over 1.5 ha, is Bangladesh, followed by Indonesia with 0.63. The highest is in Thailand with almost 4 ha, followed by Laos with 1.84. India and China are almost the same. Average farm size in Vietnam is higher with 1.63 ha but 94% of farms are smaller than 2 ha. This is similar in China and shows the effect of the earlier collectivization policies which had radically changed the feudalistic landowner system and equalized landholdings. It is different in countries where plantation agriculture by large agroindustry companies, for example in oil palm, rubber and banana, is taking place. This refers to The Philippines and Indonesia. It is also an indication of the unfinished land reform agenda, as shown by the data from India. In all countries, except in Thailand and Laos, the vast majority of farms is smaller than 2 ha, sometimes even smaller than

Table 16.3. Status and trend in farm sizes, selected Asian countries. (Authors' own table).

Country	Average Farm Size (ha)	Share of Farms Below 2 ha (%)	Past Trend in Farm Size	Agro-industry
India	1.07	70[1]	Declining	Village-based agro-industries
China	0.96	95.9	Slightly increasing	Agribusiness companies emerging
Indonesia	0.69[2]	76	Declining	Plantation sector with large-scale operations
Bangladesh	0.51	88.5	Declining	Some large companies in agribusiness but mostly small-scale
Philippines	0.83	68.3	Strongly declining	Planation crop sector slightly expanding
Vietnam	1.63[3]	94	Constant	Processing and export
Thailand	3.99	<50[4]	Constant	Agroindustry sector expanding
Laos	1.84	17.8	Slightly increasing	Plantation sector increasing

Notes: [1]estimate from Chapter 9; [2]calculated from Table 10.3; [3]calculated from Figure 5.1; [4]estimated.

1 ha. The trend in average farm size is mostly constant or declining. In China, average farm size is gradually increasing, which could be the effect of emerging agribusiness companies in some sectors. In The Philippines, however, average farm size and agricultural land is declining rapidly while the number of farm households is increasing based on reports of the 2022 Agricultural Census (PSA, 2024).

Common to all the countries is a dualistic structure of the agricultural sector. On the one hand there are agroindustry companies engaged in processing, marketing and export. They also engage in primary production in livestock and plantation crops, while rice and other field crops remain with small-scale farmers. The vast majority of them are part-time family farmers, with income from both farming and non-farm sources. Only few are full-time farmer entrepreneurs who earn most of their income from farming alone. The column 'Agroindustry' in Table 16.3 summarizes the development in the farm business sector.

From these farm size statistics one can ask whether policymakers should be worried about the persistence of small family farms that, in most cases, are not comparable to commercial farm businesses in the West. Rather they are farm household systems which combine production for home consumption and processing, marketing of surplus with income from off-farm wage employment and non-farm self-employment. So far, small farms are producing some 80% of the food worldwide. Can they continue to do so in the near future? Should part-time, multi-activity, small farm systems still receive public support or should they be rather encouraged to exit to non-farm jobs? Fan and Rue (2020) suggest that the policy should be to support only those smallholders with prospects to move up to a commercially oriented and profitable farming system. Others should be supported to give up farming altogether and move on to the non-farm sector. The philosophy of "move up or move out" is based on the principle of economies of scale and has also been a paradigm for agricultural development in the European countries. However, it did not happen that way. Although commercial farms eventually dominate, the process has taken a long time. Even in the industrialized high-income countries of the EU, part-time farming, as a multi-economic activity model, with non-farm income as a major or minor share of household income, continues to exist (Fuller, 2015; Addo and Salhofer 2022; Shahzad and Fischer, 2022). The now rich Asian economies, such as South Korea, Taiwan and Japan, show a similar picture. Farm size has remained low throughout the process of economic modernization and growth. In Taiwan and South Korea, during the course of their labor–intensive industrialization, agricultural development and technological change, with rising capital-labor substitution, factor productivity grew in both sectors and the productivity gap between the industry and agriculture sectors was narrowing (Oshima, 1986). In Japan, according to a report by the Royal Dutch Embassy (2020), by 2019, there were 3.6 million landowners out of which 2.1 million households occasionally are working on the land, while 1.1 million are part-time family farms, working on a commercial scale. Japan classifies three categories of farms, namely side-business-, semi-business- or business farm households, depending on the share of income from agriculture. Business farms are those who derive more than 50% of their income from agriculture. These are the least in numbers among the three categories. Only recently, with population aging, many farmers have no successors and thus leave their land unattended. Therefore, commercial and corporate farming is emerging, supported by policies promoting the rental of farm land.

The reasons for the persistence of small farms in both rich and less rich countries, is multi-faceted. First, demography and tradition are factors. Elder farmers are not willing to give up farming, leave their village and work as a laborer. They may earn less money on their small farm and from some occasional petty jobs in the village, but they can stay in their familiar environment and don't have to follow orders or suffer from the rudeness of a boss. Second, most farmers don't meet the education requirement for decent employment in the off-farm sector. Third, off-farm labor markets in rural areas in the developing countries are underdeveloped and accessibility is constrained by the distance to the nearest town. Fourth, in an increasingly insecure economic and political environment, livelihood security becomes a major goal for rural people, which reduces their willingness to move out of farming. In addition, further liberalization of land markets is a dilemma for governments who still have a land reform on their agenda. The democratic countries have to fear the reaction from the rural electorate and in the communist countries, within-party opposition must be expected. Above all, there need to be enough rural-based commercially minded farm entrepreneurs, who are willing to rent-in land from other farmers and undertake significant capital investment to operate as a large-scale farmer. This will not happen by itself, and it will not happen quickly. The examples of government enforced mega rice projects in Indonesia during the 1990s show that such undertakings suffer from dysfunctional management procedures with low yields and high costs and causing social disruptions among the rural communities. They leave behind a trail of degraded land and damaged biodiversity. This is likely to happen again with the current government's policy "food estate schemes" (Abdulgani and Savitri, 2025; Trisista *et al.*, 2025). Similar caution was raised about the concept of "large field farming" in Vietnam as pointed out in Chapter 5.

To conclude the farm size discussion, governments in Asian countries should give up the vision of having western-style large-scale farms within the foreseeable future. There is also no convincing evidence that small-scale farmers could not increase productivity and production to feed a growing population. We suggest that policymakers should recognize and accept the existence of multifunctional, mostly family-managed farming systems in the food crop sector. The role of governments is to guide the gradual progression towards economically and socially optimal farm sizes and farm organizations during the course of generational transitions. What is badly needed in rural Asia is a comprehensive rural development strategy, in order to transform the rural space into an attractive living and work environment. The model of EU rural development policies may serve as some useful food for thought for a start (e.g. Pelucha and Kveton, 2017). To make small rural towns and rural villages an attractive location, with people working in home offices, performing services outsourced by urban firms, offering weekend holidays on farms for stressed city people while maintaining a clean environment, and producing healthy food. The example of the Mekong villages, discussed in Chapter 13 can serve as a good example here. This can be achieved by means of well-targeted rural investment programs.

Modern Technologies

As shown by several of the policy chapters and the special topic chapters in Part III, modern technologies have been adopted by agroindustry farms in the livestock and plantation sector and by small-scale farmers alike.

In small-scale farming, use of combine harvesters of various sizes, adjusted to the structure of plot sizes, is now widespread throughout the rice areas in Asia. Private service providers and community organizations take care of harvesting and other farm activities for ploughing, planting, weeding and fertilizer application on a contract basis. Farmers choose appropriate technologies and find suitable institutional settings to optimally organize their farm. There are also no reasons why small-scale farmers should not be able to use modern information technologies and tools. Service apps as an aid to identify plant diseases and platforms to obtain marketing information for inputs and outputs are becoming more prominent. Autonomously operated planting and harvesting machines and applying fertilizer by means of drones are just the next level of mechanization (Rogelio *et al.*, 2020; Zhao *et al.*, 2020) that may soon enter the small-scale farmer world in Asia.

In corporate faming, as shown by Chapter 15 on urban farming in Singapore, automated irrigation and robotics are mandatory for the operation of such advanced production systems.

To mitigate the negative effects of climate change, smart technologies are being adopted such as zero tillage, mulching, cover cropping or intermittent irrigation and shifting to tree crop production in already disrupted ecosystems. New technologies such as remote sensing to monitor plant health, precision agriculture to optimize external input use or solar water pumps are coming to the fore. To encourage farmers to adopt climate-smart technologies, incentives and supporting institutions must be in place. For example, the economic viability of precision farming hinges on the prices of external inputs. Bottom-up approaches – anchored in local communities – play a key role, particularly in fostering climate-related education for the next generation of farmers. Public investment in land use changes, soil health improvement, and water retention infrastructure are effective when coordinated at the village level through collective action. An important point that has emerged from the policy chapters is that adaptation and mitigation strategies must look beyond agriculture, for example, through the development of a green rural and locally based economy such as renewable energy and digital services. To facilitate rapid adoption of climate-smart technologies governments need to put in place a conducive incentive system and invest in farmer education and training. This should also include management and financial literacy education to minimize the risk of over-indebtedness of rural households as a result of climate risk protection measures.

To sum up, the aim of this synthesis chapter was to highlight the main findings of the book. Excerpts of the various topics discussed in the previous chapters have shown that in order to achieve the goal of sustainable agricultural development in Asia, local, national, regional and global collaboration is needed as endorsed by Fan and Otsuka (2021). Reducing vulnerability and increasing resilience in a climate change driven and increasingly risky world remains a challenge but it can be done if decisions are guided by knowledge and understanding instead of panic and fear.

References

Abdulgani, F. and Savitri, L.A. (2025) State, capital and coercion in Indonesia's food estates. *Journal of Agrarian Change* 25, e70031.

Addo, F. and Salhofer, K. (2022) Part-time farming and scale efficiency. *German Journal of Agricultural Economics* 71(1), 16–35.

ASEAN (2022) Available at: https://asean.org/wp-content/uploads/2022/11/34-ASEAN-Master-Plan-on-Rural-Development-2022-2026.pdf

Christiaensen, L., Demery, L. and Kuhl, J. (2011) The (evolving) role of agriculture in poverty reduction – An empirical perspective. *Journal of Development Economics* 96(2), 239–254.

Devendra, C. (2007) Small farm systems to feed hungry Asia. *Outlook on Agriculture* 36(1), 7–20.

Embassy of the Kingdom of the Netherlands (2020) Land Use and Profitability of Farming in Japan. Market Report 2020. Available at: https://www.agroberichtenbuitenland.nl/landeninformatie/japan/achtergrond/marktstudies/large-agricultural-corporations (accessed 14 August 2025)

Fan, S. and Otsuka, K. (2021) Reshaping agrifood systems to achieve multiple development goals. Chapter 21, In: Otuska, K. and Fan, S. (eds) *Agricultural Development: New Perspectives in a Changing World*. IFPRI, Washington, D.C, pp. 701–714.

Fan, S. and Rue, C. (2020) The role of smallholder farms in a changing world. In: *The Role of Smallholder Farms in Food and Nutrition Security*. Springer International Publishing, Cham, pp. 13–28.

Gautam, M. and Ahmed, M. (2019) Too small to be beautiful? The farm size and productivity relationship in Bangladesh. *Food Policy* 84, 165–175.

Huang, J. (2018) Facilitating inclusive rural transformation in the Asian developing countries. *World Food Policy* 4(2), 31–55.

Kushwaha, H.L., Sinha, J., Khura, T., Kushwaha, D.K., Ekka, U. *et al.* (2016). Status and scope of robotics in agriculture. In *International Conference on Emerging Technologies in Agricultural and Food Engineering*, Vol. 12, p. 163.

Lowder, S.K., Skoet, J. and Raney, T. (2016) The number, size, and distribution of farms, smallholder farms, and family farms worldwide. *World Development* 87, 16–29.

Oshima, H.T. (1986) The transition from an agricultural to an industrial economy in East Asia. *Economic Development and Cultural Change* 34(4), 783–809.

Philippine Statistical Authority (2022) *Census of Agriculture and Fisheries 2022*. Philippine Statistical Authority, Manila, Philippines.

Ren, C., Liu, S., Van Grinsven, H., Reis, S., Jin, S. *et al.* (2019) The impact of farm size on agricultural sustainability. *Journal of Cleaner Production* 220, 357–367.

Rigg, J. (2020) *Rural Development in Southeast Asia: Dispossession, Accumulation and Persistence*. Cambridge University Press.

Rigg, J., Salamanca, A., Phongsiri, M. and Sripun, M. (2018) More farmers, less farming? Understanding the truncated agrarian transition in Thailand. *World Development* 107, 327–337.

Rogelio, J.P., Culaba, A., Dadios, E.P., Vicerra, R.R. and Bandala, A. (2020) The robotics and the role on agricultural industry in the Philippines. In: *2020 IEEE 12th International Conference on Humanoid, Nanotechnology, Information Technology, Communication and Control, Environment, and Management (HNICEM)*. IEEE, pp. 1–6.

Sen, B., Dorosh, P. and Ahmed, M. (2021) Moving out of agriculture in Bangladesh: The role of farm, non-farm and mixed households. *World Development* 144, 105479.

Thang, T.C., Khoi, D.K., Thiep, D.H., Lan, V.T., Tinh, T.V. *et al.* (2017) *Assessing the Potential of Climate Smart Agriculture in Large Rice Field Models in Vietnam*. In: *CGIAR Research Program on Climate Change, Agriculture and Food Security (CCAFS), Wageningen, the Netherlands, Copenhagen, Denmark*.

Timmer, P. (2005) Agriculture and Pro-Poor Growth: An Asian Perspective. Available at SSRN 1114155.

Trisista, R.G.M., Mintarsih, M. and Fauza, A.S. (2025) Diving deeper into Indonesia food estate policy: Its impact on local community rights and policy comparison. In: *Proceedings of the 2nd International Conference on Changing of Business Law (ICOCLB 2024), (Vol.* 902*)*. Springer Nature, p. 389.

Ullah, I., Dagar, V., Tanin, T.I., Rehman, A. and Zeeshan, M. (2024) Agricultural productivity and rural poverty in China: The impact of land reforms. *Journal of Cleaner Production* 475, 143723.

World Bank (2024) Available at: https://data.worldbank.org/indicator/NV.AGR.TOTL.ZS?locations=PH-ID

Yamauchi, F., Huang, J. and Otsuka, K. (2021) Changing farm size and agricultural development in East Asia. In: Otuska, K. and Fan, S. (eds) *Agricultural Development: New Perspectives in a Changing World*. IFPRI, Washington D.C., pp. 79–110.

Zhao, J., Yang, Y., Zheng, H. and Dong, Y. (2020) Global agricultural robotics research and development: Trend forecasts. *Journal of Physics: Conference Series* 1693(1), 012227.

Zhuang, H. (2022) Essays on the transformation of the rural economy in rubber producing areas in Southwest China. PhD thesis, Leibniz University Hannover, Germany.

17 Outlook

Hermann Waibel

This last chapter is a kind of personal outlook on Asian agriculture in the future, say in 2050, 25 years from now. I do this on two grounds. First, my own field experience and exposure to various research and development jobs in Asia. Second, the assessment of selected chapter authors of the book. I have asked them to answer four questions related to the future of agriculture (2050) in their respective countries:

(1) A key statement, on how the agricultural sector in your country will look like by 2050?
(2) What will be the major type of farm organization?
(3) What kind of technology will be used?
(4) How about the rural areas and rural villages?

In Table 17.1, the key statements of nine experts related to nine different countries, including Singapore are summarized. It is interesting to see that generally, the experts draw a positive picture of the future of agriculture in their respective countries. They expect agriculture to be modern and technically advanced, e.g. with the use of drones, robots and AI. They acknowledge that climate change effects with more frequent and more severe weather extremes will be more serious. However, they expect that farmers will have adapted to the challenges with climate-smart technologies, including ICT and modern biotechnology. They also expect large-scale farming to have increased significantly through the efforts of agro-industry corporations. Small farms will still be around, especially in rice and food crops but will mostly be organized in co-operatives and contract farming arrangements. Dominantly, farming will be more commercialized and market-oriented. The exit of aging farmers from agriculture and the lack of replacement by rural youths will foster corporate and various forms of contract farming. Deforestation and natural resource degradation will continue.

In Table 17.2, the criteria, Farm Types, Technology and Rural Areas are used to describe the experts' assessment of the future of agriculture in their countries. As regards the first criteria, the picture is mixed, some experts see corporate farming becoming dominant by 2050 (China, Thailand), while others (India, Bangladesh, Vietnam) think small-scale farming will still the major group, albeit under cooperation arrangements for technical services and marketing. The dual character of the agricultural sector which has already existed for a long time, is likely to remain in The Philippines and Indonesia, according to the experts' opinion.

It is interesting to observe that the country experts and authors of the country policy chapters overwhelmingly believe that the agricultural growth and transformation process which was going on rather steadily and for several decades after World War II, will continue, regardless of the increasing global risks, stemming from

 Asian Agriculture in a Climate Change-driven and Risky World (H. Waibel, S. Liebenehm and Ernah eds)
DOI: 10.1079/9781800627338.0017

Table 17.1. Expert key statements, Asian agriculture by 2050.

	Country	Expert	Key statement
1	China	Prof. Shi Min, Huazhong Agricultural University, China	By 2050, China's agriculture will be more technologically advanced and economically efficient, with a significant shift towards large-scale farming and high-value crops
2	India	Dr. Priyanka Parvathi, Global Market Analysist, Private Sector, India/Germany	I see farm sizes still small, but more farmer organizations and cooperatives that would strengthen bargaining power against middlemen
3	Bangladesh	Dr. Badrun Nessa Ahmed, Bangladesh Institute of Development Studies, Bangladesh	By 2050, agriculture in Bangladesh will be transformed by climate-resilient technologies, automation, and precision farming
4	Thailand	Dr. Somkid Naprom, Udon Thani Rajabhat University, Thailand	By 2050, today's farmer will be retired and will not have successors. Most young people from rural households lack agricultural knowledge and skills because they have been wage workers. They will work in large agro companies, as local entrepreneurs and/or some local farmers with capital and knowledge
5	Vietnam	Dr. Jonathan Pincus Fulbright University, USA/Vietnam	Less dependence on rice; more fish and more vegetables; fruits, higher value added commodities; capital intensive and regional trade. Mekong Delta, less rice more fish; Red River, urbanization, villages become townships; upland areas will stay much the same as now
6	The Philippines	Prof. Dr. Justurs Wesseler, Wageningen University, the Netherlands	The following 25 years will be a challenge for the Philippines because climate change is expected to increase the number and severity of tropical storms with impacts most severe for tree crops; Golden Rice will be widely adopted
7	Indonesia	Dr. Eko Ruddy Cahyadi, IPB University Bogor, Indonesia	Rural villages will suffer from changing rainfall patterns and rising temperatures, mitigation measures through climate-smart technologies like agroforestry. Some villages may engage in ecotourism and small-scale enterprises
8	Singapore	Prof. Dr. Tomoki Fujii; Singapore Management University	It's uncertain but by 2050, Singapore may have a new model of urban agriculture. It is imaginable that a sizable share of food may be produced in commercial indoor farms or from community-driven farms in line with Singapore's 30 by 30 policy
9	Laos	Phouvong Phami, MSc, Doctoral candidate, Wuhan University, China	Destruction of forests will continue; there will be more outside investors for plantation agriculture at the expense of the environment

Source: Author's own table based on expert opinion.

climate change, political and economic uncertainties. Perhaps it is normal that young scientists believe more in technological progress as a problem solver than advanced-age people like myself will do.

In the final part of the chapter (and the book) I will add my own assessment of the 2050 scenario in the Asian countries. I do this against the background of my 40-years research and development experience in the countries, covered by the book chapters. It started as an exchange student in The Philippines in 1976, with a three-month training course in rice production and continued with field research on Integrated Pest Management in the same country from 1979 to 1982 (Waibel, 1986). Then came

Table 17.2. Expert views on Asian agriculture by 2050. (Author's own table).

	Country	Major Farming Type	Technology	Rural Areas
1	China	Corporate farming dominating over individual farms	Plant breeding, biotechnology and climate-smart technologies	Abolish small and establish large villages, business initiatives by urban return migrants
2	India	Small farms still dominant under farmer organizations and cooperatives	Digitalization; molecular breeding; other biotechnologies; climate-smart technologies widespread	Better infrastructure and market access; enhanced livelihoods
3	Bangladesh	Farmland area decreasing; small-scale farming still dominating but with more cooperatives and contract farming	Small farms under cooperative setting, use of drones; gene-edited crops; vertical farming with hydroponics in urban areas	Balancing traditional rural livelihoods with modern lifestyles will remain a challenge
4	Thailand	Corporate farming will dominate commercial local farmers	Larger plots automation and climate-smart technology	No more food from natural resources; minimal social cohesion in village, isolation for elderly, gap between remote and peri-urban widened
5	Vietnam	Rice and food crops still small-scale but mostly under contract; corporate farming for fruit trees, rubber and aquaculture	Automation; drones, precision agriculture, low-methane rice technology	Heterogeneous conditions: Mekong Delta villages disappear, Red River Delta villages become small towns, Central Highlands little change
6	The Philippines	Dual structure remains: plantation sector and small-scale rice farms	Golden Rice widely adopted; Genome-edited disease-resistant banana varieties, tobacco varieties producing biopolymers for pharmaceutical industry	Bio-economy with biopolymers for bioplastics and biofuels; vitamin A deficiency among children substantially reduced
7	Indonesia	Plantation crops by corporations and larger local farmers; rice and field crops small-scale organized as cooperatives	AI-powered precision farming, drought-resistant varieties, more agroforestry and shade tree	Population 320 million, land loss in coastal areas, urbanization continues
8	Singapore	Commercial and community indoor production of vegetables	New technologies for building-integrated agriculture	n.a.
9	Laos	Small farmers decreasing; mostly with contract farming; corporate farming increasing in plantation crops	More mechanized farming, water saving technologies	Heterogeneity, if near modern infrastructures (highways, railway lines) villages become small towns, for remote villages, little change

four years as a development expert with the Extension Department in Thailand (Waibel, 1991) and two and a half years as Associate Professor at the Asian Institute of Technology (AIT) in Bangkok (Waibel and Setboosarng, 1993). After returning to Germany in 1991, first, to join the University of Goettingen and in 1994 at Leibniz University Hannover, joint research with Asian countries continued through supervision of a total of 23 PhD student theses on topics of Asian agriculture. Many of the results of these studies are in this book. A major project was the Thailand Vietnam Socioeconomic Panel (TVSEP), a rural household panel with 4400 households in 440 villages in three provinces in Thailand and Vietnam respectively (Klasen and Waibel, 2014). Another significant experience was the 10-year research consortium SURUMER (Sustainable Rubber Cultivation in the Mekong Region) in Xishuangbanna autonomous prefecture, Southern China (e.g. Jin *et al.*, 2021).

As mentioned above, my Asian students and research partners generally have a positive view of the future of agriculture. I fully understand their optimism. After all, there is no point to bury one's head in the sand. Following the words of wisdom of the late Stephen Hawkins, one might say, "looking up to the stars is always better than looking down at one's feet, as there is always something useful one can do." Although, I am not looking down at my feet, I nevertheless want to add my different, somewhat less optimistic view on three main issues, namely: (i) technology, (ii) farm size and farm type, (iii) rural areas and villages.

To begin, regarding technology, I agree with my country experts that the process of mechanization, automation, information and communication technologies in Asian agriculture will progress, regardless of whether farms will be large or small. Out of sheer necessity, and to cope with weather extremes and other shocks, many of the new technologies will have to be climate-smart. Also, I believe technological progress and advanced mechanization in agriculture will happen in all the countries covered in the book, even though their level of development, in terms of GDP per capita, differs. As regards AI, I am less optimistic about what it can do to solve the problems of climate change and global and regional political uncertainties. The experience with new "key-turn-technologies" is pretty much the same, an overestimation of the benefits and underestimation of the true costs. This has applied to genetic engineering and gene editing in plant and animal breeding, the introduction of DDT and to many other examples. *Ex ante* cost-benefit assessments should be carried out as a routine procedure, before voicing expectations.

Let me now turn to the farm size debate. Most experts expect large-scale and corporate farming to become dominant and small-scale farms to be in the minority. It is true that there were always large farms in Asia. In some countries (e.g. The Philippines, Indonesia) these have existed since colonial times. However, they mostly apply to plantation crops, managed by agro-corporations in oil palm, rubber, banana, coconut, sugarcane, coffee, tea and some fruit trees. However, in field crops the majority of farms in Asia are small. In meat production and in dairy, large commercial farms exist side by side with small farms.

To illustrate the problem, let us take the case of the largest farm in the world, Mudanjiang City Mega Farm in Heilongjiang province, north-eastern China. It is a state farm with over 100,000 dairy cows and an area for grassland and feed crops of 22.5 million acres, roughly the size of Ireland, producing 800 million litres of milk annually (DuBois, 2019). Little is known about the economics of this operation. To comment on this Chinese megaproject let me tell a little anecdote. In 1991, after I had just started my job at the University of Göttingen, I had lunch with two professors of agricultural economics (both now deceased) who had quite contradictory views about the structure of agriculture in Germany. One was Günter S., a professor of agricultural policy who believed that part-time farms are necessary and could be as efficient as large-scale entities. The other one, Wilhelm B., was a farm management economist who believed in the economies of scale and the power of average cost curves. During the course of the lunch, S. suddenly asked B., "my dear colleague, you believe in large farms, so tell me why don't all the 500,000 farms in Germany merge into one giant operation?" B. looked a bit unprepared and hesitated to answer right away. Before he could say something, S. gave the answer himself: "transactions costs my dear colleague, transactions costs!"

I do not want to question the law of economies of scale and I recognize that eventually large farms in various organizational forms will emerge in Asia. Here, in principle, I agree with my experts. But I think it will take a lot more time than some of them anticipate. And it will not happen suddenly. Furthermore, when it comes to the question of optimal farm size, there is more to consider than average cost curves. Farm families, especially when they are in a structural change and transition period, diversify their portfolio, both in farm and in non-farm activities. Most of them won't just give up farming for some low quality factory or service job with little security, right away. On the other hand, only a few of them will scale up the farm, by renting-in land from neighbors and undertaking capital investment. The business concept of "grow or go" cannot be transferred one-to-one, to family farms in Asia. The majority will adopt income combinations, whereby the share of non-farm income will rise over time. I believe that small-scale family farming will be maintained for supplementary income, food security and as a safety net. This is what the country policy chapters told us. These are also the lessons we can learn from the rich Asian countries, like Japan, Taiwan and South Korea as well as from Europe. During the course of development, various types of farms will emerge, including full-time, commercial farms, part-time farms with variable shares of non-farm income and agriculture-based enterprises like homestay, ecotourism and sales of home-made natural food products. There will also be retirement, lifestyle and hobby farms where people may keep some horses and other animals, do home gardening and nurture small fish ponds and the like.

Now, we should recall the title of the book, "climate change and a risky world." It is possible that in the near future, everything could be far less stable and much more uncomfortable than it is now. Let's just imagine a world with a growing frequency and severity of natural disasters, like storms, floods, droughts, disease and pest outbreaks and even pandemics, like Covid-19. Further let's imagine a world with numerous military conflicts, regional wars and local confrontations, for example, among ethnic and religious groups. Let's think about cyber-attacks, spam-centers like the ones in Myanmar and Cambodia, who are trying to hack accounts and personal data of innocent people. Let's also think about the crisis of global institutions, trade relations and other cross-country relations. Already we can see now that solidarity for the common goods is declining and national egoism is rising. This could mean more frequent economic downturns, more job insecurity, lower remittances, oftentimes supply chain interruptions, food shortages and inflation. Crises always mean interruptions from normal life, mental stress, pain, damage and losses, repairs and rebuilding or coping with less, sometimes for a long time. It also means higher transactions costs as people will have to spend time and money to discuss, arrange, organize and prepare for future risks. We can get a bit of a feel, what a world in crisis means, when we recall the Covid-19 pandemic, between 2019 and 2022 (e.g. Waibel *et al.*, 2020; Fan *et al.*, 2021). In most cases, shocks will hit the urban population, especially the low-income groups, harder than people living in the countryside. Against this background, rural people might be quite reluctant to give up farming. Rather, they will maintain their capacity of own food production, and making use of local resources as much as possible. Urban areas are dependent on external supply and have very limited possibility to produce food. To follow the model of Singapore with high tech, facade and roof top production (see Chapter 15), will be difficult for most of the Asian megacities. In conclusion, when taking a less optimistic view about the future, as I tend to do, the small-farmer, self-sufficiency model will still have a role to play in the future. Likewise, urban food production may become a concept that city planners can no longer put aside.

The third issue is the future of rural areas in Asia. Will rural areas be deserted villages, and home for the aged, which disappear once the old people die? Will rural areas be places that modern people just want to get out of or could these also be attractive living places for people with high paid jobs, who love to farm in their free time and make some money out of it? Let us start with a little anecdote again. Richard J. Sexton, in his book *Food Fights* (2025) articulated the tragedy of small farms in rural America. He stated that he can speak with some authority because he grew up on such a farm in Minnesota (Sexton, 2025, p. 114). He basically said that small is not beautiful at all but being small means poverty. He complained that throughout his education period, he worked at his family farm at zero wage. He called it a most futile effort trying to make his small family farm become a profitable entity. He concluded that he doesn't wish anyone to be born into a small family farm in rural America.

Now, here I put the story of my family farm in southern Germany against the story of the professor from Minnesota. To begin, I am proud that I was born into such a family farm. Assumingly, in the 11th century, the land was given as a fief by the Emperor of what was then called the "Holy Roman Empire of the German Nation." For centuries, our family farm of some 25 ha arable land plus 10 ha of forestland, was considered a large farm. Reportedly, my great-great-grandfather did not work manually himself but he had his servants and maids to do the farm and household work. However, during my teenager years, in the 1960s, the farm could no longer cope with the challenges of structural change in agriculture in Germany after World War II. At some point the farm did not generate enough money for a family and thus had to be turned into a part-time farm. My brother, who inherited the farm, took up the profession of a private tax advisor and ran the farm as a sideline, with energy crops, horses and forest wood. Tax advisor is a lucrative business in Germany because of the "jungle-like tax law" in my country. My brother's income is a multiple of that of a German university professor. Besides, he knows exactly how he can use his farm to get tax exemptions and subsidies. He modernized my parent's house, put solar panels on the huge roof of his barn and made his heating system totally reliant on renewable energy. He and his family live in a beautiful and healthy environment. If he wants to go the next city, his luxury BMW will take him there comfortably and in no time. He certainly does not regret to be born into a small family farm in southern Germany. Neither do I, although I worked there during high school vacations and semester breaks at zero wage. I just did it for my family.

Turning back to Asia, I asked four of my chapter authors, who were either raised in a rural village in their respective countries and regions or had close contact, what they think their village will look like by 2050.

The first is Suputtra, who comes from a village in Northern Thailand, co-author of Chapter 13, she wrote:

> *By the year 2050, my small village in Lamphun will become much quieter than it once was. Neighbors will know each other less, as every household is surrounded by technology – phones, the internet and smart devices that keep people connected to the world but apart from one another. The warm face-to-face conversations that once took place in front of houses or at the morning market will almost have disappeared. There will be more elderly people living in the village, since most of the younger generation have moved away to study or work in cities, other provinces or even abroad. The lively community of the past will slowly turn into a peaceful but aging society. Most villagers will still keep their small longan and mango orchards, but local labor will become scarce because there are few young people left to help with harvesting. Farmers will therefore sell their crops as "whole-orchard deals" to middlemen, which may lead to price pressure in the future. Moreover, cash will likely disappear from daily life in the village, as everyone uses digital payments and mobile banking apps instead. Life will be more convenient, yet the closeness and sense of community that once defined the village may quietly fade away.*

Next, here is the opinion of Somkid, from a rural village near Khon Kaen in Northeast Thailand and also a co-author in Chapter 13. This is what he wrote:

> *The population in my village will decline, many people will have moved to nearby towns or the provincial capital. My friends and I, for example, we work in another city, we go back home just once a year, just to visit parents and bring some money and gifts. Some of my friends who work in urban areas and cities never came back to the village, just send some money, once in a while. The natural forest to which I used to go, collecting mushrooms, has now become para rubber plantations. In 2050, I think it will still be para rubber plantation but might be with intercropping with some high-value cash crops to increase the returns to land. Our house in the village will be like a one-story concrete, in modern style and materials. Although families will still stay together in the village, interaction and interpersonal engagement will decline. We will have very little more fascinating natural food; we eat fast food, ordered with a few smartphone clicks or by voice mail. Food will be cheap but not natural and may be not as healthy as before.*

The third example is Phouvong, co-author of Chapter 11, the Laos country policy chapter.

Although he was born in a rural village in north-eastern Laos, he lost his parents at the age of 9 and was moved to an SOS children's village in Vientiane. Phouvong was repeatedly and intensively engaged in rural household surveys in many parts of Laos and he has friends everywhere in the country. So here is what he said about villages in northern Laos, an area that is increasingly becoming under Chinese influence:

> *One trend I've observed is that many rural villages in general are facing a significant decline in their young population. As they migrate to cities like Vientiane or larger urban areas, as well as to countries such as Thailand, Korea and Japan in search of better opportunities, older residents and children are left behind, leading to a demographic shift within these communities; this trend will be increased. Additionally, many villages are being removed and relocated, as the government implements a policy to consolidate smaller villages into larger ones. In the north of Laos, Chinese investors have announced plans to complete the second phase of the Vientiane-Boten Expressway by 2030, significantly transforming the region's economic landscape and improving transportation convenience. By 2050, a typical village in northern Laos is expected to harmonize traditional charm with modern amenities, featuring wooden houses, equipped with electricity. However, challenges remain. Poorly maintained roads and heavy trucks transporting natural resources and minerals complicate travel in provinces like Huaphan, Phongsaly, Xiengkuang, Xaysomboun and parts of Xayyabury. These areas are distant from the Laos-China Railway, and road conditions are often inadequate. On the other hand, large areas of land and forests have been designated for resource extraction, frequently with little consideration for local livelihoods, resulting in a sense of isolation for villagers.*

The fourth assessment comes from Eko, co-author of Chapter 10, the Indonesia chapter. He grew up in an small village in Central Java. However, his subjective assessment refers to a village in Jambi province which he studied intensively during his thesis research on smallholder oil palm farming in Indonesia. The village, named Arang-Arang, is a village with indigenous people in the oil palm areas in Jambi province. This is what he wrote:

> *By 2050, farmers in Arang-Arang village, will likely face climate-related challenges, including changing rainfall patterns and rising temperatures, affecting rubber and oil palm production. To adapt, they may apply sustainable agriculture practices, like agroforestry, or try to diversify into non-farm enterprises like home stay and ecotourism.*

What can we conclude from the assessment of my experts about rural development in the future and the four opinion statements about the rural villages in Asia? Obviously, there is no priority by governments in Asia to invest a lot in rural development. Infrastructure investments are bound to serve the cities, not the people in villages. To maintain the social structure of rural villages is not in the plan of policymakers. In short, the urban bias will persist. Such policy is based on the assumption that the future will be much like the past. However, if this assumption does not hold and instead external shocks and vulnerabilities increase, the hollowing-out of the countryside will continue. Then, countries may have to pay a high price for losing their potential safety nets and will miss opportunities. For example, destroying smaller villages with a long tradition and moving people into new larger settlements, can be costly and time consuming. There will be resistance from the local people when forced to give up their traditional culture and their ancestors' place resulting in sunk costs. Looking at the example of Europe, rural development can make rural areas become attractive and beautiful living and working places, as also mentioned in the anecdote above. Rural villages and small rural towns can be a good location for start-up business in the ICT and other sectors, with people working in home offices, away from crowded and unhealthy urban environments. Rural villages as a source of renewable energy with wind turbines, solar panels and biomass production can be the base of a bio-economy, producing biofuels, biopolymers and pharmaceuticals. I, therefore, want to make a plea for a balanced rural-urban development strategy by the Asian countries, in view of the stormy times ahead. If I am wrong and things turn out in a much more positive way by 2050, I should be happy.

References

Sexton, R.J. (2025) *Food Fight: Misguided Policies, Supply Challenges, and the Impending Struggle to Feed a Hungry World*. University of California Press.

DuBois, T.D. (2019) China's dairy century: Making, drinking and dreaming of milk. In: *Animals and Human Society in Asia: Historical, Cultural and Ethical Perspectives*. Springer International Publishing, Cham, pp. 179–211.

Jin, S., Huang, J. and Waibel, H. (2021). Location and economic resilience in rubber farming communities in southwest China. *China Agricultural Economic Review* 13(2), 367–396.

Fan, S., Teng, P., Chew, P., Smith, G. and Copeland, L. (2021) Food system resilience and COVID-19 – Lessons from the Asian experience. *Global Food Security* 28, 100501.

Klasen, S. and Waibel, H. (2014) Vulnerability to poverty in South-East Asia: Drivers, measurement, responses, and policy issues. *World Development* 71, 1.

Waibel, H. (1986) *The Economics of Integrated Pest Control in Irrigated Rice: A Case Study from the Philippines*. Springer Berlin Heidelberg, Berlin, Heidelberg.

Waibel, H. (1991) Pesticide subsidies in Southeast Asia. *FAO Plant Protection Bulletin* 38 (2), 111–120.

Waibel, H. and Setboonsarng, S. (1993) Resource degradation due to chemical inputs in vegetable-based farming systems in Thailand.

Waibel, H., Grote, U., Min, S., Nguyen, T.T. and Praneetvatakul, S. (2020) COVID-19 in the Greater Mekong Subregion: How resilient are rural households? *Food Security* 12(4), 779–782.

Index

Note: Page numbers in **bold** and *italics* refer to tables and figures, respectively.

Printed and bound by CPI Group (UK) Ltd, Croydon, CR0 4YY

16/07/2026

14922514-0001